The American Forestry Series

HENRY J. VAUX, *Consulting Editor*

FOREST AND RANGE POLICY

Its Development in the United States

The American Forestry Series

HENRY J. VAUX, *Consulting Editor*

ALLEN AND SHARPE · An Introduction to American Forestry
BAKER · Principles of Silviculture
BOYCE · Forest Pathology
BROCKMAN · Recreational Use of Wild Lands
BROWN, PANSHIN, AND FORSAITH · Textbook of Wood Technology
      Volume I—Structure, Identification, Defects, and Uses of the Commercial
          Woods of the United States
      Volume II—The Physical, Mechanical, and Chemical Properties of the
          Commercial Woods of the United States
BRUCE AND SCHUMACHER · Forest Mensuration
CHAPMAN AND MEYER · Forest Mensuration
CHAPMAN AND MEYER · Forest Valuation
DANA · Forest and Range Policy
DAVIS · American Forest Management
DAVIS · Forest Fire: Control and Use
DUERR · Fundamentals of Forestry Economics
GRAHAM · Forest Entomology
GREELEY · Forest Policy
GUISE · The Management of Farm Woodlands
HARLOW AND HARRAR · Textbook of Dendrology
HUNT AND GARRATT · Wood Preservation
PANSHIN AND DE ZEEUW · Textbook of Wood Technology, Volume I
PANSHIN, HARRAR, BAKER, AND PROCTOR · Forest Products
PRESTON · Farm Wood Crops
SHIRLEY · Forestry and Its Career Opportunities
STODDART AND SMITH · Range Management
TRIPPENSEE · Wildlife Management
      Volume I—Upland Game and General Principles
      Volume II—Fur Bearers, Waterfowl, and Fish
WACKERMAN · Harvesting Timber Crops

Walter Mulford was Consulting Editor of this series from its inception in 1931 until
January 1, 1952.

FOREST AND RANGE POLICY

Its Development in the United States

SAMUEL TRASK DANA

Dean Emeritus, School of Natural Resources
University of Michigan

McGRAW-HILL BOOK COMPANY

New York   Toronto   London

1956

FOREST AND RANGE POLICY

Copyright © 1956 by the McGraw-Hill Book Company, Inc. Printed in the United States of America. All rights reserved. This book, or parts thereof, may not be reproduced in any form without permission of the publishers.

Library of Congress Catalog Card Number 55-11168

8 9 10 11 12 13 – MP – 1 0 9

ISBN 07-015285-3

To
RUTH MERRILL DANA

Preface

"Policy," according to the dictionary, is "a settled or definite course or method adopted and followed by a government, institution, body, or individual." A single policy for an entire country in the handling of its forests, range lands, and other natural resources is therefore possible only when the central government has complete authority to dictate the courses and methods to be followed by both public and private agencies. The other extreme is found in a country like the United States where there not only may be, but are, as many policies as there are governmental units and private forest owners. There is, however, at any given time sufficient similarity among these various policies to form a somewhat indistinct but nevertheless recognizable pattern for the country as a whole.

The courses of action that constitute policies are "settled" only in a relative sense. Changing times inevitably result in changing policies, which are consequently in a constant state of flux. Some knowledge of these changes and of the forces that have controlled them is necessary both to understand the present pattern and to predict its future evolution. To present that knowledge is the object of this book.

The evolution of the policies of Federal, state, and private agencies might be recorded either by periods of time or by subjects. Here the two methods have been combined. The treatment is primarily chronological, but within the different periods, which sometimes overlap, developments relating to a particular subject are usually discussed together. Special attention is paid to Federal policies, partly because major developments in policy have until rather recently centered largely around Federal activities, and partly because limitations of space make impossible any detailed discussion of developments among forty-eight states and several million private owners.

Appendix 1 provides a brief survey of Federal policies dealing with wildlife, soil, water, and mineral resources, which are often closely related to forest and range policies. Appendix 2 presents a comprehensive, chronological summary of important events in the development of Federal policy relating to the conservation of natural resources, with dates and citations. It includes many items to which no reference is made in the text. The description of each event is made as concise as possible, with no attempt to include all details, particularly when the event is treated at some length in the body of the book. Appendix 3 brings together the references given at the end of each chapter and adds some other publications which those who desire to go deeper into the subject may wish to consult, but with no pretense of completeness.

The book is intended primarily for use as a text and for reference purposes for students of forestry, range management, and other fields of natural-resource management. It is hoped that it will also prove of service in the biological and social sciences, such as botany, zoology, geology, geography, history, economics, and political science; to professional men and executives in business and government; and to the general reader with an interest in the subject.

Grateful acknowledgment is made to the many persons who have generously assisted in supplying material and in reviewing the manuscript. Among these are C. Edward Behre and Charles E. Randall of the U.S. Forest Service; John B. Bennett, John F. Shanklin, Irving Senzel, and Walter H. Horning of the U.S. Department of the Interior; William B. Greeley, Chairman of the Board, American Forest Products Industries; Alf Z. Nelson, Forest Economist, National Lumber Manufacturers Association; Hardy L. Shirley, Dean, State of New York University College of Forestry; and Ralph S. Hosmer, Professor Emeritus of Forestry, New York State College of Agriculture. Special thanks are due to J. Willcox Brown, whose constructive criticisms were based on use of earlier drafts of the text with a class in forest policy at the University of Michigan.

The author of course assumes full responsibility for all errors and for all expressions of opinion.

SAMUEL T. DANA

Contents

CHAPTER 1

Colonists, Kings, and Forests

During the colonial period, forests constituted an important resource with respect to which American and British policies differed radically.

Four F's. Fur, fish, farms, and forests provided the economic basis for the development of the North American Colonies. Gold and silver, though destined later to play a prominent part in the building of the West, were virtually lacking on the Atlantic Seaboard. There were no wealthy Aztecs or Incas to be plundered or enslaved by another set of conquistadores. The age of metals, coal, oil, and electric power still lay in the future.

Organic rather than mineral resources furnished a livelihood for the early settlers on the American continent north of Mexico, and to a large extent molded their character. Nature was generous in her gifts of soil and water, of plants and animals. These resources, however, had to be harvested or cultivated, raw materials had to be produced and made into finished goods, before they were of value to man.

Trapping, fishing, farming, and logging were the main activities of most of the population. Their products not only supplied the basic needs of the colonists for food, clothing, and shelter but also provided a surplus for export to other countries. Manufacturing was a relatively simple process, was largely conducted in the home, and was distinctly subordinate in importance to the extractive industries.

Under these circumstances the colonial economy was essentially rural in character. People spent most of their time on the farm, in the forest, or at sea. Even the cities, which served primarily as trading centers, were hardly urban in the modern sense. Philadelphia, for example, at the middle of the eighteenth century is estimated to have had 10,000 inhabitants, Boston 7,000, and New York 5,000. Men lived much more largely by their own wits and their own muscles than they do today. Hardihood, self-reliance, versatility, adaptability, inventiveness, and ingenuity were virtues which favored the survival of their possessors.

Under All, the Land. Labor and capital were scarce. By the time of the first Federal census in 1790 the total population of the Colonies (3,924,214) was about the same as that of the present state of Indiana. Moreover, the first immigrants were for the most part from the lower and middle classes who possessed little of this world's goods. Tools, machinery, and equipment of all kinds had to be acquired painfully and gradually through the slow process of saving. Land, on the other hand, was abundant. Practically a virgin continent, with almost unlimited resources in soil, vegetation, and wildlife, was to be had for the taking.

The result of this relation between the factors of production was inevitable. Labor and capital, being scarce and therefore expensive, were used as sparingly and intensively as possible. In modern terminology, they were "conserved." Natural resources, being abundant and therefore cheap, were used as liberally and extensively as possible. In other words, they were exploited. There was nothing "ruthless" or reprehensible about this procedure. It was merely the application of sound common sense to the economic problem of making the most effective use of the productive factors at the disposal of the colonists.

The seemingly endless supply of natural resources, however, tended to change liberality into prodigality of use, to encourage unnecessary waste, and most important of all to develop an enduring belief in their literal inexhaustibility. This belief became so deeply ingrained that it still colors our thinking on the subject and greatly increases the difficulty of bringing about the more intensive management of our natural resources, which is justified by the fact that they are now relatively the least instead of the most abundant of the factors of production.

No one in those days had to be reminded that all of man's material possessions originate in natural resources—"land" in the economic sense. Firsthand experience in wresting a living from nature was the rule rather than the exception for old and young alike. Even as late as the Revolution, 95 per cent of the population was classed as rural; there were only twelve cities with a population of 5,000 or more, and none with 50,000. A largely self-sufficing agriculture provided food and most of the clothing for the family, and in the South tobacco, rice, and indigo for export to the mother country. In the absence of modern methods of fertilization, strip cropping, and contour plowing, soil depletion and erosion became sufficiently common to give serious concern to such farseeing leaders as George Washington and Thomas Jefferson.

Forests—Friend and Foe. Forests constituted the dominating feature of the colonial landscape. From Maine to Georgia, with the exception of a few mountaintops, they covered nearly 100 per cent of the land area. In variety of species, size of individual trees, and total extent, the forests of the New World were a phenomenon wholly outside the previous experience of the European immigrant. Their potential value and their apparent inexhaustibility were obvious to all who had eyes to see.

Here was an asset of inestimable worth, but here also was a local and temporary liability. The omnipresence of the forests meant that they had to be cleared away wherever it was desired to start a farm, establish a settlement, or build a road; and the very size and density which added to their value made this a difficult and time-consuming task. As a visitor from England (Isaac Weld, Jr.) expressed it: "The ground cannot be tilled, nor can the inhabitants support themselves, 'til [the trees] are removed, they are looked upon as a nuisance, and the man that can cut down the largest number . . . is looked upon as the most industrious citizen, and one that is making the greatest improvements in the country."

American Attitude. To the American colonists the forest was at the same time an impediment to be removed as rapidly as possible to make way for farms and villages, and a storehouse of materials needed for a wide variety of purposes. From the forest came wood for fuel, houses, posts, rails, poles, furniture, implements, shingles, staves, ship frames, masts, and many other products. From the forest came also tar, pitch, and turpentine for naval stores; potash for fertilizer and soap; hemlock bark for tanning. As a source of indispensable raw material, to be had for the effort involved in its harvesting, its value was outstanding.

Wood-using industries were the first form of manufacturing to be developed in the Colonies. As early as 1623, the good ship "Anne" carried a cargo of clapboards from Plymouth to England. In 1626 this was followed by a shipment of lumber from New Amsterdam. By the early 1630's several small sawmills were operating in Maine, New Hampshire, and New York. The term "lumber" to describe sawed timber was first used in 1663, when the Boston merchants were ordered "to clare the ends of all streets and wharfes that butt upon the water from all lumber and other goods." Its obvious derivation is from the "lumber," or miscellaneous junk, in the storerooms of the Lombard pawnbrokers in London. The first official use of the term in its new sense in England appears to have been in 1721.

Starting with the launching of the "Blessing of the Bay" at Medford, Massachusetts, in 1631, shipbuilding soon became a major activity, particularly in New England. Skilled craftsmen using oak for hulls and pine for masts gave American shipping an enviable reputation. At the outbreak of the Revolution, the Colonies were building 100 ships a year, and the 2,343 colonial ships then afloat constituted about a third of the total British registry.

In addition to their widespread local use, forest products formed the basis for an extensive and profitable international trade. Together with fish, fur, and tobacco, they provided the chief source of pound-sterling exchange for the purchase of manufactured goods from abroad. Lumber for construction and remanufacture and staves for cooperage were shipped in large and increasing quantities not only to England but to France, Spain, Portugal, and the West Indies. Unmanufactured material in the form of masts and spars constituted another important item of commerce. From the first shipment of fourscore masts from Virginia to England in 1609 until the outbreak of the Revolution, the mast trade was a matter of major economic and political concern to both the Colonies and the mother country.

That the forests from which these products came were literally boundless appeared self-evident to the early settlers. The farther they penetrated from the seaboard into the back country, the clearer this fact seemed to become. "We are in the midst of the greatest forests in the world," wrote Charlevoix in 1721, and "there is nothing perhaps in nature comparable to them." Why, then, should one hesitate to help himself liberally from nature's bounty without thought of the morrow, or why should he be overscrupulous in recognizing property lines?

On the other hand, it was equally clear that local depletion of the timber supply was a real possibility with the limited means of transportation then available, and also that in accessible locations and for special purposes timber did have a value which the owner would very properly wish to protect. From these opposing points of view came the actual practices and the legislative enactments which comprised the forest policy of colonial times and from which today's policy has gradually evolved.

Regulatory Forest Legislation. The first forest legislation in the Colonies came six years after the landing at Plymouth Rock and surprisingly enough was caused by anxiety concerning the timber supply. On March 29, 1626, Plymouth Colony, in order to prevent the inconveniences which might "befall the plantation by the want of timber," forbade the sale or transport of any timber whatsoever out of the colony without the approval of the governor and council. This ordinance doubtless reflects the rapidity of clearing in the colony, the importance of wood to the settlers, and the lack of roads over which supplies could be brought in from a distance. It emphasizes the significance of accessibility as an important element in the value of any natural resource.

Massachusetts Bay Colony in 1668, perhaps at British instigation, anticipated the Broad Arrow policy adopted by England in 1691, by reserving for the public all pine trees fit for masts in certain portions of the town of Exeter (later a part of New Hampshire). Under the prodding of Her Majesty's Surveyor General, New Hampshire in 1708 passed an act applying the provisions of the Broad Arrow policy to the entire colony, which with Maine was the chief source of mast timber. Curiously enough, Massachusetts in 1785, two years after the close of the Revolution, passed a similar act.

An attempt to prevent widespread clear cutting was made by William Penn, who in 1681, in the document governing the establishment of a colony in "Penn's Woods," provided that 1 acre of trees must be left for every 5 acres cleared and that special care must be taken to preserve oak for shipping and mulberry for silk.

Massachusetts in 1744 authorized any five or more proprietors of lands within Chebacco Woods in the town of Ipswich to apply for the establishment of a common woods. If two-thirds of the landowners within the proposed limits approved, all of the lands within those limits became subject to the joint control and management of the proprietors. Eleven years later similar authority was granted the contiguous proprietors of woodlands in the "Wenham Great Swamp" to form an association "for the securing of the growth and increase of a certain parcel of wood and timber." This approach to the problem of maintaining timber supplies is of interest as a forerunner of recent attempts to develop cooperative forest management through producers' cooperatives and through joint action by public and private owners of intermingled forest lands.

In 1772 New York forbade the bringing to Albany for use as firewood more than six pieces per load of wood under 6 inches in diameter at the large end for pine and under 4 inches for other species, "either for sale or otherwise." The main purpose of this act was probably to protect the purchaser from receiving inferior material rather than to safeguard young forest growth from destruction. It is, however, an interesting antecedent of modern attempts to maintain forest productivity by prescribing minimum diameter limits of cutting.

The beneficial protective influence of the forest, as well as its value as a source of timber and fuel, also received early recognition. Heavy cutting, fires, and overgrazing had resulted in such serious beach erosion and consequent ruin of neighboring meadows and other lands by drifting sand that the Massachusetts General Court in December, 1739, provided a fine of 10 shillings for each bush, shrub, or tree under 6 inches in diameter cut from the beaches and marshes of Plumb Island in Ipswich Bay and still heavier penalties for firing the beach grass, bushes, and shrubs and for the running at large of any grazing animals. This legislation had been preceded in January of the same year by an act forbidding grazing in certain portions of the town of Truro, and was followed by similar acts attempting to control cutting and grazing in other parts of Cape Cod.

These various enactments do not indicate any widespread or effective interest in the conservation of forest resources. For the most part they were aimed to meet specific and unusual situations and do not reflect any feeling of general concern as to the possibility of a future timber shortage.

Protection from Fire and Trespass. Legislation aimed at protection of the forests from fire was common during the colonial period. Massachusetts Bay Colony led off in 1631 with an order forbidding the burning of any land prior to March 1 under pain of payment of full damage and such penalty as the court might inflict. Plymouth Colony took similar action in 1633, and by 1683 all of the Colonies as far south as Pennsylvania had enacted legislation on the subject. Delaware followed suit in 1739, and North Carolina in 1777. By the time of the Revolution, fire-control legislation was therefore in existence in all of the Colonies except Maryland, Virginia, South Carolina, and Georgia.

Some of these acts recognized damage not only to merchantable timber but to young growth, soil, cattle and hogs, fences, and other improvements. Most of the acts dealt with the time of year, month, or week when fires might or might not be set and required an owner to notify his neighbors whenever he was to do any burning on his own land. Penalties included fines, imprisonment, commitment to the county workhouse, and whipping, in addition to the payment of damages for any losses caused to other owners. It was not unusual for the defendant to be required to prove his innocence; but if he was successful in doing so, he was entitled to recover costs, and sometimes double costs, from the plaintiff.

The beginnings of organized protection against forest fires are to be found in an act adopted by New York in 1743 empowering anyone who should discover a fire in the counties of Albany, Dutchess, and Suffolk and the manor of Livingston "to require and command all or any of the neighboring and adjacent inhabitants to aid & assist him" in extinguishing the fire. A fine of 6 shillings was provided for refusal, neglect, or delay of a person so commanded to comply. In 1758, the provisions of this act were extended to the entire colony of New York.

The fire-warden system started in 1760, when a special act authorized the inhabitants of the city of Albany and of each town, manor, or precinct in the counties of Albany and Ulster to elect such number of freemen as they thought necessary to act as "firemen." In 1766 these provisions were reenacted and extended to the county of Orange.

Numerous acts were also passed prohibiting the unauthorized cutting of timber on private property or common grounds. Trespassers were required to pay damages, sometimes to triple the amount involved, and in addition were often liable to fine and occasionally to imprisonment or whipping. By 1741 laws dealing with timber trespass had been enacted in all of the Colonies except Virginia, North Carolina, South Carolina, and Georgia. This geographic distribution indicates clearly the greater value attached to forests and forest products in the Northern and Middle Colonies than in the South, where agricultural products claimed more attention.

Regulation of Industry. Public regulation of industry was much more common during the colonial period than during the nineteenth century. Standard dimensions were prescribed for such products as cordwood, staves, heading, shingles, and lumber. Limited control over prices and wages was exercised in parts of New England. Inspectors, or "viewers," were appointed to see that products met specifications.

Special attention was paid to material entering into foreign trade. For example, in 1646 the General Court of Massachusetts Bay Colony, after reciting the evils to foreign trade which would result from the exportation of pipe staves of poor quality, ordered the confiscation of all staves which in the opinion of the inspectors did not meet the required standards. In addition, every master of a ship receiving substandard material was liable to a fine of £5 for each thousand staves. Connecticut in 1667 provided that all pine, spruce, or cedar boards offered for export must be either 1 full inch or ½ inch thick. Apparently the rule of *caveat emptor* applied to the local buyer.

Occasional limitations were placed on both intercolonial and international trade. Connecticut in 1714 imposed a duty on all pipe, barrel, and hogshead staves exported to New Hampshire, Massachusetts, Rhode Island, New York, or New Jersey, and the next year imposed a duty on ship timber, plank, and boards exported to any of these Colonies. Similarly, New Jersey in 1714 laid an export duty on pipe or hogshead staves shipped to any of the British colonies in America, but three years later repealed the duty on hogshead staves. However, a further restriction was imposed in 1743, when duties were levied on all logs and timber products, except firewood, exported from eastern New Jersey to any of the American Colonies.

Navigable Streams. The common practice of using rivers and even brooks to float, or "drive," logs from the woods to the sawmill led to a new conception of "navigability" which later had an important influence on forest policy. This practice soon caused difficulties under the English common law that only the parts of streams affected by tidewater are navigable and therefore open to use by the general public. On other streams the riparian owner could legally prevent their use for the transportation of logs and could also appropriate any logs that might become stranded on the adjoining land.

Connecticut attempted to meet this situation in 1752 by requiring riparian owners on the Connecticut River to file a detailed record of all "logs, timber, shingles, and staves" becoming stranded on their lands. The owner of the logs could reclaim them at any time during the ensuing six months by payment of a specified fee to the riparian owner, who was permitted to appropriate them to his own use after the expiration of that period.

New Jersey took the leadership in opening for general use streams that were in fact "navigable." An act passed in 1755 provided a fine of £5 for anyone who should "obstruct or prevent the free and uninterrupted Navigation of any River, Creek, or Stream of Water within this Colony, which is used for the Navigation of Boats or Flats, or for the transporting of Hay, Plank, Boards, or Timber," without first obtaining an act of the General Assembly for that purpose.

In 1771 the Pennsylvania Legislature declared the Delaware and Lehigh Rivers and certain parts of the Neshaminey and Lechawaxin Creeks to be "common highways . . . for vessels, boats, small craft, and rafts of any kind whatsoever" and in 1785 enacted "that the river Susquehanna shall be deemed and taken to be a public highway" in all parts of the state. The preamble to the latter act emphasized the importance both of the timber supply and of water transportation in making it accessible in the absence of roads and railroads:

The extensive countries which are watered by the river Susquehanna, and the numerous branches thereof, are stocked with immense quantities of oak, pine and other trees, suitable for staves, heading, scantling, boards, planks, timbers for ship-building, masts, yards and bowsprits, from which great profit and advantage might arise to the owners thereof, if the same could be conducted in rafts and otherwise, down the said river to the waters of the Chesapeak, which trees must otherwise perish on the lands whereon they grew.

The conception of navigability embodied in these acts gradually spread throughout the country until any stream that would float a log or a boat came to be regarded as navigable. It not only facilitated utilization of the forest but paved the way for legislation dealing with forest management on the theory that, because of their effect on runoff, forests influence the flow and therefore the navigability of streams used in interstate commerce.

British Attitude. Sir Francis Bernard, who saw long service in the Colonies as Governor of New Jersey and Massachusetts, once said:

The two great objects of Great Britain in regard to the American trade must be to oblige her American subjects to take from Great Britain only, all the manufactures and European goods which she can supply them with: 2. To regulate the foreign trade of the Americans so that the profits thereof may finally center in Great Britain, or be applied to the improvement of her empire. Whenever these two purposes militate against each other, that which is most advantageous to Great Britain ought to be preferred.

These objectives meant that the development of the Colonies was to be directed so as to make them a source of raw materials for the mother country and at the same time a market for her products. Restrictions on colonial manufactures, regulation of trade, and control of credit were the major tools used in achieving these ends. For example, woolen goods and hats could not be made for the general trade, and mills for slitting and rolling iron, and furnaces for making steel were forbidden. The *Boston Gazette* complained in 1765 that "a colonist cannot make a button, a horseshoe, nor a hobnail, but some sooty ironmonger or respectable button maker of Britain shall bawl that his honor's worship is most egregiously maltreated, injured, cheated, and robbed by the rascally American republicans." No attempt was made to curtail the output of such rough forest products as masts, lumber, cooperage, and naval stores, which did not enter into competition with British goods, except for the prohibition against the use of pine trees suitable for naval stores or other purposes.

The main trade barriers date from 1651, when the first of the navigation acts forbade the importation of goods into and the exportation of goods from Great Britain or her Colonies except in English ships or in ships of the country from which the goods came or to which they went. This legislation was stiffened in 1660 by providing that thereafter no goods should be imported into or exported from any of the British Colonies in America except in vessels belonging to Great Britain or the Colonies and of which the master and at least three-fourths of the seamen were British. These acts were supplemented by laws providing that certain "enumerated" commodities, among which were lumber and naval stores, could be exported from the Colonies only to Great Britain. That ways were often found for shipping enumerated goods to other countries can hardly be doubted.

Tariffs and Bounties. Tariffs and bounties were also used as means of directing trade into the desired channels. The former were never burdensome so far as timber products were concerned, and after 1721 no duty was imposed on such products imported into Great Britain directly from the Colonies. Although this exemption from duty was specified to be for a period of twenty-one years, the free-trade provision was actually continued until 1778.

An act of 1704 provided bounties of £1 per ton for masts, yards, and bowsprits, £3 for rosin and turpentine, £4 for tar and pitch, and £6 for hemp. The same act gave the Navy preemption on all of these articles within twenty days of their arrival in England and placed them on the enumerated list, to which lumber was added in 1721. Although somewhat reduced in 1729 except on masts and hemp, bounties remained in force until 1777. They resulted in payments by the British navy of about a million and a half pounds, more than half of which went to the Southern colonies for naval stores. Provision was made in 1765 for the payment of premiums on lumber, and in 1771 on white oak staves and heading, shipped directly from the Colonies to Great Britain in British ships.

Masts and the British Navy. By far the most important of the British policies relating to the forest resources of the American Colonies had its roots in the urgent need of the British navy and merchant marine for an adequate and continuing supply of mast timber. Prior to the seventeenth century such timber consisted almost wholly of Scotch pine from northern and central Europe, and particularly from countries with tributaries flowing into the Baltic Sea and the North Sea.

Certain disadvantages, however, attended this source of supply. The mast of a first-rate ship of the line was 40 yards long and 40 inches in diameter, while a diameter of 36 inches was required even for a third-rate ship. Trees large enough to make single-stick masts of these sizes were getting to be so scarce that "made masts" consisting of several trees fastened together were becoming the order of the day. Moreover, Great Britain was in competition with other European countries for the available supply. Shipments to England had to be made by way of the Baltic Sea and the Skagerrak, which might be closed to the British by a combination of hostile nations in time of war. Finally, trade with "Eastland" was primarily a one-way affair, since the Baltic Provinces offered little market for British manufactured goods.

Physical, economic, and political factors therefore made another source of supply highly desirable. Why should not this source be the American Colonies? American pines, and particularly white pine in the North, were far superior to Scotch pine in size and at least equal in quality of wood. There was little danger that shipments across the Atlantic could be prevented by hostile powers from reaching their destination. A two-way trade could readily be developed that would bring needed raw materials to Britain and provide an outlet for its manufactures. Such an arrangement would not only be to Great Britain's economic advantage but would strengthen the political ties between the Colonies and the mother country.

The first shipment of masts was made from Virginia in 1609, presumably of loblolly pine. The first cargo of masts went from New England in 1634, and from then on the trade centered in the Northern Colonies. Here mag-

nificent specimens of white pine occurred, not in pure stands but as individual trees often several feet in diameter and up to 200 feet tall tower-
ing above the main forest of hardwoods.

It was clear from the beginning that the supply of such trees was by no
means inexhaustible, particularly when they must be not only of adequate
size but also straight-grained and free from defect in order to be suitable for masts. In one instance, 102 out of 106 trees of sufficient size
proved to have enough decay to make them unusable. Furthermore, trees
that were in every respect ideal for masting were not infrequently sawed
into lumber.

Harvesting of mast timber was a difficult and expensive process, sometimes requiring the building of special roads and the use of as many as
twenty yoke of oxen. Full-length trees were more liable to be smashed in
river driving than shorter logs which were equally suitable for sawing into
lumber. Consequently, in spite of the much higher prices brought by
mast timber, many lumbermen did not like to bother with it and felt that
they could make more profit by manufacturing other products.

Broad Arrow Policy. Great Britain's increasing dependence on the
Colonies for mast timber, the relative scarcity of the supply, and the difficulty of making certain that all suitable trees would be used for that purpose made it imperative for the British government to take steps to safeguard the future. Accordingly, the new charter granted to the Province
of Massachusetts Bay in 1691 reserved to the Crown all trees in the
province 24 inches or more in diameter at 12 inches from the ground on
land not previously granted to any private person. Such trees were not
to be cut without specific license from the Crown under penalty of £100
for every tree illegally cut. License to cut was ordinarily granted to one
of four large contractors, who in turn arranged with the colonists for the
actual harvesting.

The policy thus inaugurated soon became known as the Broad Arrow
policy because of the practice of marking trees to be reserved for mast
purposes with three blazes representing the broad arrow, which was the
symbol of the British navy. In 1711 the original provisions were extended
to include all white pine or other pine trees fit for masts and of the specified size, not on private property in New England, New York, or New
Jersey. The same act established a penalty of £5 for unlawfully marking
a pine tree with the broad arrow. This provision was an obvious effort to
deter enterprising individuals from blazing particularly desirable trees in
order to scare off competitors and thus reserve them for their own use—a
practice that offers eloquent testimony to the attitude of the colonists
toward the entire policy.

Extension of Policy. In 1721 another amendment forbade the cutting of any white pine trees not growing within a township, from Nova Scotia to New Jersey, with penalties ranging from £5 to £50 depending on the size of the tree illegally cut. The inclusion of small trees in the prohibition against destruction makes clear the British government's realization of the necessity of protecting young growth as the only means of assuring adequate future supplies of large trees.

This act was apparently evaded by the laying out of large areas of "paper" townships. To stop this leak, it was enacted in 1729 that no one should thereafter without royal license cut or destroy any white pine trees not on private property, whether or not they were within the limits of any township already laid out or to be laid out, or on land in the Province of Massachusetts Bay that was not private property in 1691. The act also extended the area covered to include every part of America which then belonged to Great Britain or might thereafter be acquired, and provided better machinery for enforcement. The Broad Arrow policy remained in effect in this form until the Revolution.

Legal Basis of Policy. The restrictions embodied in the policy were based on the principle universally accepted in the Western world that all land originally belongs to the government. Private ownership can be established only by grant from the government, which may include in the grant such reservations as it sees fit. Even the Indians who had been living on the land for centuries before the arrival of the British were not regarded as having any title to it, but only a right of occupancy which could be extinguished by payment of just compensation.

The British were therefore on sound legal ground in the measures which they took to safeguard the vital supply of mast timber. The colonists, on the other hand, felt that they had a moral if not a legal right to the use of natural resources on public lands. After all it was they, not the British government, who were bearing the risks and hardships of subduing the wilderness and of utilizing its inexhaustible resources for the economic development of the Colonies. This conflict in attitude led to a continuing and often bitter struggle between the king's officers who were trying to enforce the laws and the colonists who were trying to evade them.

Efforts at Enforcement. As a first step toward the implementation of its colonial forest policy, the British government in 1685 appointed Edward Randolph as Surveyor of Pines and Timber in Maine. Three years later he reported, how accurately is open to question, that he had been successful in checking the devastation that had already begun in that region. More vigorous action was taken by John Bridger, who in 1705 was made Surveyor General of Her Majesty's Woods and Forests in America. In this capacity he was instructed to prevent trespass on the Queen's woods and to obtain acts by colonial legislatures in support of the Broad Arrow policy.

Efforts by the Surveyor General and his deputies to enforce the Broad Arrow policy met with vigorous opposition from the colonists. When timber illegally cut was confiscated by British officials, it would be reseized by the loggers or millmen and sawed into boards and planks just under the damning 24 inches in width. Many such boards can still be seen in New England houses. If condemned timber were offered for sale, no one would buy it. Timber agents sometimes received physical mistreatment, such as being ducked in a mill pond, and occasional threats were made against their lives.

"Pine Tree Riot." William Little in his "History of Weare, New Hampshire," gives a vivid description of the "Pine Tree Riot" which took place there in 1772. When Sheriff Benjamin Whiting called on one Ebenezer Mudgett to arrest him for making free with the king's white pine, he was told that bail would be put up the next morning. After spending a busy evening with his friends,

Mudgett went to the inn at dawn, woke the sheriff, burst into the room and told him bail was ready. Whiting rose, chid Mudgett for coming so early, and began to dress. Then more than twenty men rushed in, faces blacked, switches in their hands, to give bail. Whiting seized his pistols and would have shot some of them, but they caught him, took away his small guns, held him by his arms and legs up from the floor, his face down, two men on each side, and with their rods beat him to their hearts' content. They crossed out the account against them of all logs cut, drawn and forfeited, on his bare back, much to his great comfort and delight. They made him wish he had never heard of pine trees fit for masting the royal navy. Whiting said: "They almost killed me."

Quigley, his deputy, showed fight; they had to take up the floor over his head and beat him with long poles thrust down from the garret to capture him, and then they tickled him in the same way.

Their horses, with ears cropped, manes and tails cut and sheared, were led to the door, saddled and bridled, and they, the king's men, told to mount; they refused, force was applied; they got on and rode off down the road, with jeers, jokes and shouts ringing in their ears.

Failure of Policy. Eight of the offenders were later caught and brought to trial. Each was fined 20 shillings and the cost of prosecution. This light penalty indicated clearly the court's sympathy with the offenders, a reaction that was characteristic of colonial courts generally and that made enforcement of the law doubly difficult. When to the physical difficulty of ferreting out and apprehending offenders was added the hostility of their friends, neighbors, and the courts, the agent's task of protecting the mast timber became impossible. The situation was one that has been duplicated in every frontier region in the world.

The Broad Arrow policy was a forward-looking attempt to assure Great Britain's navy and merchant marine an adequate, continuing supply of essential mast timber. From the British point of view it was thoroughly justified on both legal and practical grounds. It failed dismally in practice because of widespread and vigorous colonial opposition over which the British were unable to exercise effective control. Concrete evidence of its failure is offered by the statement of Robert Armstrong, one-time Surveyor General, that a single survey in 1700 disclosed 25,000 logs two-thirds of which exceeded 24 inches in diameter and that during the period from 1701 to 1721 for every mast sent to England on contract for the navy 500 pines suitable for masts had been cut for other purposes.

Naval Stores and Oak Ship Timbers. Closely related to the Broad Arrow policy were the efforts of the British to increase imports of naval stores and of oak ship timber from the Colonies. In addition to the encouragement of naval-stores production by the bounties and educational activities already mentioned, the act of 1704 imposed a penalty of £5 for cutting or destroying a pitch pine or tar tree less than 12 inches in diameter at 3 feet from the ground, not within a fence or actual enclosure, in New Hampshire, Massachusetts Bay, Rhode Island, Connecticut, New York, and New Jersey. It also imposed a penalty of £10 for purposely setting fire to any woods in which there were trees prepared for the making of pitch or tar, without first giving notice to the person who had prepared the trees. The yield from the northern pines, was, however, insufficient to justify intensive efforts to obtain naval stores from that region, and reliance was soon placed on bounties to yield adequate supplies from the South.

British shipbuilders and the British navy always had a strong preference for native oak. Although timber from the Continent and from America had to be used to some extent to supplement the local supply, it was regarded as of inferior quality. Nevertheless, in 1696 the Navy Board sent John Bridger with three other commissioners to New England to investigate among other things the possibility of increasing the supply of ship timber. One of his first acts was to dispatch a cargo consisting largely of oak to Deptford, where the dockyard inspecting officials reported that "the wood in general is of very tender and 'frow' substance mingled with red veins and subject to many worm holes with signs of decay." Another shipment in 1700 also met an unfavorable reception, and both oak timber and American ships made of oak continued in bad repute in England.

This situation existed in spite of the fact that shipbuilding was one of the chief industries in the American Colonies and that American ships, in which oak played a prominent part, gave complete satisfaction to their owners. Perhaps the explanation of the paradox lies in the fact that the colonists kept the best material for their own use and that ships built for sale were often constructed hastily of green timber, while they used better wood and built more carefully for themselves.

Forests and Independence. Timber may well be added to "tea, taxes, and tyranny" as one of the causes of the American Revolution. For nearly a century the irritation caused by the Broad Arrow policy and attempts at its enforcement had been rankling in the breasts of the colonists. There can be no doubt that the timber agents were among those whom the signers of the Declaration of Independence had in mind in the complaint that "he has erected a multitude of new offices and sent hither swarms of officers to harass our people, and eat out their substance." The forest policy of the British was one of the factors that had conditioned the Colonies for revolt.

That policy not only helped to bring on the Revolution but also to contribute to its successful outcome. With the outbreak of war, trade between the two countries was interrupted, and the last shipment of masts from the Colonies reached England in July, 1775. At that time the navy had on hand approximately a three-year supply of mast timber, for which it was then almost wholly dependent on the American Colonies. In spite of the obvious wisdom of taking immediate steps to rebuild the trade with the Baltic countries, this was not done, presumably because of the mistaken belief that the war would not last long enough to be of serious consequence.

As a result of this failure to replenish the mast supply, the British navy soon found many of its ships in disrepair and failed to give the usual excellent account of itself in its encounters with the French navy. In the summer of 1778 the fleet under Admiral John Byron, which was trying to intercept an expedition under D'Estaing bound for America, was dispersed by a severe storm and many of its ships badly mauled. The damage to the "Invincible" as described by her captain may be taken as typical. "The mainmast was sprung close to the gundeck, so much that it was expected to go with every roll. We cut it away on the quarter deck to prevent the upper deck from being torn up. . . . The foremast went in three pieces, all of which fell on the forecastle and wounded several men but only one killed. The bowsprit being sprung some time before and now so bad, we had to cut part of that away." In contrast with the English experience, D'Estaing brought his fleet safely across the Atlantic.

Trouble resulting from inadequacy of the mast supply continued to dog the British throughout the war. Admiral Augustus Keppel, on being court-martialed for retreating before a superior French fleet in 1778, offered as his defense the complete lack of masts and naval stores in the dockyards. The climax came in 1781 when Admiral Graves, because of the many sprung masts and leaky hulls in his fleet, failed to force an entrance to Chesapeake Bay through the French fleet and returned to New York for repairs. When he finally sailed a month later, with 7,000 troops under General Clinton for the relief of Cornwallis, the surrender at Yorktown had already taken place.

Forests also helped to prepare the Colonies for economic as well as political independence. Ships, lumber, staves, heading, and shingles were among the products of some of the first and most important manufacturing industries in the Colonies and formed the basis for much of its international trade. From pine, oak, and other trees, and from the industries dependent on them, came in large measure the economic strength that enabled the Colonies to stand on their own feet both before and after the Revolution.

Summary. To the American colonist, the omnipresent forest came to be regarded as an inexhaustible resource that was both an asset and a liability. It provided abundant raw material for the building, equipping, and heating of his home and for the manufacture of goods that constituted an important part of his international trade; but it was also a nuisance that had to be cleared away with much effort to make room for farms, settlements, and roads. In spite of its general inexhaustibility, most of the forest was inaccessible because of lack of transportation facilities, and numerous acts were accordingly passed to protect it from fire and trespass. That these laws were generally enforced or affected materially the utilization of the forest is unlikely.

To the British government, the American forest was a source of lumber, masts, and other timber products to be imported in exchange for exports of manufactured goods. Controls over shipping, tariffs, and bounties were used to direct trade into channels that would be most profitable from the British point of view. Special attention was paid to maintaining a continuous supply of mast timber from the Colonies. The Broad Arrow policy which was adopted in 1691 for this purpose not only failed to save much mast timber but aroused great irritation on the part of the colonists. It was in part the cause both of the outbreak of the Revolution and its successful outcome.

REFERENCES [1]

Albion, Robert G.: "Forests and Sea Power," chaps. 6–7, Harvard University Press, Cambridge, Mass., 1926.

Cameron, Jenks: "The Development of Governmental Forest Control in the United States," chaps. 1–2, Johns Hopkins Press, Baltimore, 1928.

Kinney, Jay P: Forest Legislation in America Prior to March 4, 1789, *N.Y. (Cornell) Agr. Expt. Sta. Bul.* 370, pp. 358–405, 1916.

Lillard, Richard G.: "The Great Forest," part II, Alfred A. Knopf, Inc., New York, 1947.

---

[1] The references at the end of each chapter apply specifically to the material contained in that chapter. A more general, but still selective, list of references on the broad subject of forest and range policy is given in Appendix 3.

CHAPTER 2

Acquisition and Disposal of the Public Domain

The United States began its career as an independent nation with a population of less than four million people confined chiefly to the Atlantic Seaboard. West of the Allegheny and Appalachian Mountains lay a vast expanse of primeval forest, untouched by the plow or the ax, largely unexplored, and uninhabited except by a relatively few Indians. The first century of the new nation's existence had to be devoted to the subduing of an untamed wilderness and the conversion of its natural resources into usable wealth. Once again, the forests were both a liability and an asset—an obstacle that had to be removed to make way for settlement, but increasingly a valuable source of indispensable raw materials as population, industry, and trade expanded.

Cessions by the States. After the Revolution, the thirteen states claimed ownership of all the lands not already in private ownership to which the British had formerly held title. Massachusetts, Connecticut, New York, Virginia, North Carolina, South Carolina, and Georgia laid claim to large areas of "western lands" on the basis of their colonial boundaries. These boundaries were generally vague and often overlapping.

Even before the end of the war, the other six colonies, under the leadership of Maryland, started a movement to force the cession of these western lands to the central government. The objective was threefold: to bring about greater equality in land resources among the several states, to provide tangible assets with which to meet the overwhelming national debt, and to strengthen the Union by the increased feeling of political and economic solidarity which it was believed would be fostered by common ownership of such an extensive and potentially valuable territory. The Continental Congress in 1780 resolved that lands ceded to the United States should be used for the common benefit of all the states. Promises of land bounties to induce enlistments in the Continental Army had been made as early as 1776 in the expectation that land for the purpose would be available.

The proposed cessions were made by all seven states during the period from 1781 to 1802. New York, which had by far the smallest area at stake, was the first state to take action and Georgia the last. Four of the states accompanied their cessions by certain reservations, of which Connecticut's "western reserve" and Virginia's "military lands," both in Ohio, are the best known. North Carolina's cession of the area which is now the state of Tennessee was so encumbered with reservations and claims that the United States in 1806 and 1846 relinquished whatever rights it might have had under the original grant.

Altogether the cessions comprised a total area of 233 million acres and embraced the present states of Ohio, Indiana, Illinois, Michigan, and Wisconsin, that part of Minnesota lying east of the Mississippi River, and all of Alabama and Mississippi lying north of the 31st parallel of latitude. They made the Federal government the owner of all the land not already in private ownership within this territory.

Other Accessions. The next addition to the public domain came in 1803 with the Louisiana Purchase from France. In view of the strategic importance and natural wealth of the territory involved, this purchase of 523 million acres at a cost of less than 5 cents per acre was a bargain of the first water. In 1819 came the Florida purchase of 43 million acres from Spain for about 16 cents an acre. The "Oregon compromise" of 1846 added without cost another 181 million acres in the Pacific Northwest and settled by peaceful means an issue that had threatened to lead to war with Great Britain. In 1848 the treaty ending the war with Mexico transferred to the United States nearly 335 million acres in the Pacific Southwest in return for a payment of $16,300,000, or about 5 cents per acre.

In 1850, Texas, which had retained title to all the public lands within its borders on its admission to the Union in 1845, sold to the United States an irregularly shaped piece of land comprising nearly 79 million acres in what are now the states of Oklahoma, Kansas, Colorado, Wyoming, and New Mexico for nearly 20 cents per acre. Another 19 million acres on the southern borders of Arizona and New Mexico were acquired in 1853 by the Gadsden Purchase from Mexico at a cost of about 53 cents per acre. Finally came the purchase from Russia in 1867 of more than 365 million acres in Alaska for a little over 2 cents per acre.

Original Public Domain. These various cessions and acquisitions brought into existence the "original public domain," which has played such an important part in virtually every phase of the development of the United States. It includes all land, exclusive of that already in private ownership, acquired by the United States by the cessions, purchases, and treaties described above. It thus includes all land subject to disposal under the general land laws, but does not include land acquired for specific purposes such as post-office or customhouse sites, military reservations, or national forests purchased under the Weeks Act. Lands obtained in exchange for public lands or timber thereon are subject to legislation applying to the original public domain.

The public domain has always been subject to complete control by the Congress of the United States under Article IV, Section 3, Paragraph 2 of the Constitution: "The Congress shall have Power to dispose of and make all needful Rules and Regulations respecting the Territory or other Property belonging to the United States." This paragraph has repeatedly been interpreted by the Supreme Court as giving Congress full power to sell, give away, or retain the public domain for such purposes and by such methods as it sees fit.

Only about 411 million acres, or 29 per cent of the original public domain in the states, is still in Federal ownership. This area is divided into the two classes of "unreserved" areas, which are still open to disposal under the general land laws, and "reserved" areas, which are not currently open but the status of which may at any time be changed by Congress, or in most cases by the Executive. "Public domain" and "public land(s)" are commonly used synonymously, and both terms are occasionally limited to unreserved lands. Their meaning often has to be determined from the context in which they are used.

Table 1 presents the major facts concerning the location, area, and initial cost of the original public domain. It shows that in seven decades the United States has swept across the continent, so that by the middle of the nineteenth century it controlled the entire 1,904 million acres of land lying between the Atlantic and the Pacific, Canada and Mexico. More than three-fourths of this vast area was at one time or another in Federal ownership as part of the public domain—a veritable empire acquired by an expenditure of about the present appraised value of the real estate in an average city of 50,000 inhabitants. The public lands contained agricultural, forest, range, wildlife, and mineral resources of incalculable value. The forest and range policy of the country, like so many of its other policies, is inextricably interwoven with their disposal and management.

Disposal of the Public Domain.[1] What to do with the public lands was one of the most difficult and most persistent of the many thorny problems faced by the young nation. For years it was the occasion for more oratory and more legislation than any other single subject. On one point only was there general agreement—that title to the lands should not be

[1] This book deals chiefly with public lands in the continental United States exclusive of Alaska. The Alaskan lands have for the most part been handled under special legislation differing in many respects from that applicable elsewhere.

retained permanently by the Federal government but should be passed to states and to private owners as rapidly as was consistent with their orderly development. Nearly a century was to elapse before there was any substantial advocacy of the modern practice of combining Federal ownership and private utilization.

On all other points there was heated controversy in which broad national policies and interregional rivalries played a prominent part. Decisions with respect to the handling of the public domain became deeply involved with such issues as states' rights, Federal aid, slavery and free soil, free land, the tariff, and relations between labor and capital.

The early debates were concerned chiefly with the purpose and terms of disposal. For example, should the lands be given away to promote settlement, perhaps at the expense of the thirteen states, or sold to provide revenue; or could both objectives be attained? Should survey proceed or follow disposal? Should it be by the rectangular system or by metes and bounds? What price, if any, should be charged, and should sales be made only for cash or on credit? Should there be any maximum or minimum limit on the area which one person could acquire? Should sales be made at the seat of the government or at local land offices?

Table 1. Acquisition of the Original Public Domain

					Per cent of total	
Origin	Year	Land area, million acres	Total cost, million dollars	Average cost, cents per acre	Public domain	Land area of United States
State cessions 1781–1802		233	$ 6.2*	2.7	16	12
Louisiana Purchase ..	1803	523	23.2	4.4	36	27
Red River Basin †		29	....	...	2	2
Florida purchase	1819	43	6.7	15.6	3	2
Oregon compromise ..	1846	181	....	...	13	10
Mexican cession	1848	335	16.3	4.9	23	18
Texas purchase	1850	79	15.5	19.6	6	4
Gadsden Purchase ...	1853	19	10.0	52.6	1	1
Totals		1,442‡	$77.9	5.4	100	76
Alaska purchase	1867	365‡	$ 7.2	2.0		

* Payment to Georgia.

† Drainage basin of the Red River of the North, south of the 49th parallel. Authorities differ about the method and date of its acquisition. Some hold that it was a part of the Louisiana Purchase; others, that it was acquired from Great Britain.

‡ Inland waters comprise an additional area of 20 million acres in the continental United States and 10 million acres in Alaska.

SOURCE: U.S. Department of the Interior, Bureau of Land Management.

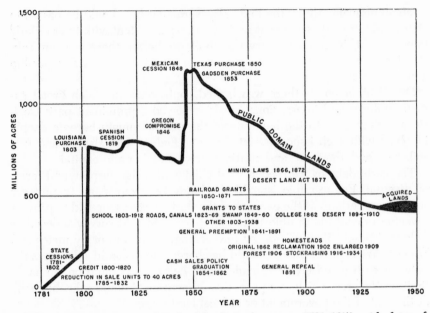

Fig. 1. Approximate area of Federal lands in the states, 1781–1950, with dates of important acquisitions and laws providing for their disposal. (*Courtesy U.S. Department of the Interior, Bureau of Land Management.*)

General Sales. The first decision on these points came with the enactment by the Continental Congress of the ordinance of 1785. That ordinance established the rectangular system of public-land surveys with townships 6 miles square divided into 36 sections of 640 acres each. It provided that after survey the lands would be sold at auction for cash to the highest bidder at not less than $1 per acre. Half of the townships were to be sold entire, the other half in sections, with the two types alternating in checkerboard fashion. Section 16 was reserved for common-school purposes, and four other sections were reserved for later disposal by the government. Reservation was also made of one-third of all gold, silver, and copper mines, to be sold or otherwise disposed of as Congress might direct. Administration of the ordinance was placed in the Board of Treasury.

This legislation aimed both to promote settlement and to provide revenue. Unfortunately the financial hopes which attended its passage were doomed to disappointment. The first patent was issued in 1788, but sales were slow. Even $1 per acre was high in comparison with the price at which lands could be bought from the states, and in addition few immigrants could raise the $640 in cash which was the minimum needed to purchase a piece of government land. A provision in the ordinance of

1787 dealing with the government of the Northwest Territory that one-third of the sale price could be paid in cash and the balance in three months did little to relieve this situation. Immigrants were, however, "squatting" on the land in considerable numbers, without payment and of course without obtaining legal title, in spite of the prohibition against the practice.

Another means of disposing of the public lands during the early period was by private sale in large blocks to companies and individuals. The theory behind this procedure was that substantial immediate income could be obtained by selling several townships to a single buyer at considerably reduced rates and that the buyer would then actively promote the resale of the land to actual settlers. The practice, however, involved so much speculation and other difficulties that it was soon abandoned,

Credit Sales. The system of land disposal embodied in the ordinance of 1785 proved so unsatisfactory in practice that it was completely revamped by acts passed in 1796, 1800, and 1804. These acts raised the minimum price of public lands to $2 per acre; provided for the payment of the purchase price in installments over a four-year period; reduced the minimum area that would be sold to 160 acres; limited lands reserved from sale to Section 16 and to sections containing salt springs, both of which were reserved for educational purposes; established the system of local land offices; and declared all navigable streams within the territory covered by the acts to be public highways.

Endless difficulties followed the adoption of the installment plan of payment for public-land purchases. Delinquency was the rule rather than the exception. Theoretically the government could evict a delinquent and confiscate his land, but frontier sentiment made this step a practical impossibility even if the government had been disposed to adopt so drastic a course of action. Many relief acts were required to take care of settlers whose intentions were honest and who were obviously an asset to the country but who simply could not meet the obligations they had assumed. Then there were others with less honest intentions whose sole purpose in making a first payment was to keep off competitors and to legalize their presence on the land for a few years while they could exhaust the fertility of the soil or harvest the timber for commercial purposes before abandoning it and moving on to repeat the process elsewhere. Sales increased materially, especially in Ohio, but payments lagged. By 1820 delinquent purchasers owed the government more than 21 million dollars.

Return to Cash Sales. The next revision, which came in 1820, established the general sales policy in almost its final form. The installment plan was abolished, and both the minimum price and the minimum **area** were reduced. Thereafter public lands were to be sold at auction to the highest bidder in half-quarter sections of 80 acres at not less than $1.25 per acre, with full payment at time of sale. The minimum area was further reduced in 1832 in private sales to 40 acres, which has since remained the standard unit of measure in the sale of public lands, just as $1.25 per acre remained the standard base price until 1934. No classification was made of the lands offered for sale, and no restriction was placed on the maximum area that might be purchased by one person.

Sales, which had been running well over a million acres a year during the period from 1814 to 1819, fell off considerably during the ten years following passage of the act of 1820. This was doubtless due in part to the requirement of full payment at the time of sale, but probably even more to the financial situation resulting from the panic of 1819. During the 1830's sales again boomed, largely as a result of speculative buying, until in 1836 they reached an all-time high of more than 20 million acres with receipts of more than 25 million dollars.

Effect of Specie Circular. In that year the Treasury Department issued its famous Specie Circular instructing local land-office officials to accept only gold and silver in payment for public lands, except for actual settlers buying not more than 320 acres. President Jackson later stated that the order was necessary to stop wildcat bank inflation in the West on which speculation was feeding, and recommended that the public lands be sold only to actual settlers for immediate settlement and cultivation. The bursting of the bubble of inflation and speculation led to the severe panic of 1837.

Sales of public lands fell off by more than 70 per cent to 5,600,000 acres in 1837 and by more than 90 per cent to 1,165,000 acres in 1841. They never again ran as high as 5,000,000 acres a year except for the three years following the passage of the Graduation Act of 1854, which materially reduced the price of land that had been on the market for ten years or more. There was a brisk but short-lived demand for land at the reduced prices, presumably because the best bargains were snapped up promptly. Although the act attempted to limit sales to actual settlers, this provision was easily evaded, and it is probable that the bulk of the purchases were made by speculators. Of the 77,600,000 acres subject to sale under the act, some 25,700,000 acres were actually sold at an average price of about 32 cents an acre. The act was repealed by an act of June 2, 1862, which logically put a stop to sales at reduced prices after passage of the Homestead Act.

"Squatting." "Squatting" on public lands was a practice that originated in colonial times and continued with increasing frequency after the Revolution. Its justification was found in the slowness of public owners to open resources to utilization and in the apparent inexhaustibility of these resources.

Although a few members of Congress championed the practice from the very beginning, it was slow to receive legislative approval. The majority felt that it should be forbidden, in the interest of orderly procedure, to obtain maximum prices for the land, and to avoid the complications certain to arise if Indian lands were taken up prior to extinguishment of the Indians' right of occupancy. When it became evident that little attention was being paid to the provisions in the early land legislation prohibiting settlement prior to survey and sale, Congress in 1807 adopted an antitrespass law with teeth in it. This law authorized the President of the United States to direct the marshal to take such measures, including the employment of military force, as he judged necessary and proper to remove anyone who should take possession of or attempt to settle on the public lands until authorized by law.

In practice the law did little to stop squatting. Surveys and sales proceeded so slowly that the Western lands had to be occupied illegally if they were to be occupied at all. Although the act of 1807 was never formally repealed, Congress recognized the facts of frontier life by repeatedly passing special laws which in effect legalized preemption in specific cases. On the frontier, the squatters were not classed with other lawbreakers but were referred to as "a very respectable class of citizens," "a sturdy class of pioneers," "the hardy yeomanry," and "meritorious and industrious citizens."

In 1828 the Public Lands Committee of the House of Representatives expressed the view that squatting was both inevitable and desirable. It pointed out that the pioneer settler, although technically a trespasser, was actually a benefactor, not a malefactor, whose enterprise and contribution to the development of the country should be rewarded by permitting him to buy without competition the land on which he had settled. An illustration of this enterprise was the presence in 1838 in what is now Iowa of an estimated twenty to thirty thousand squatters on land that had never been offered for sale.

Claim Associations. So strong was the sentiment in favor of squatters that "claim associations" were often organized for their protection. These associations had their own rules and regulations for seeing that justice was done as between legitimate claimants and intruders and also as between rival claimants. When a sale was to be held, the association would register on a plat the location of each claim and the name of the claimant and would appoint a "bidder" to bid off the lands so registered in the names of the respective claimants. The Commissioner of the General Land Office in 1836 complained that receipts from sales of public land had been cut down by some millions of dollars by these "unlawful organizations."

Violators of "claim law" received harsh treatment, often being subjected to physical violence when moral suasion failed to induce repentance. B. H. Hibbard, in "The History of Agriculture in Dane County, Wisconsin," tells the story of one claim jumper on whom a committee of the local claim association

exhausted their verbal arguments in vain, then putting a rope around the waist of the culprit, led him to a pond, cut a hole in the ice, and immersed him. He was soon drawn out, but, being still in a combative and profane frame of mind, was treated to another ducking and on his second coming out was unable to continue his side of the debate, so the negative was declared closed, and after returning to the house, the dripping defender of that side set his signature to the papers and with uplifted right hand swore that it was his "voluntary act and deed."

Experience in the settlement of the agricultural lands, and also in the utilization of the other resources of the public domain, has demonstrated conclusively that laws must have popular support to be effective. A vigorous people bent on occupying a virgin territory of fabulous richness will not be deterred by legislation unsuited economically, socially, or psychologically to the conditions with which it attempts to deal. An editorial in the *Chicago Democrat* of June 4, 1835, referring to an approaching sale of the public lands, voiced the spirit of the times:

"Public opinion" is stronger than the law, it has been well said, and we trust it may prove so in this case, and that the strangers who come among us, and especially our own citizens, will not attempt to commit so gross an act of injustice as to interfere with the purchase of the quarter section, on which improvements have been made by the actual settler. We trust for the peace and quietness of our town that these local customs, to which long custom has given the force of law . . . which have been repeatedly sanctioned by the general government . . . and which are so strongly sustained by the principles of justice and equity, will not be outraged at the coming sale.

It is a safe guess that they were not.

Preemption Sales. Growing pressure to legalize preemption led in 1830 to passage of an act granting preemption rights for one year to settlers on the public lands. Such settlers on proof of settlement or improvement might purchase not more than 160 acres at the minimum price of $1.25

per acre. The act was not to delay the regular sale of the public lands, and preemption rights were not transferable.

Approval of preemption as a basic policy came with passage of the act of September 4, 1841, dubbed by Senator Benton of Missouri the "Log Cabin Bill." It authorized every head of a family, widow, or single man over twenty-one years of age, who was a citizen of the United States or had declared his intention to become a citizen, to settle upon and purchase at $1.25 per acre not more than 160 acres of surveyed, unoccupied, unreserved, nonmineral public lands, subject to certain restrictions. Preemptors must inhabit and improve the land and erect a dwelling thereon. They must swear that the land was being taken up for their own exclusive use and benefit, and any assignment of the preemption right prior to the issuance of patent was null and void.

As might have been anticipated, it proved impossible to limit preemption to surveyed lands. In 1853 and 1854 the privilege was therefore extended to unsurveyed lands in six states, and in 1862 in all states.

The Preemption Act was a victory for the West,[1] which had consistently struggled to curb speculation and to facilitate actual settlement of the public lands. It was a defeat for the East, which feared that preemption would reduce government receipts, would lower the price of lands already sold, and by attracting immigrants to the West would deplete the supply of labor and increase wages in the East. The act placed the emphasis on settlement rather than revenue as the primary objective in the disposal of the public domain. It also aimed to favor the settler as against the speculator and to encourage the establishment of many small farms.

Unfortunately, widespread use was made of the act for the fraudulent acquisition of lands not primarily valuable for cultivation. In 1882, for example, the Commissioner of the General Land Office stated that it was "one of the causes of frauds in land entries which have approached great magnitude. . . . A material proportion of the preemption entries now made are fraudulent in character, being chiefly placed upon valuable timber or mineral lands, or water rights, and made in the interest and by the procurement of others, and not for the purpose of residence and improvement by the professed preemptor." In response to repeated urgings, Congress in 1891 finally put a stop to preemption, the need for which had disappeared with the passage of the Homestead Act in 1862.

Demand for Free Land. Efforts to obtain public lands free of charge date back to 1797, when settlers along the Ohio River asked for a grant of 400 acres per family in return for an agreement to remain on the land for three years before receiving title. From then on, petitions for donations reached Congress irregularly but with increasing frequency. In 1828

---

[1] The "West" at this time included the country west of the thirteen original states.

the Public Lands Committee of the House, in the same report in which it urged legalization of preemption, recommended the grant of 80 acres "to the heads of such families as will cultivate, improve, and reside on the same for five years." Four special "donation" acts, passed between 1842 and 1854, resulted in the grant of about 3,100,000 acres to frontier settlers in Florida, Oregon, Washington, and New Mexico, with certain restrictions as to residence and cultivation.

Free land had always been favored in the West, but at first was generally opposed in the East, particularly in the South. As the population of the public-land states increased, the pressure on Congress also increased, so that during the twenty years preceding the outbreak of the War between the States the controversy developed into one of the hottest political issues of the day. The West now received strong support from the Northeast, where increased Federal receipts from the tariff and an increased labor supply had largely removed the region's former objections to cheap, and still more to free, land. Rapid development of the West now became a desirable objective as a means of expanding the market for Eastern manufactures. Labor leaders favored free land because it offered every man an opportunity to make a living by his own efforts and because it would encourage westward migration and thereby tend to avoid an oversupply of labor in the industrial East. The South, on the other hand, remained unalterably opposed because of the well-founded conviction that the plantation system, which was dependent on slave labor, could not compete with the system of small farms operated by their owners.

Although the terms "free land" and "free soil" are not synonymous, the ideas for which they stand became inextricably interwoven. The Free Soil Party, in its 1848 platform, resolved that "the free grant to actual settlers . . . is a wise and just measure of public policy which will promote, in various ways, the interest of all the states of the Union." In 1852, the Free Soil Democrats declared "that the public lands of the United States belong to the people, and should not be sold to individuals nor granted to corporations, but should be held as a sacred trust for the benefit of the people, and should be granted in limited quantities, free of cost, to landless settlers."

At the same time the South was urging the annexation of Cuba as a means of expanding the area under the sovereignty of the United States open to slavery. This situation led to an interregional clash in Congress, with the North and West seeking "land for the landless" and the South seeking "slaves for the slaveless." Free land won out, but only after a bitter fight. Land reformers, free-soilers, and Horace Greeley (who was as strong for homesteading as he had been against preemption) failed to obtain favorable action from Congress until 1860, when the Republicans succeeded in forcing through a measure that was promptly vetoed by

President Buchanan. Public-land policy became an important issue in the political campaign of 1860, and Representative Owen Lovejoy of Illinois later declared that Lincoln would not have been elected without the pledge of the Republican party to support the homestead bill.

Homestead Grants. The new Congress, in the absence of most of its members from the Southern states, promptly and overwhelmingly passed the Homestead Act of May 20, 1862. That act provided that any person who was the head of a family or over twenty-one years of age and who was a citizen of the United States or had declared his intention to become a citizen might enter upon not more than 160 acres of nonmineral land subject to sale at a minimum price of $1.25 per acre or not more than 80 acres subject to sale at a minimum price of $2.50 per acre. The latter category included alternate sections retained by the government in grants to assist in the construction of railroads and other permanent improvements. Free patent could be secured by the settler on payment of certain fees and on proof that he had resided upon and cultivated the land for five years. Commutation, or purchase of the land at its regular price, was possible at any time after six months from the date of filing.

Several later acts granted Union soldiers more favorable treatment than other citizens in the exercise of their homestead rights, but land bounties for veterans, although vigorously advocated, were not approved. The War between the States thus marked the end of the military land-bounty policy.

The Homestead Act has sometimes been characterized as embodying the most progressive land-disposal policy ever adopted anywhere in the world. Whether or not it deserves this high praise, it was certainly effective in speeding up the development of the West on terms that opened its agricultural resources to poor and rich alike. As long as the supply lasted, it was literally true that anyone with the necessary skill and stamina could get himself a farm. The act was, however, based on experience in the humid parts of the country and was not suited to the semihumid conditions in the prairie states and the Far West, where 160 acres was inadequate to support a family.

Liberalization of Homestead Act. This defect was tardily remedied by two acts passed in 1904 and 1909. The first act increased the maximum size of homesteads in western Nebraska to 640 acres and at the same time required the construction of permanent improvements to the extent of not less than $1.25 per acre. Then in 1909 the Enlarged Homestead Act made it possible to acquire homesteads of 320 acres anywhere in the nine Western states and territories of Arizona, Colorado, Montana, Nevada, New Mexico, Oregon, Utah, Washington, and Wyoming, but forbade commutation. The lands entered must be nonmineral and nonirrigable and must contain no merchantable timber. The provisions of the act were later extended to Idaho, Kansas, North Dakota, South Dakota, and California.

Experience also showed that in the semiarid regions five years constituted a pretty long period of residence to require of the settler. Critics pointed out that the Homestead Act was in effect a wager in which the government bet a man 160 acres that he could not live on it for five years and that the government was too often the winner. Congress finally recognized this situation by passage of the Three-Year Homestead Act of June 6, 1912, which provided that patent could be obtained after three years of residence covering at least seven months per year. The act also set minimum cultivation requirements.

Two other acts dealing with homesteading should be mentioned here, but will be discussed in detail in later chapters. These are the Forest Homestead Act of 1906, which opened up certain areas in forest reserves to homesteading, and the Stockraising Homestead Act of 1916, which permitted the homesteading of areas primarily valuable for grazing rather than for cultivation.

Abuses. One of the most abused features of the Homestead Act of 1862 was the commutation provision which permitted an entryman to purchase his claim at any time after six months for $1.25 per acre. This provision was commonly used, particularly after 1880, for the acquisition of lands primarily valuable for timber, minerals, or forage, and also for the consolidation into single ownerships of agricultural lands suitable for the production of crops such as wheat, which can most advantageously be grown in large holdings.

The general revision act of 1891 extended the period that must elapse before commutation could take place to fourteen months, but this was not long enough to stop the practice. For example, in the single state of North Dakota, in the ten years between 1900 and 1910, 5,781,000 acres were commuted as against 5,614,000 acres on which final proof was made. The problem of how to prevent abuse of the privilege, without being unfair to the actual settler who had a legitimate reason for wishing to purchase before the end of the period required to obtain free title, was never solved.

As in the case of the Preemption Act, fraud was often perpetrated in the acquisition of public lands under the Homestead Act. A "twelve-by-fourteen" house might be a dry-goods box, with the dimensions measured in inches instead of feet; a "shingle roof" might consist of two shingles; a house that from a little distance looked habitable might have no floor or a wooden chimney. "Dummy" entrymen could be used to block up holdings of considerable size. There is no way of knowing how many million acres of valuable timber and mineral land were acquired on the basis of affidavits swearing that they had been lived on and improved for agricultural purposes, but the figure must be a sizable one. Land-office officials were too few and far between for effective enforcement of the laws, and public opinion commonly tolerated or even supported the illegal practices. It was not until the 1890's and early 1900's that the timber thief came to be generally regarded as a criminal.

Results. Some 285,000,000 acres were patented under the Homestead Act. This area constitutes nearly a fifth of the original public domain and more than a fourth of the total area sold or given to individuals, corporations, institutions, and states. The homestead policy was on the whole one of the few really bright spots in legislation concerning the public domain and did much to develop the country in a democratic way.

Timber-culture Grants. Settlers on the fertile soils of the Middle Western plains and prairies were seriously handicapped by lack of wood and water. Trees were scarce except along the river bottoms, so that practically all of the wood needed for construction purposes had to be imported, largely from the Lake states, and precipitation was often so low as to make farming without irrigation a hazardous enterprise. Tree planting was advocated as a means of meeting both needs. Forests would obviously produce wood, and at that time it was generally believed that they would materially increase rainfall.

Commissioner Wilson of the General Land Office, in his annual report for 1866, suggested that tree planting be required of all homesteaders in localities where there is a scarcity of timber. Two years later, he stated: "If one-third the surface of the great plains were covered with forest there is every reason to believe the climate would be greatly improved, the value of the whole area as a grazing country wonderfully enhanced, and the greater portion of the soil would be susceptible of a high state of cultivation." "Multipurpose" forestry, about which we hear so much today, seemed the way to get more wood and more water.

Congress rose to the occasion by passing the act of March 3, 1873, which offered to donate 160 acres of public land to any person who would plant 40 acres of it to trees, not more than 12 feet apart each way (302 per acre), and keep them in a growing and healthy condition by cultivation for a period of ten years. Neither residence nor other improvements were required. Tree planting by homesteaders was also encouraged by permitting them to receive patent at the end of three years on the submission of satisfactory proof of having had under cultivation in this way for two years 1 acre of trees for each 16 acres in the homestead claim. In other words, timber culture for two years was accepted in lieu of the final two years of residence and cultivation normally required under the Homestead Act.

Modification of Requirements. Amendments to the act were almost immediately found necessary. The offer to grant land to "any person" was too liberal, while the requirement that the entire 40 acres must be planted in one year was too strict. In 1874 Congress therefore limited the privilege to heads of families or persons over twenty-one years of age who were citizens or had declared their intention to become citizens; extended the period of planting to four years; and reduced the period of cultivation to eight years. The total area that could be acquired by any one person was limited to 160 acres.

A final amendment in 1878 reduced the area that must be planted to one-sixteenth of the area entered. It also increased the number of trees to be planted to 2,700 per acre, of which 675 had to be living and thrifty at the time of final proof. An extension of one year in the time of cultivation and planting was allowed for each year the trees were destroyed by grasshoppers or drought.

Failure to Attain Objectives. Although there was no geographic limitation in the act of 1873, its title, "To encourage the growth of timber in the Prairie States," indicates the region where Congress expected it to be used. Tree planting here is a difficult and expensive undertaking which the early settlers had neither the knowledge nor the money to handle successfully. Parts of the region are too dry for forest growth, and even where conditions are favorable successful planting demands technical skill of high order, as the experience of the Federal government during the last twenty years in its shelterbelt program amply demonstrates. The policy was, therefore, doomed to failure from the outset.

Even if the plantations had thrived, it is doubtful whether they would have had any significant effect on precipitation. On the other hand, there is no question as to the beneficial effect of tree growth in reducing wind velocity, checking evaporation, providing shelter for man and beast, and in general adding to the "livableness" of a region. Had the establishment of small groves and windbreaks been practicable at that time, they would have been an asset from the point of view of comfort and crop production as well as of the wood supply.

As with most of the other land laws, both the terms and the spirit of the timber-culture acts were widely violated. Entries were made for speculative purposes, with a view to selling relinquishments at a substantial profit to actual settlers. Dummy entrymen were used to acquire large tracts for wheat and cattle ranges. As Commissioner Sparks of the General Land Office put it in 1885: "Within the great stock ranges of Nebraska, Kansas, Colorado, and elsewhere, one quarter of nearly every section is covered by a timber-culture entry made for the use of the cattle owners, usually by their herdsmen who make false land office affidavits as a part of the condition of their employment." He estimated that 90 per cent of the entries were fraudulent. Instances were even known where the law was used to obtain title to areas already wooded.

Repeal. In 1882 Commissioner McFarland of the General Land Office called attention to the abuses under the act and the next year recommended its repeal.

Continued experience has demonstrated that these abuses are inherent in the law, and beyond the reach of administrative methods for their correction. . . . My information leads me to the conclusion that a majority of entries under the timber-culture act are made for speculative purposes and not for the cultivation of timber. Compliance with the law in these cases is a mere pretence and does not result in the production of timber. . . . My information is that no trees are to be seen over vast regions of country where timber-culture entries have been most numerous. . . . I am convinced that the public interests will be served by a total repeal of the law, and I recommend such repeal.

Repeal was finally effected on March 3, 1891, in the omnibus revision of the land laws entitled "An Act to repeal the timber-culture laws, and for other purposes." Altogether some eleven million acres were patented under these laws, mainly in the Dakotas, Nebraska, and Kansas. Although a few remnants are still to be seen of the many plantations established in good faith, the results were extremely meager from the point of view of forest establishment, while fraud was widespread. One of the best-intentioned of the land laws had proved one of the most complete failures.

Desert-land Sales. An attempt to promote settlement of the arid region, to which the Homestead Act obviously was not adapted, was made in 1877 through the Desert-Land Act. This act provided for the sale at $1.25 per acre of 640 acres of land unfit for cultivation without irrigation to settlers who would irrigate it within three years after filing. Why the precedent of a free grant of 160 acres set by the Homestead Act should not have been followed is unclear. Possibly a charge was thought to be justified because of the greater productivity of irrigated land, but in that case the larger size of the grant is difficult to explain.

As usual, the law was more honored in the breach than in the observance. Speculation was rife and collusive entries were common. Large areas primarily valuable for grazing or for the production of hay without irrigation were taken up. Also, control over much grazing land to which title was not acquired was obtained by making desert entries on land including or bordering on water. In 1889 the United States Surveyor General for Idaho, after commenting on the folly of permitting private parties to file on water almost at will, made this constructive proposal: "The irrigation of the arid lands of the West should be undertaken by the government or the lands be granted to the respective states and territories upon such terms and conditions as will insure the construction of necessary canals and reservoirs for reclaiming all of the lands possible."

President Hayes in his annual message of December 3, 1877, referring to the lands west of the 100th meridian, had suggested that "a system of leasehold tenure would make them a source of profit to the United States, while at the same time legalizing the business of cattle raising which is at present carried upon them." Many years were to elapse before these eminently sane recommendations were approved by Congress in the Reclamation Act of 1902 and the Taylor Grazing Act of 1934. Alienation, not reservation, of the public domain was still the order of the day.

Criticism of the act resulted in its amendment in 1890 and 1891. That these amendments effected little improvement in the situation is indicated by the remark some years later of an official of the General Land Office: "I may properly say that I have traveled through localities in which more or less extensive areas of land had been disposed of under the desert land law, and a most critical inspection of these lands failed to reveal the slightest evidence that they had ever received any treatment different from that to which the yet unappropriated public lands in the same localities had been subjected." Some ten million acres have been patented under the act, chiefly in Montana, Wyoming, Colorado, Idaho, and California. In spite of repeated recommendations for its repeal, the law is still on the statute books.

Timber, Range, and Mineral Lands. Although prior to 1891 there was very little legislation relating specifically to the public timberlands and none relating to range lands, both classes of land played an important part in the settlement of the West. So important, indeed, is their role in the development of forest and range policy as to justify devoting a separate chapter to them. Their treatment during this period is accordingly discussed in Chapter 3.

The disposal and reservation of mineral lands are treated in Appendix 1, which surveys Federal policy with respect to natural resources other than forest and range lands.

Grants for Education. From 1785 on, Section 16 in every township was reserved from disposal under the various land laws for later donation to the several states for the support of education. The first grant was made

to Ohio on its admission to the Union in 1803. In 1848, when Oregon was
first organized as a territory, Section 36 was also reserved for the same
purpose. Still later, Sections 2 and 32 were added to the grants made to
Utah, Arizona, and New Mexico. Altogether, twelve states received grants
of one section per township, fourteen states received grants of two sections, and three states received grants of four sections. When any of the
granted sections were not available because of previous occupation or
reservation, the state was ordinarily allowed to make lieu selections of
any other available public lands.

Sale of the school lands by the states was prohibited until 1826, when
Ohio was permitted to sell them, with the requirement that the sale must
be approved by the township concerned and that the proceeds must be
permanently invested in some productive fund. The same privilege was
later extended to the other states, often with a limitation as to the minimum price at which sales could be made. The proceeds from these
grants, which totaled nearly 78 million acres, assisted in the support of
the common schools in all of the public-land states, in some of which they
are still a significant item.

Land-grant Colleges. A substantial expansion of Federal aid to education took place with the passage of the Morrill Act of July 2, 1862,
making liberal grants of public land for the establishment of colleges of
agriculture and the mechanic arts. Such action had been under consideration for some time and had been urged by several state legislatures. It
was generally favored by the Northern states as giving them an opportunity to obtain some direct benefit from the Western lands; was opposed
by the Southern states as inexpedient and unconstitutional; and was
fought by the Western states because of the anticipated evils of land
speculation and absentee landlordism. A similar measure had been vetoed
by President Buchanan in 1858.

Grants under this act totaled about 10 million acres and provided
important support for the "land-grant colleges," as the institutions bene-
fiting from it came to be called. The Second Morrill Act, in 1890, provided further support in the form of continuing cash grants to be paid
out of the proceeds from the sale of the public lands. An amendment in
1903 authorized the use of other funds in case these proceeds were inadequate to meet the full amount of the appropriation—an indication of the
extent to which the public-land business had declined by the beginning
of the present century.

Other Grants for Education. Numerous other, much smaller, grants of
land were made for the support of individual universities, normal schools,
and other educational institutions. In addition, the proceeds from grants

to finance the construction of permanent improvements were sometimes used by the states for educational purposes. On the whole, the public lands have made a significant contribution to the development of both common-school and higher education in the United States.

Grants for Permanent Improvements. The public lands and receipts therefrom have also been used in liberal measure to assist the states in the construction of transportation facilities and other permanent improvements. Acts of 1802 and 1803 providing for the establishment of the state of Ohio granted the state 3 per cent of the net proceeds from all sales of public lands within the state for the construction of roads.[1]

The precedent thus set gradually led to the distribution to the public-land states of 5 per cent of the net proceeds from the sale of public lands within their borders. States admitted since 1889 have been required to add this amount to the common-school fund. All states on their admission to the Union agreed not to tax lands in Federal ownership—a situation that has led in recent years to much agitation in favor of payments in lieu of taxes.

Many grants of public land to assist in the construction of specific internal improvements were made from 1823 to 1871. These include donations to various states of 3,300,000 acres for the construction of wagon roads, 4,600,000 acres for the construction of canals, and 1,500,000 acres for river improvement. In 1827, donations to aid Indiana and Illinois in canal construction inaugurated the alternate-section system of land grants which was later followed on a much larger scale with the railroads. The practice of charging a "double-minimum" price of $2.50 for land in the sections retained by the government started in 1846 with a grant for the Fox River Canal in Wisconsin.

Railroad Land Grants. Land grants to aid in the construction of railroads greatly overshadowed all other grants for permanent improvements not only in area but in economic and political importance. From 1835 on, Congress frequently granted railroads free right of way through public lands, and this privilege was made general in 1852.

The need for better transportation facilities to connect the rapidly developing Middle Western and Far Western states with the Eastern states was generally recognized, but there was much opposition to large grants of public lands to aid in their construction. This opposition came largely from the South, which feared that the proposed railroads would link the agricultural West and the industrial East more closely together

[1] An additional 5 per cent was set aside by the Federal government for construction of roads leading to and through the state.

to its own detriment. It further feared that the increased accessibility of Western lands would strengthen pressure for the enactment of homestead legislation, to which it was also vigorously opposed.

The first large grant was made in 1850 to assist in the construction of the Illinois Central Railroad from Chicago to Mobile. The act granted to Illinois, Alabama, and Mississippi (the three public-land states through which the road would pass) a right of way not over 200 feet wide; free use of construction material; and every alternate section of land designated by even numbers for six sections in width on each side of the road, with the right to make lieu selections in place of alienated lands to a distance of not more than 15 miles from the road.

The alternate sections retained by the government were to be sold at not less than $2.50 per acre. This, it was argued, would double the receipts that would otherwise be received from the lands, so that the donation of the other sections would actually result in no loss to the government—a plain case of eating one's cake and having it too.

Property and troops of the United States were at all times to be transported over the railroad free, and the mails at such rates as Congress might fix. Although the grant was made to the states, it was understood that the road might be built by a private company which would be subsidized by the proceeds from the sale of the lands. This was the procedure followed.

The first transcontinental grants were made in 1862 after the South was no longer in a position to object. The Union Pacific and the Central Pacific Railroads were then given alternate sections of public land, at first to a distance of 10 miles and later of 20 miles, on each side of the road. Mineral lands were excluded, but an amendment in 1864 specified that coal and iron were not to be classified as minerals. In 1864 the Northern Pacific Railroad received a grant of alternate sections to a distance of 40 miles on each side of the road in the territories through which it passed and of 20 miles in the states, again with the exclusion of mineral lands. Similar grants were made in 1866 to the Atlantic and Pacific Railroad (now the Atchison, Topeka, and Santa Fe), the Southern Pacific Railroad, and the Oregon and California Railroad.

Discontinuance and Relinquishment. The liberality of these grants removed large areas of accessible land from the operation of the Homestead Act and thereby aroused hostility on the part of the settlers. As a result of mounting opposition, the last railroad grant was made in 1871. An act passed in 1890 provided that all grants adjoining rights of way on which railroads had never actually been constructed should be forfeited to the United States and restored to the public domain. The total area received by the railroads amounted to more than 128 million acres, of which about 91 million acres went direct to railroad corporations and the other 37 million acres to states for the benefit of the railroads.

All the railroad land grants were made on condition that the recipient transport government shipments at reduced rates. With the passage of time and the growth of government, this requirement imposed such an increasingly heavy burden that the railroads sought relief. Congress recognized the situation by including in the Transportation Act of 1940 a section authorizing the payment of full tariff rates to land-grant railroads which would waive all further claims under their grants. Lands already patented were not affected.

The seventy-two land-grant railroads promptly filed waivers of all unsatisfied claims, but only six of the roads had claims that were not already completely satisfied. These claims aggregated about 8,300,000 acres, of which some 3,600,000 acres were within national forests and 2,100,000 acres within Indian reservations. Nearly 77 per cent of the claims were in Arizona, Montana, and California. Railroad land grants are now a closed book so far as further litigation between the railroads and the government is concerned.

Swampland Grants. Efforts to have swamplands in the public domain donated to the states date back to 1826. They bore fruit in 1849 when Congress granted to Louisiana "the whole of those swamps or over-flowed lands, which may be, or are found unfit for cultivation." Proceeds from the sale of the lands were to be used exclusively, as far as necessary, for the construction of levees and drains. In 1850 a similar grant was made to Alabama, Arkansas, California, Florida, Illinois, Indiana, Iowa, Michigan, Mississippi, Missouri, Ohio, and Wisconsin, and in 1860 to Minnesota and Oregon.

These acts resulted in much fraud and relatively little reclamation. Most states made their own determination of the public lands within their borders which were "swamp and overflowed," commonly through agents employed for the purpose. Much excellent agricultural land and timberland found its way into the hands of the states, and later of private owners. In Florida whole townships classified as swamp proved on investigation to be substantially dry land, and at two places in California irrigation works were found on areas claimed as swamp.

Commissioner Sparks of the General Land Office in 1887 commented that "states have disposed of the lands in large quantities for a small consideration, or granted them to railroads and other corporations. But a small proportion of the proceeds has been applied to the reclamation of the land, and the purposes of the grant have generally been ignored and defeated." Some 65 million acres were donated to the fifteen states to which grants were made. Another piece of public-land legislation, however attractive in theory, had proved a failure in practice.

500,000-acre Grants. During the first half of the last century the proposal to cede the public lands to the states was almost continuously before Congress, particularly after 1826. President Jackson favored the proposal, which was commonly linked with states' rights, and in 1837 Senator Calhoun of South Carolina introduced a bill providing for the cession to the states of all the public lands within their borders.

Although no such general cession was ever approved, its advocacy was undoubtedly influential in bringing about the inclusion in the Preemption Act of 1841 of a provision granting 500,000 acres to each of the existing states and to such new states as might later be admitted to the Union. Grants already received from the Federal government were to be deducted from the 500,000 acres. The lands were to be sold by the states for not less than $1.25 per acre and the proceeds used only for the construction of internal improvements. In practice they were also used for many other purposes. The total area passing to the states under this grant was approximately 8 million acres.

Distribution of Public-land Receipts. The same act also provided for the distribution to the several states and territories of the net proceeds from the sale of the public lands. This provision had long been the cause of heated political controversy. It had the support of most, but not all, of the non-public-land states as a means of giving them a direct return from a resource which was the property of the entire nation. It was favored by the manufacturing interests as a means of reducing the income of the Federal government from this source and thereby making it necessary to rely more heavily on the tariff, and was opposed for the same reason by those who wanted a low tariff.

One argument in favor of distribution which has a strange ring today was that it would help to avoid the danger of building up a surplus in the Federal Treasury. The public debt had been extinguished in 1835, and there were those who feared that this situation might lead to the accumulation of surplus funds which public officials would be tempted to use wastefully or even dishonestly.

An argument against the plan was that it would make the states dependent on the Federal government and thus weaken their power. President Jackson in 1833 vetoed the first bill providing for distribution largely for this reason.

It appears to me [he stated] that a more direct road to consolidation cannot be devised. Money is power, and in that government which pays the public officers of the states will all political power be substantially concentrated. The state governments, if governments they might be called, would lose all their independence and dignity. . . . It is too obvious that such a course would subvert our well-balanced system of government, and ultimately deprive us of all the blessings now derived from our happy union.

The act as finally approved provided that 10 per cent of the net proceeds from the sale of public lands in the nine existing public-land states would be distributed to the state concerned. The remainder of the proceeds from these states and all of the proceeds from the other public lands were to be distributed quarterly to the 26 states, the District of Columbia, and the territories of Wisconsin, Iowa, and Florida according to their population as shown by the latest census. Each state could use the amounts which it received for such purposes as the legislature might direct.

Distribution was to cease in case of war, if the minimum price of public lands were increased above $1.25 per acre (except in alternate sections where the "double-minimum" price applied), or if the duties fixed by the act of March 2, 1833, were raised above 20 per cent. The latter situation arose the very next year (1842), when an act dealing primarily with the tariff suspended indefinitely the distribution of proceeds from the sale of public lands. Only three quarterly installments amounting to $691,000 were paid, and distribution was never resumed. Several states refused to accept their share of the fund, which South Carolina referred to as a "bribe."

Military Land Bounties. Military land bounties were at first offered to encourage soldiers to enlist and later used to reward them for service. It was hoped that land bounties would help to populate the frontier with a particularly vigorous class of settlers. Most of the soldiers, however, were not interested in migrating to the Western wilderness and preferred to sell their warrants for whatever they could get. This led to lively speculation in military-bounty warrants, which at one time were quoted on the New York stock exchange.

In 1855 the bounty system was liberalized so as to give 160 acres to every soldier who had participated in any war from the Revolution to date. Service need not have been longer than fourteen days. Total lands acquired have amounted to about 61 million acres.

Sale of Isolated and Mountainous Tracts. In 1846 Congress authorized the Commissioner of the General Land Office to sell isolated tracts of unappropriated and unreserved public land at not less than $1.25 per acre. In 1895 this authority was specifically limited to not more than

160 acres. It was extended by later legislation, including the Taylor Grazing Act of 1934, which permitted the sale of any isolated tract of not more than 760 acres at public auction for not less than its appraised value and not more than three times that value, with a preference right to adjoining owners.

The same act continued the Secretary's authority, granted in 1912, to sell not more than 160 acres of land which is mountainous or too rough for cultivation, whether isolated or not, on application of adjoining owners. In 1947 the maximum size of isolated tracts that the Secretary could sell was increased to 1,520 acres, and of mountainous tracts to 760 acres.

Sale and Lease of Small Tracts. A new method of land disposal was inaugurated in 1938 when the Secretary of the Interior was authorized to sell or lease not more than 5 acres of certain public lands which he may classify as home, cabin, health, convalescent, recreational, or business sites, subject to a reservation to the United States of all oil, gas, and other mineral deposits. Present regulations under the act provide for leases up to five years, with or without an option to purchase.

This act and the acts providing for the sale of isolated and mountainous tracts permit the disposal of scattered pieces of public domain which do not fit into any of the major systems of Federal reserves, such as national forests and national parks. They have largely displaced the Homestead Act as the popular method for obtaining public land. In 1952, for example, 25,397 small-tract leases covering 114,333 acres (including Alaska) were in force, while there were 478 small-tract sales comprising 2,058 acres. Public-auction sales of isolated and mountainous tracts numbered 637 and covered 88,482 acres. The 357 final homestead entries approved included only 41,965 acres, of which 32 per cent were in Alaska.

Extent of Disposals. Although this chapter deals primarily with alienation of the public lands prior to 1891, the greater portion of the disposals had been completed by that time. It is therefore not inappropriate to conclude the present discussion with a summary of the total area disposed of by the various methods adopted by Congress. Table 2 presents this information as of 1953.

Altogether, title to 1,030 million acres, or more than 70 per cent of the original public domain, has been transferred from the Federal government to other owners. Of this total, some 69 per cent has gone to individuals and institutions, 22 per cent to the states, and 9 per cent direct to railroad corporations. Public, private, and preemption sales and homestead grants comprise nearly three-fifths of all disposals. Grants to aid in the construction of permanent improvements and for the support of education also occupy a place of major importance. The rise and fall of the public domain (illustrated in Figure 1) have been of the utmost significance in the development of the nation.

Reservations and Withdrawals. Policies with respect to public lands that have been reserved and withdrawn for specific purposes will be discussed in subsequent chapters and in Appendix 1. Such lands include national forests, Oregon and California Railroad revested lands, national parks, national monuments, grazing districts, wildlife reserves, power and water reservations, and mineral reserves. The approximate areas involved are shown in Table 3 and in Figure 2.

Table 2. Area of Public Lands Disposed of under the Public-land Laws, as of 1953

Method	Area, million acres	Per cent of Total disposals	Per cent of Original public domain
Public, private, and preemption sales, mineral entries, and miscellaneous disposals	300.0	29	20
Homestead grants and sales	285.0	28	19
Grants to states:			
For support of common schools	77.5	8	5
For reclamation of swampland	64.9	6	4
For construction of railroads	37.1	4	3
For support of miscellaneous institutions	20.6	2	1
For purposes not classified elsewhere	14.3	1	1
For construction of canals	4.6 ⎫		
For construction of wagon roads	3.3 ⎬	1	1
For improvement of rivers	1.5 ⎭		
Totals	223.8	22	15
Grants to railroad corporations	91.3	9	6
Grants to veterans as military bounties	61.0	6	4
Confirmed private-land claims	34.0	3	2
Sales under timber and stone laws	13.9	1	1
Grants and sales under timber-culture laws	10.9	1	1
Sales under desert-land laws	10.0	1	1
Totals	1,029.9	100	70

SOURCE: U.S. Department of the Interior, Bureau of Land Management.

Table 3. Area of Federal Lands in Continental United States by Type of Reservation

	Public domain, acres	Acquired lands, acres
National forests (1953)[a]	139,762,332	20,735,949
Oregon and California Railroad and Coos Bay Wagon Road lands (1953)	2,611,309[b]	
National parks, monuments, etc. (1949)	11,908,241	2,047,397
Grazing districts (1953)	148,846,347[c]	249,809
Wildlife reserves (1950)	1,666,303	2,462,481
Indian lands (1950)	55,608,363	1,671,366[d]
Reclamation withdrawals (1950)	9,284,007	643,553
Soil Conservation Service (1950)	400,737	7,014,347
Military and naval reserves[e]	13,000,000	8,458,455
Other reserves (1950)	844,006	999,923[f]
Unreserved lands, outside grazing districts (1953)	27,417,248	
Totals	411,348,893	44,283,280

[a] Includes all lands administered by the Forest Service except the formerly controverted O. and C. lands.
[b] Includes 462,731 acres of formerly "controverted" lands.
[c] Includes stock driveways, public water reserves, and other miscellaneous withdrawals within grazing districts.
[d] As of 1945.
[e] Estimated from various reports.
[f] Adjusted to total.
SOURCE: U.S. Department of the Interior, Bureau of Land Management (based on *U.S. Dept. Agr. Cir.* 909).

Shortly after passage of the Taylor Grazing Act, President Roosevelt by Executive orders dated Nov. 26, 1934, and Feb. 5, 1935, withdrew from entry, pending classification, practically all of the remaining unreserved and unappropriated public domain in the states. The objective was to determine the most useful purpose to which the lands may be put in furtherance of the land program and conservation and development of natural resources.

It was anticipated that lands found suitable for cultivation would be opened to entry under the Homestead Act, while those primarily valuable for programs involving grazing, timber, wildlife, or park purposes would be placed in the appropriate form of reservation. The lands were later fully opened to prospecting and patenting for minerals. With this exception, land is available for private acquisition only after being classified as suitable for the purpose for which it is sought.

The extent to which times have changed is illustrated by the fact that in 1953 the area of unreserved land outside of grazing districts (27 million acres) was about equal to the area covered by final homestead entries during the three years from 1913 to 1915. Out of 357 tracts for which application for listing was made in the first five years following the Roosevelt withdrawals, only nineteen were classified as suitable for agriculture. Truly the public domain that our forefathers knew has become

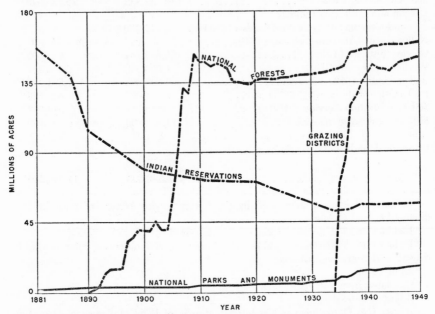

Fig. 2. Net area of major Federal reservations in the states, 1881–1949. (*Courtesy U.S. Department of the Interior, Bureau of Land Management.*)

a matter of history from the standpoint of area, character, and availability for private acquisition.

Summary. The century following the American Revolution was marked by the occupation and development of a virgin continent at a rate unparalleled in previous history. This development was dependent in large measure on the rich agricultural, timber, and mineral resources contained in the billion and a half acres of public domain. The apparent inexhaustibility of these resources made it inevitable that they should be used freely, often wastefully by present standards.

During this period the basic policy was to transfer public lands to private ownership. Sometimes this was done by grants to the states, chiefly for the support of education and the construction of permanent improvements. More often the transfer of title was direct to the individual or corporation, with general sales, preemption sales, homestead grants, and railroad grants the principal means of disposal. In retrospect, much of the legislation was unsound and administration ineffective.

The public domain had a profound influence on the economic, social, and political life of the nation. Free and abundant natural resources of every description promoted material prosperity, broke down barriers between classes, and bred a sturdy individualism that rebelled at governmental restraints. Soil, timber, forage, wildlife, minerals were to be had for the taking, evasion of laws attempting to control their disposal and exploitation was easy and respectable.

Not until late in the nineteenth century was the laissez-faire philosophy of the frontier seriously challenged. Then came the establishment of forest reserves, the crusade for "conservation," and the retention in public ownership of other natural resources. Gradually the national policy with respect to the public domain changed from one of almost complete disposal to one of virtually complete reservation. Yet the legacy of individualism and of unconcern for the future bequeathed to us by free land still colors our thinking and creates an inertia which strongly influences current attitudes toward natural-resource policies.

REFERENCES

Clawson, Marion: "Uncle Sam's Acres," chaps. 1–3, Dodd, Mead & Company, Inc., New York, 1951.

Hibbard, Benjamin H.: "A History of the Public Land Policies," The Macmillan Company, New York, 1924.

Peffer, E. Louise: "The Closing of the Public Domain," chap. 1, Stanford University Press, Stanford, Calif., 1951.

Robbins, Roy M.: "Our Landed Heritage," Peter Smith, New York, 1950.

U.S. Department of the Interior, Bureau of Land Management: "Brief Notes on the Public Domain," by Irving Senzel, 1950.

————: "Graphic Notes on the Public Domain," 1950.

U.S. Department of the Interior, General Land Office: "Land of the Free," 1940.

CHAPTER 3

Exploitation of Timber and Forage

For nearly a century after the close of the Revolution there was no general legislation dealing with the enormous area of forest land included in the public domain. Until 1878, the only legal way to acquire either timber or timberlands from the Federal government, except incidentally in the case of lands primarily valuable for agriculture, was under the various acts providing for auction sales.

Naval-timber Reserves. The need to build up a strong navy resulted in special legislation to assure adequate supplies of timber for this purpose. Following authorization in 1794 for the construction of six frigates, and the establishment of the Navy Department in 1798, Congress in 1799 appropriated $200,000 for the purchase of timber or of lands on which timber suitable for naval construction was growing and for the preservation of such timber for future use.

At this time the Navy Department was particularly interested in live oak, which was regarded as the best species for the construction of hulls and the supply of which on lands under the control of the United States was becoming scarce. Two islands aggregating 1,950 acres and containing fair amounts of live oak were promptly purchased off the coast of Georgia at a cost of $22,500. No further purchases were made, presumably because the Louisiana Purchase, and more particularly the Florida purchase, placed under Federal ownership large areas containing considerable quantities of live oak. Concern continued to be felt, however, for the adequacy of the future supply.

Congress reacted in 1817 by authorizing the Secretary of the Navy, with the approval of the President, to reserve from sale public lands containing live oak and red cedar for the sole purpose of supplying timber for the Navy of the United States. The unauthorized removal of any timber from such reservations or of any live oak or red cedar from any other public lands was punishable by a fine of not more than $500 and imprisonment for not more than six months. Administration of the reserves was placed under the Navy Department, while administration of the other public lands remained in the Treasury Department.

Thus did the United States embrace the principle of Great Britain's hated Broad Arrow policy. The kind of trees and the regions involved were different, but the underlying philosophy that government has the right to exercise absolute control over the use of public lands was identical. The American government went even further in its restrictions than the British by reserving not only particular species but entire areas from unauthorized use. White pine could not have been reserved, even if there had been fears as to the adequacy of the mast supply, since the entire stand in the Northeast was on state or private lands; but all live oak and red cedar were specifically protected from illegal cutting on all lands which the Federal government did own and all kinds of trees on areas set apart as reservations. These provisions materially strengthened the general prohibitions against entry on the public lands embodied in the anti-trespass act of 1807.

Ineffectiveness of Reserves. Nevertheless the frontier spirit in the South differed little from that of pre-Revolutionary days in the North. Government, whether its headquarters were in London or in Washington, was regarded as a distant and unreasonable despot whose edicts were to be evaded whenever they conflicted with the interests of those who were undergoing the dangers and hardships of conquering the wilderness. That the mere placing of a law on the statute books would effectively put a stop to illegal cutting was not to be expected.

In a letter to the House Committee on Naval Affairs in 1827, Secretary of the Navy Southard stated that on February 29, 1820, the President had reserved 19,000 acres of live oak lands; that efforts to protect these and other public lands from trespass had been ineffective; and that the most valuable and accessible timber had been removed from the whole coast. The Secretary declared that the value of this kind of timber made it necessary "to use all the means in our power to obtain, preserve, and increase it." He therefore recommended the purchase of particularly valuable tracts of timber; the creation of additional reservations in Florida and Louisiana; the planting of trees on government land; and, in order to assure an adequate immediate supply in case of war, the purchase and storage of large quantities of timber.

Santa Rosa. As a result of the Secretary's letter, Congress on March 27, 1827, reinforced the act of 1817 by authorizing the President to take proper measures to preserve the live oak timber growing on lands of the United States and to reserve from sale such public lands as might be found to contain live oak or other timber in sufficient quantity to render it valuable for naval purposes. Several additional reservations were soon established in Louisiana, Alabama, and Florida. Of these, the most interesting both silviculturally and politically was the Santa Rosa Reserve adjacent to Pensacola, Florida.

Henry M. Brackenridge, Federal judge for West Florida, occupied a piece of land within the boundaries of this reserve and had ambitions to supplement his judicial activities by becoming its superintendent. In support of his candidacy, he indicated his technical qualifications for the position in a letter of June 1, 1828, to Secretary Southard which was later published under the title "On the Cultivation of Live Oak." No doubt Judge Brackenridge's obvious familiarity with the subject was in part responsible for his receiving the desired appointment, which was accompanied by instructions, apparently originating with President Adams, himself an enthusiastic tree planter, to undertake experimental plantings of live oak acorns. The Santa Rosa Reserve thus became America's first forest experiment station.

Now it happened that within the area was a tract of some 1,400 acres owned by Judge Brackenridge and Colonel Joseph M. White, Florida's delegate to Congress, which they desired to sell to the government. A somewhat delicate situation, with obvious political implications, was created by the attempt of a Federal judge and a member of Congress to sell the United States a piece of land within a reservation for the existence of which they were largely responsible. Secretary Southard consequently took the precaution of presenting the facts to Congress, which in March, 1828, authorized the expenditure of not more than $10,000 for the purchase of such lands as the President might think "necessary and proper to provide live oak and other timber for the use of the Navy of the United States." Of this amount $9,000 was actually spent to buy the Brackenridge-White lands and a few other small parcels.

Early in 1829, Secretary of the Navy John Branch, Secretary Southard's successor, directed Judge Brackenridge to suspend all operations on the plantation until further orders, and about a year later instructed him to terminate reforestation activities on the reserve by January 18, 1831. In his annual report to Congress in December, 1830, he called attention to the orders, which he justified on the ground that the natural supply of live oak was adequate to make its artificial propagation or culture unnecessary.

This action brought a vigorous reaction from Colonel White, who succeeded in having the House Naval Affairs Committee instructed to investigate the entire naval-timber-reserve program. The resulting report expressed doubt as to the necessity of growing oaks from acorns, but declared extensive reservation to be imperative, with accompanying measures to encourage and facilitate growth of the young live oaks. The one positive recommendation of the Committee was that existing legislation for the preservation of live oak and red cedar timber should be strengthened by extending its penalties to timber on any lands acquired or to be acquired for naval purposes. Congress went even further in this respect than the Committee had recommended by passing the important antitrespass act of March 2, 1831, which will be discussed later in this chapter.

In 1831 Secretary Branch was succeeded by Levi Woodbury of New Hampshire, who soon found not only Santa Rosa but the whole naval-timber-reserve policy on his desk. As to Santa Rosa, he came to the conclusion that it would be wise for a few years to continue to destroy the annual growth of other wood injurious to young live-oak trees and to trim and train the thriftiest new ones appearing—in other words, to practice improvement cuttings and prunings. By December, 1833, the "nursery" so treated covered 225 acres and included over 60,000 trees. Judge Brackenridge was relieved of his duties as superintendent of the reserve in 1832, and these duties were assigned to the commander of the Pensacola Navy Yard, under whose direction the enterprise languished.

Growth of the Reserves. With respect to the naval-timber-reserve system as a whole, Secretary Woodbury viewed the situation seriously but without alarm. Because of the "great abundance of our public lands, on which the live oak tree spontaneously thrives," he could see no occasion to make additional purchases of private lands or to carry further the artificial cultivation of live oak on any of the public lands. He also felt that the economic law of supply and demand would induce private owners to protect and even to produce adequate supplies of live oak whenever it became profitable to do so. Somewhat incidentally, he called attention to the fact that the Navy used three times as much wood of other species such as white oak, pine, larch, cedar, locust, and elm as it did of live oak, which was used only in the frames of ships, and suggested the possible desirability of more general reservation of public lands—a suggestion embodied nearly sixty years later in the Forest Reserve Act of 1891.

Despite his optimism, Secretary Woodbury believed in playing safe. He therefore recommended not only continuing but expanding the naval-timber reserves. In order to determine the area needed he went through much the same calculation that a modern forest owner would use in finding the area required to support a plant of given capacity. On the basis of past experience he estimated the Navy's needs for live oak at 62,286 cubic feet per year. At 50 cubic feet per tree this meant an annual consumption of 1,245 trees. With an average of 20 trees per acre, these could be harvested from 62 acres. A rotation of seventy-five years would then give 4,640 acres as the total area required to put the supply on a sustained-yield basis. Finally, to make assurance much more than doubly sure, he multiplied this figure by a safety factor of 35 and came out with 160,000 acres as the total area that should be set aside as reservations. To reach this goal would require nearly doubling the existing reserve area of 86,000 acres.

The total area actually reserved from the public domain was reported by Secretary of the Navy Gideon Welles in 1868 as follows:

State	Acres
Florida	210,540
Alabama	221
Mississippi	25,849
Louisiana	27,840
Total	264,450

Decline of the Reserves. Somehow the Navy continued to obtain the timber it needed in spite of heavy cutting, much of it in trespass by "the unprincipled of all nations," and interest in the reserves gradually waned. Legislation in 1843 opened to settlement certain reservations of live oak lands in Louisiana, and from then on the area continued to dwindle. Iron and steel began to come into use for shipbuilding in the 1850's, and the destruction of the wooden ships of the Federal fleet by the Confederate ironclad, the "Merrimac," in the battle of Hampton Roads in 1862, marked the end of wood as an indispensable material in ship construction.

In 1879, the bulk of the reservations in Florida, where most of them were located, were restored to the unreserved public domain, and in 1895 the same action was taken with the remaining area in Florida, Alabama, and Mississippi. Reservation of the last remnant of the reserves— 3,000 acres in Louisiana, almost completely covered by private claims— took place in 1923.

The naval-timber episode, transitory as it was, constituted an important landmark in the development of this country's forest policy. It emphasized the then indispensability of wood in the vital task of building up an effective navy and merchant marine. It recognized the value of the public lands as a source of timber supply and the right of Congress to exercise complete control over those lands in the national interest. It demonstrated the impossibility, under frontier conditions, of effectively preventing trespass on a natural resource publicly owned and generally regarded as inexhaustible. It initiated the first Federal experiment in reservation, timber culture, and forest management. It paved the way for the present system of national forests.

Timber Trespass. So far as legislative pronouncements are concerned, the prevention of trespass on the public lands was always the official

policy of the American government. Administrative and judicial enforcement of those laws was quite another matter.

Question early arose as to whether the general antitrespass act of 1807
forbidding anyone to settle or occupy the public lands until authorized
by law applied to timberlands. In 1821 the Attorney General of the United
States ruled that it did and that the government could remove timber
wasters by military force and subject them to fine and imprisonment. He
added that quite aside from that statute the government passed the same
common-law remedy as individuals for the protection of its property. A
few weeks later the Commissioner of the General Land Office instructed
all registers, receivers, and district attorneys to take every proper means
to prevent trespass and to advertise in the local papers that "those lawless
persons who are guilty of intruding on lands of the United States and of
committing waste of public timber" would be "prosecuted to the utmost
rigor of the law." The old Broad Arrow policy was now enlarged to
include every species on every acre of the public lands.

Looting of Naval-timber Reserves. The agents employed to stop trespass on the naval-timber reserves were for the most part men who owed
their appointment to political pull rather than to ability and who would
have been able to accomplish little even under more favorable circumstances. "Irish Apothecaryism," as Cameron picturesquely terms the
method of selecting public officials on the basis of political influence without regard to their fitness for the job, was as wasteful of public funds and
as ineffective in serving the public interest then as it is today.

The situation finally became so malodorous that the Senate in 1843
passed a resolution asking for

the evidence . . . that depredations of a most ruinous kind are being daily committed on the navy timber . . . and also . . . proof . . . that any foreign government has, by contracts with any of our citizens, obtained supplies of live oak
cut from the public land.

Secretary of the Navy Upshur in reply to this resolution supplied ample
evidence that depredations were actually taking place, particularly in the
more accessible timber. Among the supporting documents was a letter
from Captain Hughes that illustrated the way in which legislation intended to promote settlement was fraudulently used to obtain timber:

The cutting of live oak from the public domain has been, for many years,
an illicit occupation; but recently it has assumed a legal form, under the action
of the pre-emption law, and the evil has been greatly extended. Persons squat
on the *cheniers*, as the live oak groves are called, and cut off the timber as
fast as possible, paying no regard even to its economic removal but destroying it in the most wasteful manner; and as soon as the oaks, which constituted

the only value of the land, have disappeared, they renounce their "rights," squat somewhere else . . . and repeat the same plundering operation, to the manifest injury of the United States, and to the serious annoyance of the *bona fide* settlers, whose interests are intimately connected with those of the Government.

He recommended modifying the preemption law so as to exempt live oak from its operation or to require the land to be paid for prior to removal of the timber. Secretary Upshur supplemented this recommendation by suggesting condemnation proceedings against property cut in trespass, since he saw little hope of convicting trespassers by jury trials. Congress took no action, and annual reports from the Navy Department contain no further reference to the naval-timber reserves until 1851.

During the 1850's, the reports of Captain Joseph Smith, Chief of the Bureau of Yards and Docks, were surprisingly optimistic as to the effectiveness of the live oak agents in reducing timber depredations. He did, however, remark on the handicap of having to try trespassers before local juries which would impose penalties of "one cent," and added that "the woodsmen of pioneer life, in too many instances, do not regard their obligations to public law when private interests can profit by its transgression."

"Round Forties." One method of acquiring government timber which he described was also widely used on unreserved public lands in other parts of the country. The timber operator would establish title or claim to a small piece of land and then extend his cutting on adjacent land without regard to the legal boundaries of his own property. The center of operations was known as a "round forty" because the owner would instruct the cutters to cut "all around the forty," or as a "rubber forty" because its boundaries were so elastic.

Local reaction to this practice is well illustrated by a case in Alabama, in which an enterprising lumberman who had extended his "round forty" to include some naval timber was convicted and fined. Thereupon the Congressman from the gentleman's district presented to the Secretary of the Navy a huge petition from the local citizenry pointing out that the timber in question had previously been so inaccessible as to be worthless and that the alleged trespasser had really performed a public service by opening it up for utilization. The governor of the state described him as one of the finest citizens and stated that such use of public timber had been "common practice in Alabama for thirty years." To which the Chief of the Bureau of Yards and Docks replied that if this were the case, it was high time to do something about it. Some years later the Federal district attorney for Alabama reported that the usual failure of juries to convict trespassers—possibly for the alleged reason that timber agents failed to present evidence of the trespass, possibly for other reasons—had led Federal judges to instruct juries to render verdicts of acquittal, and thus give tacit approval to the plundering of government timberlands.

First Appropriations for Timber Protection. In 1871 Congress appropriated $5,000 for the fiscal year 1872 for the protection of timberlands in naval-timber reserves. The next year it appropriated $10,000 for the protection of public timberlands in general from trespass and fraud. These acts constituted the first specific appropriations by Congress for the purpose, since timber agents were previously paid out of general funds at the disposal of the department concerned.

Appropriations for the protection of the naval-timber reserves continued for only a few years, that for 1876 being the last for this purpose. Their discontinuance indicated lack of interest in the reserves rather than any decrease in the depredations on them.

Antitrespass Law of 1831. On paper, the government's authority to control trespass on the public lands generally was much strengthened by passage of the act of March 2, 1831, which had resulted from the congressional investigation of the Santa Rosa Reserve. Although the title of the act referred to naval-timber reserves only, its text provided penalties not only for trespassers on these reserves but also for anyone who should cut or remove any live oak, red cedar, or other timber from any other public lands of the United States without written authorization or with intent to export it or use it for any other purpose than for the Navy of the United States. Congress thus gave its official stamp of approval of the Attorney General's interpretation of the act of 1807 that *all* kinds of trees on *all* public lands were to be protected from trespass.

The discrepancy between the title and the body of the act led one Ephraim Briggs, who had been convicted of timber stealing in Michigan, to test the validity of that part of the act applying to unreserved public lands. The case went to the United States Supreme Court, which in January, 1850, decided that the enacting clause governed and that the penalties provided by the act applied to any trespass on any public lands.

Additional antitrespass legislation was provided in 1859 by an act forbidding the unlawful cutting of timber on lands of the United States reserved or purchased for military or other purposes. This provision became Section 5388 of the Revised Statutes of 1878 and is a part of the present Criminal Code.

Creation of Department of the Interior. In 1849 Congress created the Department of the Interior, transferred to it the General Land Office established in 1812 in the Treasury Department, and made it responsible for administering all of the public lands except those under the direction of another department, such as the naval-timber reserves. The new Department also took over the Office of Indian Affairs from the War Department and the Patent Office from the Department of State. This change in administration, coupled with the 1850 decision of the Supreme Court, resulted in a flurry of action against timber trespassers in the early 1850's.

Drama in the Lake States. Action against timber trespassers centered largely in the Lake states, where the exploitation of the enormously rich pineries that was to continue for the next half century was just getting under way on a scale to attract national attention. Congressman Fuller of Maine, in a speech in the House of Representatives in 1852, spoke in glowing terms of "the stalwart sons of Maine" who were "marching away by scores and hundreds to the piny woods of the Northwest" and who "like the renowned men of the olden time . . . are famous for lifting up the axes upon the thick trees." The mission of these lusty woodsmen and of the forests which they were harvesting was eloquently expressed at about the same time by Representative Eastman of Wisconsin.

Upon the rivers which are tributary to the Mississippi, and also upon those which empty themselves into Lake Michigan, there are interminable forests of pine, sufficient to supply all the wants of the citizens in the country, from which this supply can be drawn for all time to come. A few years since, and that whole country was unknown to the white man. The planter and farmer of Missouri, Illinois, and Iowa had exhausted the limited supply of timber which the groves upon the streams in the prairie lands had afforded, and the demand for pine timber with which to fence their farms and build their houses, called into activity a class of persons who fear no danger, and whose courage and enterprise are equal to any emergency. At great risk, and with intense labor, these lumbermen . . . penetrated this forest, and erected mills . . . which at the same time supplied the settlers upon the vast prairies . . . with lumber, and opened a market for the corn, flour, pork, beef, and cattle of the farmers, which they took in exchange. By this internal and domestic commerce two objects were effected, both highly advantageous to the interests of the Federal Government. Without lumber with which to erect houses and make fences, the vast prairies of the lower country could not be settled and farmed profitably, or with success, and there were no supplies from which to draw this necessary article, but from the forests of the North.

Mr. Eastman was incensed that government agents should be "persecuting" the enterprising individuals whose efforts were supplying the immense quantities of wood needed for the building of the nation, warned that "open revolt" might result, and argued that the inexhaustible forests of the public domain should have been thrown open for use without restriction. His views were identical with those of Representative Sibley of Minnesota, who a few weeks before had declared that the laws under which the "outrages in Wisconsin and Minnesota" were perpetrated should be repealed as "a disgrace to the country and to the nineteenth century." Senator Dodge of Iowa similarly inveighed against governmental harassment of the timber cutters in the new states of the Northwest. His declaration that he would never again vote "to extend the power of this government" has a familiarly modern ring.

Agent Willard Forces the Issue. Agent Willard was a particularly obnoxious timber agent whose activities had far-reaching results. His demand that the "trespassers almost innumerable" whom he found should cease their illegal activities was met by a flat refusal. Local feeling reached a high pitch. "Threats were openly made that none of the property seized should be removed from the place of seizure except by the original claimant or trespasser, and under his control; and that any attempt to remove the same on the part of the agent, or any third party to whom it might be sold by him, would be resisted by force and violence."

Thus challenged, Willard in 1853 seized and offered for sale a large quantity of stolen timber at Manistee, Michigan, most of which was either burned or restolen. He then obtained warrants for the arrest of the leading offenders, who promptly escaped or were violently rescued by mobs of their friends. Not until he called for assistance from armed bluejackets from the warship "U.S.S. Michigan" did he succeed in holding his prisoners in custody. Most of those finally brought to trial in Detroit the following year were convicted; but the penalties of small fines and jail sentences of from one day to one year seem hardly commensurate with the gravity of the offense, which had included not only trespass but armed resistance and the death of Federal officers.

Administrative Adjustments. The outbursts in Congress and episodes such as that at Manistee were doubtless responsible for a series of events relating to the administration of the timber-trespass laws that occurred in 1854 and 1855. On January 19, 1854, the Secretary of the Interior, who previously had himself handled the matter of timber protection and whose signature appeared on Mr. Willard's appointment, turned the whole business over to the "sound judgment and discretion" of the Commissioner of the General Land Office "as the public officer who from position and experience in such matters is most properly chargeable therewith."

Nine days later the Commissioner issued a circular to the timber agents informing them that the main object of their appointment was the "prevention of waste or trespass on the public lands and the destruction or carrying away of the public timber" and instructing them not to

harass actual settlers and not to enter into any compounding with trespassers. On June 27, Secretary McClelland, in response to a resolution
moved by Mr. Eastman, presented to the House of Representatives a
comprehensive statement on the situation with a decidedly apologetic
ring: "It is no part of the policy of the department to oppress or in any
manner to embarrass the citizens of the States in which depredations
are committed." The most interesting feature of this document is the
account of their experiences submitted by Agent Willard and Agent
Estes, kindred souls who were irritating despoilers of the public timber
in Michigan and Minnesota.

In June, 1855, John Wilson, the immediate superior of these two gentlemen, was removed from his position as Commissioner of the General Land
Office and assigned to assist in certain negotiations with the Indians in
the West. This action, however, was apparently not enough to mollify
opponents of the enforcement policy. On December 24, 1855, his successor notified all registers and receivers of land offices that "the Secretary of the Interior has concluded to change the present system of timber
agencies, and to devolve the duties connected therewith upon the officers
of the local land districts."

In other words, timber agents were to be done away with and their
duties taken over by the local land officers, already fully occupied with
other activities, with no increase in salary. Delegation of authority could
be made to special deputies on pressing emergency, but the fact of such
delegation must be reported "instanter" with an explanation of its necessity. When the emergency was not pressing, the facts were to be referred
to the Washington office for consideration and instructions.

Triumph of Compromise. For the next twenty years trespassers on the
pineries of the Lake states and other public timberlands led a relatively
peaceful life so far as government interference was concerned. The Commissioner's circular of 1855 had told the local land officers that "under no
circumstances will you compound or compromise," and these orders were
never officially revoked. In practice, however, compromise soon became
the order of the day, with the open approval of the Washington office.

Commissioner Edmunds of the General Land Office, for example, in his
annual report for 1864, after pointing out that his office had taken repres-
sive measures against timber spoliation on the public lands from Michigan to California, had this to say on the subject:

Wherever trespass has actually taken place, but found not to be willful but
through ignorance, it has not been the policy of the department to pursue the
offenders in a vindictive spirit, but, where the lumber has been taken from

offered land simply to require the actual entry of the premises and payment of costs. In case of unoffered or unsurveyed lands, we have enforced the payment of a liberal stumpage.

In his next report (1865) the same official commented on the failure of the timber-agent system to stop trespass in the face of "the pressing demand of settlers and the avarice of capitalists" and pointed with pride to the effectiveness of the new "compromise system."

Experience has taught us that when community interests conflict with law, and public opinion is in conflict with its enforcement it becomes virtually inoperative. Hence, by other means equally effective, ends unattainable by legal exactions may be accomplished and public and private interest secured. The department by a civil procedure, and avoiding the criminal courts, has legitimately converted waste timber into a productive fund, and is gradually suppressing an evil hitherto commensurate with the timber domain of the West. The present laws, discreetly administered, are ample for protection, unless Congress should deem proper by express enactment to give direct sanction to the authority of the Commissioner, now regarded by this office as legitimately incidental, of relaxing or enforcing the penalty imposed by the act of March 2, 1831, on such conditions as shall seem meet to him in all cases involving the spoliation of public timber.

Collection of Stumpage. In passing judgment on the lenient policy followed by the Department of the Interior during this period, it must be remembered that there was no way by which lumbermen could legally acquire either public timber or timberland except at auction sales. These sales were designed primarily for the disposal of agricultural lands and became increasingly infrequent after the passage of the Preemption Act of 1841 and the Homestead Act of 1862. Nevertheless, timber had to be cut from Federal lands to meet the needs of the rapidly expanding lumber industry, which was performing an indispensable service in providing a product that was essential for both the agricultural and the industrial development of the nation.

In order, therefore, to meet the exigencies of the case [said Commissioner Edmunds] this office proposed a *compromise*, substituting a uniform tariff of fees, in lieu of selling the timber seized, mitigating thereby the penalty in consideration of the peculiar local necessities of the settlers.

The arrangement proposed rests on the principle of treating the parties as offenders under extenuating circumstances, and releasing them on conditions ample to meet the exactions of justice—a principle applicable as well *before* as *after* conviction. Hence, while the law is not evaded, nor its violation countenanced, the wants of new settlements are gratified so far as consistent with sound policy and the necessity of the case.

Edmunds' successor, Commissioner Joseph S. Wilson, in his annual
report for 1867, stated that

where extenuating circumstances exist [the procedure of charging] a reasonable
tariff as stumpage for timber cut . . . while the criminal prosecution is dismissed [has] operated beneficially. Instead of mulcting the Government in
heavy costs, after long and unsuccessful prosecution according to the old system, doubtful cases have been compromised, and considerable revenue thereby
placed in the treasury.

The "stumpage" thus collected differed in several important ways from
the stumpage later collected in sales of timber from national forests. No
attempt was made to appraise its true value; there was always danger
that not all of the amounts collected would actually reach the United
States Treasury; the operator, not the government, decided where and
how much to cut; and no supervision was exercised to make sure that the
cutting was done without unnecessary waste and in a way to perpetuate
the forest. Under these conditions, and with "vindictiveness" succeeded
by a spirit of reasonableness on the part of the General Land Office, it
is no wonder that there was a long period of quiet on the Western front.

Compromise Questioned. Commissioner S. S. Burdett, in his annual
report for 1874, went straight to the heart of the timberland problem:

I fail to find, from the beginning of the Government to the present time, a
single enactment of Congress providing any distinctive method for the disposal
of that vastly extensive and proverbially valuable class of lands known as "pine
lands." These lands are notoriously unsuited to general agricultural uses, but
have been held subject only to preemption and homestead entry.

The product of these lands is of universal use, and forms the staple of commerce of no inconsiderable portion of the Nation. The difference between the
Government price and the actual value thereof is large, yet Congress provides
that these lands shall be disposed of under the preemption laws at $1.25 per
acre, or under the homestead laws by commutation . . . at the same rate.

He regarded the sale of the timber without the land, under careful
supervision, as the ideal method of disposal, but feared that this would
be ruinously expensive and would afford "opportunities for fraudulent
collusion and unjust exactions" on the part of corrupt public servants.
He therefore recommended that the pine and fir lands be removed from
the operation of these laws, carefully appraised, and sold for cash
at not less than their appraised value.

I am strongly of the opinion that the wisest policy the Government can pursue
in respect to this class of lands is that which will most speedily divest it of title
to the same for a fair consideration, for the reason that depredations to an
enormous extent are constantly occurring, which existing laws are powerless
to prevent and seem legally powerless to punish.

Apparently despairing of effective governmental action, he expressed the view that

the greatest protection to the timber of the country, now rapidly decreasing, will be found in placing it under private guardianship.

Two years later (1876), Commissioner J. A. Williamson heartily endorsed these recommendations.

If the pine lands were appraised at their full value and depredations prevented, they would sell as fast as the Government would desire to see them denuded of their forests. The timber would be more carefully husbanded in the hands of men whom it had cost a fair price than in the hands of the lawless trespasser or the bogus homesteader or preemptor. It is an anomalous fact that the Government is giving away the rich alluvial soil of Iowa, Nebraska, Kansas, and Minnesota to any citizen who will plant a few acres of cottonwood or other inferior timber, while under the provisions of the preemption and homestead laws it is granting a license to destroy millions of acres of pine forests of almost incalculable value, which should be preserved as a nation's heritage.

He declared vigorously that "no compounding with the offenders should be allowed, as is now the custom," favored resumption of the special-agent system, and instructed registers and receivers of the local land office to employ no further deputies without specific authority from his office.

Schurz Enforces the Law. Commissioner Williamson received strong support from Carl Schurz, who became Secretary of the Interior under President Hayes in the spring of 1877. On May 2 of that year he informed all local land officers that they were relieved from their duties connected with timber and that these duties would hereafter be performed by clerks or employees acting under direct instructions from the Commissioner. This action marked the beginning of a new campaign of law enforcement unprecedented in the previous history of the Department.

Secretary Schurz's German upbringing led him to approach the public-land problem with strong conviction as to the value of forests in the national economy and the importance of rigorous enforcement of the law. His views on the latter point were expressed with admirable clarity in his first annual report, for the fiscal year 1877:

That the law prohibits the taking of timber by unauthorized persons from the public lands of the United States, is a universally known fact. That the laws are made to be executed, ought to be a universally accepted doctrine. . . . There may be circumstances under which the rigorous execution of a law may be difficult or inconvenient, or obnoxious to public sentiment, or working particular hardship; in such cases it is the business of the legislative power to adapt the law to such circumstances. It is the business of the Executive to enforce the law as it stands.

In the same report he took cognizance of the fact that the new enforcement policy had

called forth remonstrance from several parts of the country where seizures were made. Lumber-merchants, saw-mill owners, and timber operators in some of the timber districts complained that private property had been or was apt to be seized together with logs wrongfully taken from the public lands of the United States, and that, by the proceedings carried on, business in certain localities would be severely injured and many laboring people put out of work. The agents of this department are instructed to use the utmost care in respecting private property; and, as far as the department is informed, those instructions have, a very few trifling and promptly corrected mistakes excepted, been strictly obeyed. As to the injury done to business, if that business consists in wrongfully taking timber from the public lands of the United States and manufacturing it into lumber and selling it, it is just the business which it is the duty of this department to suppress for the protection of the public interest.

Action and Reaction. For the first time the timber-agent system brought substantial results, in spite of the constant complaint of the Department that inadequate amounts for effective prosecution of the campaign against timber trespassers were appropriated by Congress. In his final report for the fiscal year 1880, Secretary Schurz called attention to the fact that during the nearly twenty-two years from December 24, 1855, to April 5, 1877, only $248,796 had been recovered and turned in to the Treasury as a result of timber trespass; while during the three years from April 5, 1877, to June 30, 1880, $242,377 had been actually collected, with uncollected judgments for about as much more.

Even more important than this favorable showing was the marked reduction in trespass on the public timberlands which he believed had been effected by the enforcement program. He warned, however, that

the evil will spring up again if the efforts of the government to arrest it should be in the least relaxed in the future, or if Congress should fail, by leaving the laws of the country in their present condition, to show an active sympathy with this policy.

He also renewed the recommendations which he had made repeatedly but without avail for the enactment of legislation providing for the sale of timber without the land under regulations to prevent waste and indiscriminate destruction and prescribing a severe penalty for the willful, negligent, or careless setting of fires upon the public lands.

Although Williamson reported in 1880 that "requests are received from every direction that special agents should be sent to stop the plundering upon the public timber," the reaction in Congress was hardly enthusiastic. Admitting that timber stealing could not be officially condoned, many members nevertheless felt that there was an element of unfairness in the drastic and unheralded change from lax to strict enforcement of the law. This feeling resulted in passage of the act of June 15, 1880, relieving timber trespassers on the public lands prior to March 1, 1879, from both civil and criminal prosecution on payment of $1.25 per acre for the total acreage involved. Opponents of this legislation from the East and the Lake states referred to it sarcastically as "the bill to license thieves on the public domain" and "the bill to condone crime and invite trespass and encourage theft." While perhaps generous in its terms, it is significant that the act by implication warned trespassers after the date set to expect no further leniency from Congress. Stealing of public property apparently was no longer quite as respectable as it long had been.

Following the departure of Secretary Schurz and Commissioner Williamson from their posts in the Department of the Interior in 1881, the pendulum again swung in the other direction. Commissioner McFarland in 1882, requesting additional appropriations for suppressing the extensive inroads still taking place on the public lands, stated that he did

not apprehend that it is the purpose or desire of the government to make the public timber a source of revenue or profit. . . . Accordingly, in all cases where there are extenuating circumstances surrounding the act of trespass, and the party shows a desire to compromise, and makes proposition of settlement before suit has been instituted, the practice has invariably been to recommend the same to the Secretary of the Interior for his acceptance.

Compromise, however, was not back for long. William Andrew Jackson Sparks, who in 1885 became Commissioner of the General Land Office under President Cleveland and Secretary Lamar, proved to be another crusader.

The policy of this department prior to the commencement of the present administration [he wrote in his first annual report] appears to have promoted rather than checked timber depredations, the laws having been construed with a liberality of license in favor of persons and corporations committing trespass, and damages, when claimed, being compromised at nominal rates.

Sparks would have none of this policy and undertook the prosecution of the "bold, defiant, and persistent depredators on the public domain" so vigorously as to elicit from the Senate a demand to know by what authority the timber involved was being seized and sold. Worse still, he suspended from final action all entries in regions where fraud had run rampant and undertook a vigorous campaign to put a stop to the "widespread, persistent public land robbery committed under guise of the various forms of public land entry."

The moral and political climate was not yet favorable for such drastic action. Although Cleveland supported the reform program, he was unable to save Sparks, who was dismissed from office in November, 1887; and the next year Secretary Vilas revoked the entire reform program.

Depredations on Indian Lands. Timber trespass was not confined to the public domain but extended also to Indian reservations, which contained much excellent timber. In the 1850's Timber Agent Estes had reported extensive depredations on the lands of the Chippewas and Menominees, and in 1877 a Chippewa chief complained to an Indian commissioner: "The whites are encroaching upon my reservation; they are stealing my timber."

Two years later the government brought suit to enforce the penalties provided by the acts of 1831 and 1859 against a lumberman who was making free with timber on lands of the Cherokee Nation, with unsuccessful results. The United States Circuit Court regretfully ruled that those acts applied only to "lands of the United States and did not therefore afford protection to reservations which were the property of the Indian." The judge pointed out that it was the manifest duty of Congress to protect these Indians from unlawful intrusion from without and from violation of their rights by any and all persons and stated that if Congress would enact the necessary legislation, the court would see that it was vigorously enforced.

After nearly ten years, during which the legislation urged by the court was also recommended by the Public Land Commission and in four special messages by President Arthur and President Cleveland, Congress got around to taking action. In 1888, it amended the Revised Statutes so as to prohibit specifically trespass on timberlands in all Indian reservations. A few months later the act of January 4, 1889, dealing only with the Chippewa Indian Reservation in Minnesota, provided for the cession to the United States of certain lands in the reservation which were to be disposed of so as to bring maximum prices for both land and timber, with the proceeds going into a permanent fund for the benefit of the tribe. These are the lands for the management of which more detailed provisions were made in the Morris Act of 1902.

Alternative Policies. Question of course arises as to whether the government could not have adopted a more constructive policy of handling the public timberlands than forbidding trespass on them and then virtually ignoring such trespass except for occasional sporadic and generally ineffective attempts at law enforcement. Certainly a complete locking up of the timber, even by the use of armed force, was impossible and undesirable. Wood in liberal amounts, which could come only from the public lands, was an essential item in the march of the American people across the continent.

Complete abandonment of any pretense of protecting the government's timber property was not impossible but was certainly undesirable. It would have led to even more confusion than actually attended the utilization of this property and would have greatly increased the difficulty of reserving it from appropriation and placing it under conservative management when in the fullness of time these actions became feasible.

Another alternative would have been to exempt public lands primarily valuable for the production of timber from the operation of the laws intended to promote settlement and to provide for the disposal of land and timber, or of timber only, at their estimated value. This procedure would have materially increased governmental receipts from timberlands and, if the timber only had been sold, would have permitted a later choice as to whether the lands involved should be transferred to private ownership or retained permanently in Federal ownership. While forest management of a high order would have been out of the question because of lack of technical knowledge and the temper of the people, it might have been possible to avoid some of the wasteful and destructive cutting, followed by even more destructive fires, that actually took place.

Such a policy would probably have been impossible of adoption until well after the middle of the nineteenth century because of the belief that most forest lands in the Eastern United States were suitable for agriculture and that the timber supply was inexhaustible. But by the 1870's or 1880's there is at least the possibility that it might have been accepted without too serious opposition. In retrospect it seems unfortunate that Congress did not take earlier action on the many recommendations it received from high public officials and other foresighted citizens that public land of chief value for timber production be so classified and the timber thereon sold separately from the land.

Free Timber. The Schurz-Williamson attack on timber trespass focused public attention on the fact that large areas of public lands were primarily valuable for timber production and that no distinctive method had been provided by which title either to these lands or the timber thereon could legally be obtained. Congress reacted by passing simultaneously two laws on the subject—the Free Timber Act and the Timber and Stone Act. These acts did little to improve the situation and in some ways made it worse.

The Free Timber Act of June 3, 1878, provided that residents of Colorado, Nevada, New Mexico, Arizona, Utah, Wyoming, Dakota, Idaho, and Montana, who are citizens of the United States or had declared their intentions to become citizens, might cut timber on mineral lands for building, agricultural, mining, or other domestic purposes, subject to such regulations as the Secretary of the Interior might prescribe. On its face the act was a well-intentioned attempt to provide without charge the timber needed by bona fide settlers and miners in the development of their farms and mineral claims. Yet it was also so broad and so restrictive in its provisions as to make its effective administration impossible. For example, cutting was limited to mineral lands, but as the Public Land Commission pointed out in 1880, "perhaps not one acre in 5,000 in the states and territories named is mineral and not one acre in 5,000 of what may be mineral is known to be such."

Among the detailed regulations for administration of the act issued by the Secretary of the Interior were a prohibition against the cutting of any tree less than 8 inches in diameter and a requirement that the brush and tops must be disposed of in such manner as to prevent the spread of forest fires. The difficulty of enforcement is illustrated by the statement of Commissioner Williamson in 1880 that "hundreds of charcoal burners consume in aggregate 1,200 cords of wood daily and use chiefly the small trees from 3 to 6 inches in diameter, it being more easily handled and quickly charred. This is done in direct violation of law and departmental regulation for the protection of undergrowth of timber on public lands."

Almost immediately after passage of the act he expressed the opinion that it was equivalent to a donation of all the timberlands to the inhabitants of the states named. Secretary Schurz promptly called attention to the basic defects in the act, stated that the machinery of the land offices was utterly inadequate to accommodate the object in view, and recommended its replacement by a law providing for the sale of timber without the land.

Congress not only turned a deaf ear to repeated recommendations by Commissioners of the General Land Office and Secretaries of the Interior that the act be repealed but materially extended its scope. The act of March 3, 1891, in effect permitted the free cutting of timber anywhere on the public domain in all of the Rocky Mountain states except Arizona and New Mexico not only for mining, agricultural, and domestic purposes but also for manufacturing. In 1893 these liberalized provisions were extended to Arizona and New Mexico, and in 1901 to all of the public-land states. Greater generosity would be hard to imagine.

Sale of Timber and Stone. The Timber and Stone Act of June 3, 1878, provided for the sale in Washington, Oregon, California, and Nevada of 160 acres of unoccupied, surveyed, nonmineral land, chiefly valuable for timber or stone and unfit for cultivation, which had not been offered at public sale, for not less than $2.50 per acre. The purchaser had to swear that he did not apply to buy the land on speculation but in good faith for his own exclusive use and benefit.

The act also forbade the unlawful cutting or wanton destruction of timber on any public lands or its removal for export or other disposal, with less severe penalties than those prescribed by the antitrespass acts of 1831 and 1859. This provision, which in 1892 was extended to all the public-land states, forms the basis for the section in the present Criminal Code relating to trespass on unreserved public lands. Further provisions of the act granted permission to miners and farmers to clear and use such timber from the public lands as was necessary for improvement, relieved trespassers who had not exported the stolen timber from the United States from both criminal and civil prosecution on payment of $2.50 for the timber, and covered all moneys collected as the result of trespass into the treasury of the United States.

The intent of the best-known part of the act—that relating to the sale of timber and stone—is far from clear. If the primary object was to provide actual settlers on the Pacific Coast with a supply of wood in addition to that on their preemption or homestead claims, it is odd that they were not simply offered free timber as was done by the act of the same date applying to the Rocky Mountain states, or that residents of the latter states were not given the opportunity to acquire wood lots. Possibly Congress felt that the large timber on the Pacific Coast was so much more valuable than that in the Rocky Mountains that a charge should be made for it, in return for which the purchasers should be allowed to obtain title to the land. On the other hand, if the object was to make timber legally available to the lumber industry for manufacture and sale in the open market, the maximum area of 160 acres that could be acquired by any one person was entirely too small for the purpose. The lumping of stone with timber in the act is also a minor mystery.

Whatever the intent, bona fide settlers made little use of the act, since they were usually more concerned with clearing off the heavy stands of timber already on their claims than in acquiring more. As might have been anticipated, the act was used chiefly by commercial timber operators to build up large holdings of valuable timberlands, with fraud playing a conspicuous part in the process. Commissioner Williamson immediately expressed the view, in which Secretary Schurz concurred, that under the provisions of the law timberlands would speedily be taken up and passed into the hands of speculators.

Their fears were amply justified. Speculation and fraud, which would undoubtedly have been rife under any circumstances, were greatly facilitated by a Supreme Court decision that a person who planned the immediate sale of his timber claim to another party was nevertheless taking it up "for his own exclusive use and benefit," unless collusion to sell prior to the making of the entry could be proved—obviously an exceedingly difficult task. The original purchaser from the government did not even have to await the issuance of patent. He could sell immediately upon receipt of a certificate of entry, and thereafter any number of subsequent transfers were of course possible.

Public Land Commission. General dissatisfaction with the operation of the land laws led Congress in 1879 to create a Public Land Commission to codify existing legislation and to recommend the best methods for the disposal of the public lands. The Commission visited all the Western states and territories and in 1880 submitted a constructive report which included the draft of a comprehensive bill "To provide for the survey and disposal of the public lands of the United States." One of its major recommendations was that the public domain be classified by government surveyors into arable lands, mineral lands, irrigable lands, pasturage lands, and timberlands, and that land in each class be disposed of under laws and regulations suited to its particular character.

The proposed bill provided that "all lands, excepting mineral, which are chiefly valuable for timber of commercial value for sawed or hewed timber, shall be classified as timber lands." The phrase "chiefly valuable for timber of commercial value" was intended to exclude arable, irrigable, and pasture lands. The Commission pointed out that

timber lands, as thus defined, comprise a comparatively small area. In Arkansas, Louisiana, Mississippi, Alabama, and Florida, but little, if any, lands will fall under the classification. There may be small areas in Michigan, Wisconsin, and Minnesota, which should come under the classification, but the existence of such tracts is a matter of doubt, and consequently the classification will probably only include lands in the Territories and Pacific States.

This analysis of the situation reflected the prevailing view that forest lands in the Lake states and the South were essentially agricultural in character.

All timberlands were to be withdrawn from sale or other disposal. Timber more than 8 inches in diameter on even-numbered sections was to be sold for cash in lots of not less than 40 or more than 640 acres, with no sales on odd-numbered sections. Settlers would be allowed free use of timber for building, agricultural, mining, and other purposes, but not for sale, commerce, or export, subject to such regulations as the Secretary of the Interior might prescribe. The Commission expressed the view that these provisions would "result in the maintenance and reproduction of the forest. The experiment is at least well worth trying."

Congress took no immediate action on any of the recommendations of the Commission, but they probably had an indirect influence on subsequent legislation such as the General Revision Act of 1891, including the provision for the establishment of forest reserves. One extremely valuable by-product of the work of the Commission was the preparation by one of its members, Thomas Donaldson, of an exhaustive history of the public domain. The third edition of this volume, published in 1884, constitutes an unequaled storehouse of information on legislation and regulations relating to the public lands prior to December 1, 1883.

Fraud and Collusion. Commissioner McFarland in 1884 emphasized the prevalently illegal character of entries under the Timber and Stone Act and stated that

the result of the operation of the act is the transfer of the title of the United States to timber lands practically, in bulk, to a few large operators. The preventive measures at the command of this office have proven wholly inadequate to counteract this result. The requirements of the law are slight and easily evaded, and evidence of fraudulent proceedings rests so much within the knowledge of interested parties that specific testimony can rarely be obtained. Thus, while results are observable, easily demonstrated, and of common notoriety, the processes by which they are reached are difficult to trace in a legal proceeding.

Two years later (1886) Commissioner Sparks got down to cases. One company in the Olympia district in Washington procured a number of persons, including several of its employees, to make nine timberland entries and eight preemption entries on 2,700 acres of valuable timberland. The timberland entrymen were given descriptions of their tracts by the company's agent, who paid the purchase money and land-office fees and obtained the certificates of entry, which were promptly conveyed to representatives of the company.

A speculator in the Vancouver district in Washington had an agent hire forty-five men to make timber entries on 7,000 acres of the finest timberland, most of which was valuable for agricultural purposes. Up to twenty claimants at a time were taken to the land office by the agent, who paid for the lands and arranged for their immediate conveyance to the speculator. In the Oregon City district in Oregon, 4,200 acres of agricultural lands containing valuable stands of timber were entered by persons who had never seen the tracts and knew nothing of their character.

Skulduggery in the Redwoods. Considerably more spectacular was the attempt to obtain title to some 100,000 acres of choice redwood land in the Humboldt district in California. Investigations by an agent of the land office showed that well-informed timber locators and expert surveyors had been employed to locate and survey the best lands.

Others were then hired to go upon the streets of Eureka and elsewhere and find persons who could be induced to sign applications for land and transfer their interests to the company, a consideration of fifty dollars being paid for each tract of one hundred and sixty acres so secured. The company's agent received five dollars for each applicant obtained. No effort seems to have been made to keep the matter secret, and all classes of people were approached by agents and principals of the company and asked to sign applications.

Sailors were caught while in port and hurried into a saloon or to a certain notary public's office and induced to sign applications and convey the lands to a member of the firm. Farmers were stopped on their way to their homes, and merchants were called from their counters and persuaded to allow their names to be used to obtain title to the lands. The company's agent presented the applications to the register and receiver in blocks of as many as twenty-five at one time; paid the fees; had the proper notices published; hired men to make the proofs; paid for the lands and received the duplicate receipts; yet the register and receiver and some of the special agents appear to have been the only persons in the vicinity who were ignorant of the frauds.

The ramifications of this particular case involved not only the local land office but penetrated into the headquarters of the General Land Office in Washington. On the strength of a letter from one of his subordinates recommending the entries for approval, the Commissioner directed the issuance of patents to 22,000 acres of land, although evidence of the fraudulent character of the claims was at the time on file in a detailed report from the special agent who had been assigned to investigate the case. This agent, who stated that he had declined a proffered bribe of $5,000 to suppress the facts and abandon the investigation,

was subsequently suspended from duty and afterwards dismissed from the service at the instance, as understood in this office, of great influence brought against him from the Pacific coast and in Washington.

The special agent who followed him was soon spotted by representatives of the company and given a lively time.

Some of the witnesses were spirited out of the country; others were threatened and intimidated; spies were employed to watch and follow the agent and report the names of all persons who conversed with or called upon him; and on one occasion two persons who were about to enter the agent's room at his hotel for the purpose of conferring with him in reference to the entries were knocked down and dragged away.

In spite of all difficulties, sufficient evidence was eventually obtained to result in cancellation of both patented and unpatented claims.

Results of Timber and Stone Act. The outcome in the case of the attempted redwoods frauds was, however, unusual. In most instances fraud was either not detected or could not be legally proved. As Commissioner Sparks realistically remarked, "It is possible that this grab is only an atom compared with what they may have secured by similar methods in other States and Territories." His statement with reference to public-land legislation in general that "it is idle to expect that public lands can be saved from fraudulent appropriations if laws under which fraud is made easy are permitted to remain, and if no adequate means are provided by Congress for suppression of fraud" expressed the sentiments of many a Commissioner of the General Land Office and Secretary of the Interior.

Some 15 million acres were sold under the Timber and Stone Act for a tithe of their true value. Although the area is not large compared with that disposed of under several other laws, the values involved were tremendous and the methods of acquisition generally unsavory. The only response by Congress to numerous and repeated recommendations that the law be repealed was to extend its provisions in 1892 to all of the public-land states. One improvement in administration was effected in 1908, when President Roosevelt and Secretary Garfield decided that not "less than" meant what it said. Land was thereafter sold for its appraised value instead of for the flat minimum of $2.50 per acre that had previously been charged.

Neglect of Range Lands. No legislation providing specifically for the disposal of public lands primarily valuable for grazing was passed by Congress prior to the Stockraising Homestead Act of 1916, which will be discussed in a later chapter. Until that time stockmen either obtained patent under legislation dealing with other classes of land, notably the preemption, homestead, timber-culture, and desert-land laws, or made free use of the range with no attempt to establish title except to the water holes that were the key to its successful utilization. For the most part the extensive areas of range land in the semiarid and arid West were looked upon as a public commons, the use of which under frontier theory was open to anyone, but which in practice was limited chiefly to those who controlled the limited water supply.

Although such use was unauthorized by law and constituted trespass against the government, no steps to prevent it were taken until some of the larger owners in effect turned immense areas of open range into private preserves by the construction of hundreds of miles of fences. This procedure led to complaints by the smaller owners and prospective settlers and to remedial action by the government.

In 1882 Commissioner McFarland of the General Land Office reported the widespread fencing of public lands

where no pretext of ownership or of legal claim to any part of the land exists. The usual routes of travel are also cut off by these inclosures, and the inhabitants of the country are in many instances compelled to go a great way around or to tear down the fences, thus incurring the risk of disturbance and perhaps bloodshed . . . in some cases State laws have provided for a nominal tax upon "possessory rights," the effect of which is represented to be to locally legalize this infringement upon the laws of the United States. . . .

It is undoubtedly true that the vast plains and mountain ranges west of the Mississippi River must be relied upon for an important proportion of the sheep and cattle husbandry required by the necessities of national consumption, but it does not therefore follow that this industry should be the subject of individual or corporate monopoly any more than that other agricultural pursuits should be so controlled.

The next year Commissioner McFarland, by direction of Secretary of the Interior Henry M. Teller, gave notice that the Department would

interpose no objection to the destruction of these fences by persons who desire to make *bona fide* settlement on the inclosed tracts, but are prevented by the fences, or by threats of violence, from doing so.

He added that the government would take proper proceedings against persons unlawfully enclosing the public lands. Experience soon showed, however, that it is one thing to issue orders and another to get them executed. Secretary Teller consequently urged Congress to pass legislation that would remove any doubt as to the authority of the Department and would enable it to remove illegal fences without the expense and delay of a suit in equity.

Control of Illegal Fencing. Two years later (in 1885) Congress passed an act making the enclosure of the public domain by fences illegal and authorizing their destruction. Since passage of the act by itself had little effect, President Cleveland on August 7 issued a proclamation calling attention to violations of the act and enjoining upon all persons obedience to it. More effective was the employment of two inspectors who proceeded to make life uncomfortable for the fence builders, chiefly in New Mexico and Wyoming, fortunately without encountering serious physical resistance. The law was catching up with the frontier.

The area included in fenced enclosures is unknown. Secretary Lamar in his annual report for 1887 stated that the General Land Office had records showing the existence of 465 illegal enclosures aggregating nearly 7 million acres. Of this area, nearly 5 million acres had already been opened up, and proceedings had been instituted for the destruction of fences enclosing an additional 3,275,000 acres.

That the total area involved was considerably larger is indicated by the following extract from a letter addressed by a stockman at Santa Fe to an inspector in the Department of the Interior:

The mania for fence-building began in 1878 and culminated in 1881, when a vast number of inclosures of public lands, ranging from 10,000 to 2,000,000 acres, existed all along the Rocky Mountain slope.

Every one engaged in the cattle business who could raise money enough had as large a pasture as he could build. The Government made no objection to such inclosures and a tacit understanding seemed to exist among ranchmen that all could build as many pastures as they saw fit, so they had the water rights in fee.

Rather unexpected, in view of the entire history of the public domain, is Secretary Lamar's surprise at the existence of such a situation:

How the illegal occupation and possession of the public domain could have grown to such enormous proportions is beyond my comprehension.

In its social, moral, and political aspect, not less than in its economic results, the existence and growth of such lawless combinations resting upon unlawful inclosures of the public domain, and protected by the acquirement of titles, through devious or more openly fraudulent methods, to the controlling sources of water supply, constitute to-day a phenomenon of American life and a problem of political import that demand the intelligent attention and action of the Congress of the United States. When it was first represented that substantially the entire grazing country west of the one hundredth meridian had been fenced in by cattlemen, I could not give credence to the statement of such lawlessness and rapacity. But the testimony of the residents of the Territories, who have no motive for misrepresentation, corroborated by reports based on personal inspection of agents and surveyors now on record in the Department, establish the fact. Indeed, it is publicly announced, by the trespassers themselves, who not only acknowledge that they have unlawfully fenced in the public lands without any limit except their own pleasure and power, but they justify their action, and complain against the execution of the law as an oppressive and unjust hardship.

Although removal of the illegal fences was pretty well completed by 1890, no attempt was made to prevent or to regulate grazing on the open range, which became the scene of many a lively conflict between stockmen and settlers ("nesters") and between cattlemen and sheepmen. Repeated recommendations that the lands be managed under a leasing system failed to get action from Congress until passage of the Taylor Grazing Act in 1934. The implied right to free use of the public lands for grazing purposes apparently conferred by the policy of noninterference was later to raise important questions concerning the administration of range lands included in the national forests.

Summary. A large part of the public domain consisted of land primarily valuable for timber, which was urgently needed for innumerable essential uses. Yet for a century Congress passed no legislation dealing specifically with the disposal of timber or timberlands. This situation inevitably resulted in extensive trespass on government timberlands and in the fraudulent acquisition of these lands under laws intended for other purposes. The Free Timber Act and the Timber and Stone Act of 1878, which for the first time recognized the existence of a timber problem, were so poorly framed as to make protection of the government's interests even more difficult than before.

In spite of the fact that they were so long ignored by Congress, the forests on the public lands played a major part in promoting the amazing advance of agriculture and industry in the United States and in molding the character of the people. The story has its splendid as well as its sordid side. Much of America's forest wealth was unquestionably plundered, legally and physically, and that plundering left an unwelcome legacy of depleted lands. It also helped mightily to conquer a continent and to build a nation. The many difficult problems created by the destructive utilization of the forest on both public and private lands gave serious concern to a steadily increasing number of farsighted persons both in and out of government. The steps taken toward their solution will be discussed in the next chapter.

Range lands in the public domain were even more completely ignored by Congress than were timberlands. Utilization of their rich forage resources for the production of livestock was nevertheless essential for the development of the Western country. Trespass and fraudulent acquisition were therefore almost universal. Except for the passage in 1885 of an act forbidding illegal fencing, Congress did nothing to remedy the situation until well into the twentieth century.

REFERENCES

Cameron, Jenks: American Forest Influences—Sea Power, *Amer. Forests*, 32: 707–711, 767–768, 1926.

————: President Adams' Acorns, *Amer. Forests*, 34:131–134, 1928.

————: An Anchor to Forestward, *Amer. Forests*, 34:199–201, 235, 1928.

————: Who Killed Santa Rosa? *Amer. Forests*, 34:263–266, 312, 1928.

————: "The Development of Governmental Forest Control in the United States," chaps. 3–7, Johns Hopkins Press, Baltimore, 1928.

Greeley, William B.: "Forests and Men," chaps. 2–3, Doubleday & Company, Inc., New York, 1951.

————: "Forest Policy," chap. 12, McGraw-Hill Book Company, Inc., New York, 1953.

Hibbard, Benjamin H.: "History of the Public Land Policies," chaps. 21, 23, The Macmillan Company, New York, 1924.

Ise, John: "The United States Forest Policy," chap. 1, Yale University Press, New Haven, Conn., 1920.

CHAPTER 4

Forestry in the Offing

For nearly a century following the outbreak of the American Revolution, interest in the forest resources of the country as expressed in legislative enactments was decidedly on the wane.

New Attitudes Appear. Prior to 1873 Congress passed only five laws dealing specifically with timber—the naval-timber purchase act of 1799, the two naval-timber-reserve acts of 1817 and 1827, and the timber-trespass acts of 1831 and 1859. These laws were occasioned primarily by concern for an adequate supply of timber for the Navy, although the trespass acts of 1831 and 1859 were broad enough in their terms to apply to all timber anywhere on the public domain.

During the same period most of the states and territories enacted laws aimed at the control of forest fires, but little attempt to enforce them was made. There was almost a complete dearth of state legislation dealing with trespass and the regulation of industry, which had received so much attention during the colonial period.

The change in attitude was due to three main factors. First was the growing belief in the inexhaustibility of the forest. Until about the middle of the last century a small population, chiefly agricultural in character, had made only insignificant inroads on the more than 600 million acres of virgin forest. In the second place, the building of roads, canals, and presently railroads was constantly opening up new bodies of timber. As the development of transportation facilities pushed back the frontier, the supply of accessible timber increased faster than that in the older settlements was being cut out. Finally, there was an outburst of the spirit of *laissez faire.* The new belief in economic as well as political freedom is well expressed in a resolution adopted by the Continental Congress in 1788: "It hath been found by experience that limitation in the price of commodities is not only ineffective for the purpose proposed, but likewise very productive of evil consequences, to the great detriment of the public service and the grievous oppression of individuals."

The same causes that reduced the interest of the legislator in the forest increased the interest of the businessman. Apparent inexhaustibility, greater accessibility, and rugged individualism encouraged heavier and heavier cutting to meet the insatiable demand for wood created by a rapidly growing population and an expanding industry. Sawmills grew in size and output as the center of the lumber industry moved from the Northeast through the Central Atlantic states to the Lake states, then to the South, and finally in the present century to the Pacific Coast. There was no lack of economic interest in the forest in a country where practically the entire population relied on wood for houses, vehicles, fences, boxes, railroad ties, and fuel.

Early Literature. Scientific interest in the forest was decidedly limited but keen. It was at first concerned chiefly with research in dendrology and dealt with determination of the characteristics and distribution of the amazing number and variety of trees that composed the American forest. Foremost among early students of the subject were two Frenchmen—André Michaux and his son, François André Michaux. The explorations of the elder Michaux covered the forests of North America west to the Mississippi River and north toward Hudson Bay during the period from 1785 to 1796. They were made under the auspices of the French government and were concerned chiefly, though by no means exclusively, with the oaks—a reflection of the governmental concern of the day with oak for naval construction. André Michaux was the author of "Histoire des Chênes de l'Amérique," published in Paris in 1801, and of "Flora Boreali-Americana," published in Paris in 1803, the year after his death.

François André Michaux extended and intensified the studies initiated by his father. He is the author of several important works, of which the best known is "The North American Sylva: A Description of the Forest Trees of the United States, Canada, and Nova Scotia, with a Description of the Most Useful European Forest Trees." This three-volume book, originally written in French, was translated into English by Augustus L. Hillhouse and published in Paris in 1819. It was later revised and enlarged by Thomas Nuttall.

The revision was published in three volumes in 1849 under the title "North American Sylva" and remained for many years the standard work in its field. Michaux, in the original text, had expressed his concern over the failure of either the Federal government or the states to establish forest reserves.

An alarming destruction of the trees proper for building has been the consequence, an evil which is increasing and which will continue to increase with the increase of population. The effect is already very sensibly felt in the large cities, where the complaint is every year becoming more serious, not only of excessive dearness of fuel, but of the scarcity of timber. Even now inferior wood is frequently substituted for the White Oak; and the Live Oak so highly esteemed in shipbuilding, will soon become extinct upon the islands of Georgia.

The younger Michaux in 1855 made bequests of $12,000 to the American Philosophical Society and of $8,000 to the Massachusetts Society of Agriculture and Arts in order to contribute "to the extension and progress of agriculture, and more especially of sylviculture, in the United States." The bequests did not become available until 1870, following the death of Michaux's wife, by which time their value in American money had been reduced approximately 50 per cent. The grant to the American Philosophical Society was used for the establishment of the Michaux Grove of oaks in Fairmount Park in Philadelphia and for the endowment of the Michaux lecture course at the University of Pennsylvania. Dr. James T. Rothrock, the "father of Pennsylvania forestry," made a signal contribution to sound thinking with respect to forest policy in the annual lectures which he gave in this course during the period from 1877 to 1894. The legacy to the Massachusetts Society of Agriculture and Arts was used for the support of the Botanical Garden and the Arnold Arboretum at Harvard University and for the publication of a pamphlet on forest culture.

Three other books attest to the interest in the dendrological aspects of forestry during the first half of the nineteenth century. In 1837 the Massachusetts legislature ordered a special survey of the state's forest resources. The result was *A Report on the Trees and Shrubs Growing Naturally in Massachusetts* by George B. Emerson, a comprehensive treatise on the subject which was published in 1846 as a state document. This classic was republished in 1875 by Little, Brown & Company in a two-volume edition with 624 pages and 124 plates. In it Emerson commented on the fact that the little state of Massachusetts had more species of timber trees than any country of Europe and made many suggestions with respect to their cultivation. Books with a more extensive geographic coverage are J. D. Brown's "Sylva Americana," published in 1832, and R. U. Piper's "The Trees of America," published in 1855. Both men viewed with alarm the current "devastation" of the forests.

Encouragement of Tree Planting. Sporadic attempts were made to encourage tree planting and experimental activities. For example, in 1804 the Massachusetts State Society for Promoting Agriculture offered prizes for the best plantations of certain hardwoods "in the proportion of 2,400 to the acre"; and in 1818 the state authorized agricultural societies to offer premiums to encourage the growth of oaks and other trees of value for shipbuilding material. Mention has previously been made of Brackenridge's essay on the live oak and of the experiments in the planting and pruning of live oak at the Santa Rosa naval-timber reserve. At about this same time (1827) the Secretary of the Treasury informed certain consuls of the desire of President Adams to introduce into the United States all such trees and plants, including "forest trees useful for timber," as gave promise of being of value in this country and requested their cooperation (at their own expense) in obtaining such material.

This first official attempt at foreign-plant introduction brought at least one response—from Dr. Henry Perrine, a consul at Campeachy, Mexico. He succeeded in getting Congress in 1838 to grant him and his associates a township of public lands at the south end of the Florida peninsula on condition that the lands be occupied by bona fide settlers actually engaged in the propagation or cultivation of useful tropical plants; but this enterprise was cut short by his murder two years later by Seminole Indians. Another proposed experiment was nipped in the bud when the Senate Committee on Public Lands in 1830 turned down the request of several citizens for a grant of a township of land in Missouri for the purpose of experimenting in the raising of forest timber in the prairies. The refusal was based on the ground that "it is only necessary to keep out the fire, to cover those prairies with timber by the operation of nature."

The efforts of the Federal government to encourage tree planting through the Timber Culture Act of 1873 and later amendments are discussed in Chapter 2.

Warnings of Timber Shortage. The clearest evidence of the growing interest in forests is the repeated warning voiced by men both in and out of government from about the middle of the century on, with respect to the dangers faced by the country as a result of the mounting destruction of forests. In 1849 the Commissioner of Patents stated: "The waste of valuable timber in the United States will hardly begin to be appreciated until our population reaches 50,000,000. Then the folly and shortsightedness of this age will meet with a degree of censure and reproach not pleasant to contemplate." Eleven years later a special article in the annual report of the Patent Office discussed at length the influence of forests on climate and health. In 1858 the Georgia Legislature requested Congress to appoint a commission to inquire into the limits and extent of the southern pine belt and its probable duration under the present rate of depletion.

One of the most thoughtful articles of the day, "American Forests, Their Destruction and Preservation," was written by the Rev. Frederick Starr of St. Louis and published in the annual report of the Commissioner of Agriculture for 1865.

It is feared it will be long [wrote Mr. Starr], perhaps a full century, before the results at which we ought to aim as a nation will be realized by our whole country, to wit, that we should raise an adequate supply of wood and timber for all our wants. The evils which are anticipated will probably increase upon us for thirty years to come, with ten-fold the rapidity with which restoring or ameliorating measures shall be adopted.

In 1869 a committee appointed by the Michigan Legislature to look into the matter of forest destruction reported:

The interests to be subserved, and the evils to be avoided by our action on this subject have reference not alone to this year or the next score of years, but generations yet unborn will bless or curse our memory according as we preserve for them what the munificent past has so richly bestowed upon us, or as we lend our influence to continue and accelerate the wasteful destruction everywhere at work in our beautiful state.

This sentiment found its echo many years later in Gifford Pinchot's admonition:

According as we accept or ignore our responsibility as trustees of the nation's welfare, our children's children for uncounted generations will call us blessed, or will lay their suffering at our doors.

A few years earlier (1864), George P. Marsh had published his famous book "Man and Nature," which was republished in 1872 under the title "The Earth as Modified by Human Action." This book dealt in great detail with the evil effects of forest destruction on climate and the water supply, particularly in the Mediterranean region, and had great influence on current thinking on these subjects in the United States. It is still a classic in its field.

Attitude of Interior Department. Annual reports of the Secretary of the Interior and the Commissioner of the General Land Office show increasing recognition of the threat posed by forest destruction to the economy of the country. Commissioner Wilson in 1866 felt that the supply of timber in the Lake states was already "so diminished as to be a matter of serious concern." In 1868 he predicted that in forty or fifty years our own forests would have disappeared and those of Canada would be approaching exhaustion.

Commissioner Williamson in 1876 expressed the belief that "a national calamity is being rapidly and surely brought upon this country by the useless destruction of the forests." Nine years later (1885) Commissioner Sparks stated: "The importance and necessity of preserving our remaining forest and woodland is urged upon the attention of the legislators and the public by thoughtful persons, scientific bodies, and patriotic associations throughout the country."

During the four years that he was Secretary of the Interior (1877–1881), Carl Schurz emphasized the importance of forests and the dangers of forest destruction with characteristic vigor. In his annual report for 1877 he declared:

The rapidity with which this country is being stripped of its forests must alarm every thinking man. It has been estimated by good authority that, if we go on at the present rate, the supply of timber in the United States will, in less than twenty years, fall considerably short of our home necessities. How disastrously the destruction of the forests of a country affects the regularity of the water supply in the rivers necessary for navigation, increases the frequency of freshets and inundations, dries up springs, and transforms fertile agricultural districts into barren wastes, is a matter of universal experience the world over. It is the highest time that we should turn our earnest attention to this subject which so seriously concerns our national prosperity.

In subsequent reports he elaborated on the idea that "a provident policy, having our future wants in view, cannot be adopted too soon." His final warning was decidedly to the point:

Our forests are disappearing with appalling rapidity, especially in those parts of the country where they will not renew themselves when once indiscriminately destroyed. Like spendthrifts, we are living not upon the interest but upon the capital. The consequences will inevitably be disastrous, unless the Congress of the United States soon wakes up to the greatness of the danger and puts this ruinous business upon a different footing by proper legislation. . . .

From Talk to Action. During the 1870's several events occurred which helped to translate from mere talk into action the growing awareness of the essential part played by forests in both the temporary and the permanent prosperity of the country. In 1872 Congress established the Yellowstone National Park in Wyoming and thereby took the first step toward the present extensive system of public-land reservations. The Hot Springs Reservation in Arkansas (renamed the Hot Springs National Park in 1921) had, it is true, been created as early as 1832 by the reservation of four sections of public land containing the well-known hot springs, but the area involved was so small and its use so completely restricted to medical purposes as to place it in a class by itself.

Great credit is due to Cornelius Hedges, Nathaniel P. Langford, and their associates, who first really explored the Yellowstone wonderland of geysers, hot springs, lakes, falls, and forests, for their fight to have it preserved for the enjoyment of all the people, unspoiled and free of toll, instead of stretching one of the public-land laws to acquire it for their own use and profit, as they might readily have done. Congress was, however, so slow about taking any action for the protection and administration of the park that in 1877 Secretary Schurz reported: "No appropriation has been made for the pay of a superintendent or the survey of the park, and no revenues have been received, nor have any leases been granted by the department."

Congress was slow also to take the next step in the expansion of the national-park system. In 1879 Secretary Schurz recommended "that the President be authorized to withdraw from sale or other disposition an area at least equal to two townships in the coast range in the northern, and an equal area in the southern part of the State of California" for the preservation of samples of the virgin forests of redwoods and big trees, "the noblest and oldest species of trees in the world." Not until 1890 was any protection given the big trees, and California later had to assume the task of creating a park in the redwoods by purchasing at great expense lands once in the public domain that had been allowed to pass into private hands.

Arbor Day and Arnold Arboretum. The first Arbor Day was celebrated in Nebraska in 1872 at the instance of J. Sterling Morton, a prominent citizen who was later Secretary of Agriculture. The purpose was to stimulate interest in tree planting, and the idea proved so popular that the celebration of Arbor Day, proclaimed in each case by the governor of the state concerned, soon became a national habit. The doubt subsequently felt by many as to the effectiveness of Arbor Day in arousing interest in forest planting as contrasted with shade-tree planting was expressed by B. E. Fernow: "I am not sure but this otherwise interesting and beautiful idea has had a retarding influence on practical forestry by misleading people into thinking that tree planting was the main issue instead of a conservative management of existing forests." Fernow was, nevertheless, active in urging not only state but national observance of Arbor Day, and there can be little question that its net effect has been helpful from the standpoint of forestry as well as of arboriculture.

Still a third event of importance in 1872 was the bequest of $100,000 to Harvard University by James Arnold for the purpose of establishing in the Bussey Institution a professorship of tree culture and of creating and maintaining on the Bussey estate a collection of all the trees, shrubs, and herbs, both native and exotic, that could be grown out of doors in that climate. This bequest led to the establishment of the famous Arnold Arboretum, which has contributed materially to our knowledge of the taxonomic and silvical characteristics of the trees and shrubs that have gradually been included in it.

Mention has already been made in Chapters 2 and 3 of the appropriations provided by Congress in 1871 and 1872 for the protection from timber trespass of lands in naval-timber reserves and on the unreserved public domain; of the Timber Culture Act of 1873; of the Free Timber Act and the Timber and Stone Act of 1878; and of the Public Land Commission of 1879. These various measures represent the transition from talk to action during the 1870's so far as timber on the public lands is concerned. The action programs embodied in the establishment of the Division of Forestry in 1876 and the passage of the Forest Reserve Acts of 1891 and 1897 will be discussed later, together with the educational activities that played such an important part in the development of forest policy during this period.

A.A.A.S. Takes a Hand. At the twenty-second meeting of the American Association for the Advancement of Science (A.A.A.S.) held at Portland, Maine, in 1873, Franklin B. Hough of Lowville, New York, presented a brief but forceful paper that was to have notable consequences. The association was organized in 1849 for the purpose of promoting progress in all fields of science—physical, biological, and social. It is a popular organization in the sense that membership is open to anyone interested in its objectives, but leadership is naturally exercised by persons who are themselves engaged in scientific work and who are usually members of the many affiliated scientific societies. Effective support in the development of a sound national forest policy has been rendered by the association on several occasions.

Hough was a medical doctor, naturalist, historian, and statistician who had developed a keen interest in forestry. In 1872 he served on a New York State Park Commission which recommended legislation stopping further sale of state lands. His paper before the A.A.A.S. was entitled "On the Duty of Governments in the Preservation of Forests." In it he stressed the need to educate both the general public and timberland owners as to the importance of forests in relation to timber supply, climate, erosion, and runoff and indicated several ways in which such educational activities might be conducted. At his suggestion the association "*Resolved,* That a committee [shall] be appointed by the association to memorialize Congress and the several State legislatures upon the importance of promoting the cultivation of timber and the preservation of forests, and to recommend proper legislation for securing these objects."

The committee had a distinguished membership which included Franklin B. Hough, chairman, and such eminent natural scientists as George B. Emerson, Asa Gray, William H. Brewer, and E. W. Hilgard. Hough and Emerson, acting in behalf of the committee, formulated the views of the association in a statement which was transmitted to Congress by President Grant in February, 1874. After pointing out the great public injury likely to result from the rapid exhaustion of the forests of the country, with no effectual provision against waste or for the renewal of the supply, the memorial requested the creation of a commission of forestry which would report directly to the President, as was the case with the recently created Commission on Fish and Fisheries.

Establishment of Division of Forestry. After a public hearing, the House Committee on Public Lands reported favorably a bill introduced by Representative Herndon of Texas providing for the appointment of a Commissioner of Forestry. No further action was taken on the bill at that session of Congress, apparently because of pressure of other business rather than of any outright opposition. The bill was again introduced in the next Congress by Representative Dunnell of Minnesota, but the prospect of favorable action seemed so slim that Mr. Dunnell changed his tactics and attempted to achieve much the same end by adding an amendment to the section of the sundry civil appropriation bill for 1877 providing for the free distribution of seeds for experimental purposes.

This rider, which remained in the act as approved by the President on August 15, 1876, provided

that two thousand dollars . . . shall be expended by the Commissioner of Agriculture as compensation to some man of approved attainments, who is practically well acquainted with methods of statistical inquiry, and who has evinced an intimate acquaintance with questions relating to the national wants in regard to timber to prosecute investigations and inquiries, with the view of ascertaining the annual amount of consumption, importation, and exportation of timber and other forest products, the probable supply for future wants, the means best adapted to their preservation and renewal, the influence of forests upon climate, and the measures that have been successfully applied in foreign countries, or that may be deemed applicable to this country, for the preservation and restoration or planting of forests, and to report upon the same to the Commissioner of Agriculture, to be transmitted by him in a special report to Congress.

Thus did the promotion of forestry become a function of the Department of Agriculture through a rider attached to a minor item in an appropriation bill without debate and probably without even the knowledge of most members of Congress. The Commissioner of Agriculture, Frederick Watts, lost no time in complying with these instructions. On August 30, 1876, Hough was appointed to undertake the comprehensive task outlined by Congress, with an appropriation of $2,000 to cover all expenditures for salary and other costs. The Federal government embarked on its first venture in the field of forestry with one man and a financial shoestring. From this modest beginning has evolved the present U.S. Forest Service.

The Division under Hough. Hough tackled his encyclopedic assignment with energy and enthusiasm. In the absence of funds for travel or the employment of assistants, he had to compile the facts for his report by reading and correspondence. He was a tireless letter writer and succeeded in collecting an amazing amount of information from persons having some knowledge of the subject of his inquiries both at home and abroad. The result was the monumental *Report upon Forestry* (650 pages) prepared in 1877 and published in an edition of 25,000 copies in 1878. This report constitutes a comprehensive collection of information of the most heterogeneous character. The material varied in quality from the trivial to the highly significant but is all of historical interest as representing current information and thinking on the subjects treated.

The second *Report upon Forestry* was published in 1880 and ran to 618 pages, of which nearly 500 pages were devoted to United States exports and imports of forest products and to the timber resources and timber trade of Canada. The remainder of the report dealt mainly with the Timber Culture Act of 1878, with timber on the public lands, with recent state and territorial legislation relating to forestry, and with miscellaneous topics.

Hough's third report, published in 1882, shrank to a mere 318 pages. As usual it covered a wide variety of subjects, but with major attention to forest fires, the importance of which was strongly stressed. With respect to timber on the public lands, he expressed his agreement with the reservation policy proposed by Schurz and Williamson:

We would therefore earnestly recommend *that the principal bodies of timber land still remaining the property of the government . . . be withdrawn from sale or grant under the existing modes for conveying the public lands, and that they be placed under regulations calculated to secure an economical use of the existing timber, and a proper revenue from its sale, the title being retained by the government, and the young timber, in all leases for cutting, being reserved and protected for a future supply.*

The report also urged vigorously the establishment of forest experiment stations in various parts of the country and discussed the importance of meteorological observations as a means of determining the effect of forests on climate. Many of the views with respect to forest research and forest influences expressed in this report were doubtless influenced by a trip which he had made to Europe in 1881.

The general character of Hough's three reports, because of which he was awarded a diploma by the Vienna International Congress of 1882, was well summarized by Fernow in 1899 as follows:

The appropriations being extremely limited, special original research was excluded, and Dr. Hough being acquainted with the subject as an interested layman and not as a professional forester, these reports, while valuable compilations of facts from various sources, naturally did not contain any original matter, except such suggestions as Dr. Hough could make with regard to the duties of the Government with reference to the forestry interests of the country and especially of the public domain.

Certainly he was one of the outstanding leaders of the day in bringing about public appreciation of the importance of the country's forest resources and of the need for their conservative management. In addition to his official reports, Hough wrote a book entitled "The Elements of Forestry," published in 1882, which constituted the first American textbook on the subject.

The Division under Egleston. By 1881 the studies and other activities of the Department of Agriculture relating to forestry had attained sufficient scope and stability to justify their recognition by the Commissioner for administrative purposes as the Division of Forestry. Two years later (1883) Hough was replaced as chief of the Division by Nathaniel H. Egleston of Massachusetts, a former minister with little previous experience in the field. Hough continued his services as an "agent" in the Division. Whatever the reasons for the change, they were not based on the relative competence of the two men.

In his first annual report (for 1883) to the Commissioner of Agriculture, Egleston went all out in his emphasis on the indispensability of forests as a basis for urging Federal action in their protection:

The Government cannot interfere with or regulate the use or consumption of forests which belong to individuals, corporations, or the separate States. That must be left to the influence of increased and diffused knowledge and enlightened self-interest. But nothing seems clearer than that the Government should take care of its own property and use it for the general welfare. And today it has no property so valuable as its forests. Its mines, its forts, its ships, the coined money in its vaults, taken together, are hardly comparable with them. These might all be lost without essential or permanent injury to the nation, while the loss of the forests would threaten desolation and national decay and destruction.

In addition to his advocacy of governmental action to protect the timber on the public lands, he recommended establishment by the government of forest schools and of forest experiment stations at the capital and in other parts of the country, including "that peculiar region, the Pacific."

The fourth and final special *Report on Forestry* was published in 1884 under Egleston's name. It was less heterogeneous than its predecessors and consisted mainly of reports by agents of the Division, four of them by Hough, on tree planting, timber culture, forest conditions, and the utilization of forest products. One of its most interesting parts was a county-by-county presentation of the results of tree planting in the Prairie states, with a list of the trees that had proved successful and unsuccessful. Most prominent among the reasons for failure were drought, freezing, insects, prairie fires, negligence, and lack of knowledge and experience.

Fernow Takes Over. In 1886 Congress granted statutory recognition, previously lacking, to the Division of Forestry. That same year, Bernhard Eduard Fernow became its chief, a Republican chosen by a Democratic administration as the only man in the country with the requisite professional qualifications for the office.

Fernow was a German who had received professional training in forestry at the well-known school of forestry at Muenden in the Province of Hannover in western Prussia, followed by several years of practical experience in the diverse forests of Silesia, Brandenburg, and East Prussia. He had come to the United States in 1875 at the age of twenty-five ostensibly to attend the Centennial Exposition at Philadelphia, but actually to marry an American girl to whom he had become engaged while she was on a visit to Germany. He became a citizen of the United States on December 14, 1883.

Prior to his appointment in the Federal service, Fernow had spent most of his time in the management of iron furnaces for Cooper, Hewitt and Co. in Pennsylvania. This position he owed to his friend and mentor Rossiter W. Raymond, who as United States Commissioner of Mines and Mining had observed and warned of the destruction of the forests. Fernow's work included the management of 15,000 acres of hardwoods to supply charcoal for the iron works, and he made the most of the opportunity to study American forest conditions on the ground. He had first come to the attention of the growing group of men interested in forestry matters at the American Forestry Congress at Cincinnati in 1882, at which he read a paper and of which he later served as secretary for several years.

With Fernow's advent the work of the Division of Forestry took a new slant. While the Division of necessity continued to be primarily a bureau of information and investigation, the fact that its activities were now under the direction of a professional forester gave them a different tone. In particular Fernow stressed forest management as a much broader and more fundamental aspect of forestry than forest "culture," which had come to mean little more than reforestation by tree planting. To him the real objective was to protect and harvest the forest so as to avoid the need for planting. In season and out, he preached the doctrine that forestry should start with the forest, not with bare land. Like his predecessors he favored the establishment of forest experiment stations as a means of providing the basic information on which sound forest management must rest.

Research and Education. Under Fernow's direction, the research work of the Division became more intensive and more professional in character. It covered the entire scope of forestry, which he described as resting on three main bases: (1) scientific basis, including forest biology, timber physics, and soil physics and soil chemistry; (2) economic basis, including statistics, technology (applied timber physics), and forest policy; and (3) practical basis, including organization of the forest, management of the forest, forest regulation, and harvest. Although he believed that forests have an effect on precipitation, he succeeded in getting excused from spending $2,000 appropriated by Congress in 1890 for conduct by the Division of experiments in the artificial production of rain.

An important innovation in the work of the Division was the inauguration of research in "timber physics," which Fernow defined as comprising "not only the anatomy, the chemical composition, the physical and mechanical properties of wood, but also its diseases and defects and a knowledge of the influences and conditions which determine structural, physical, chemical, mechanical, and technical properties." The new field, he pointed out, had economic as well as scientific utility since "the properties upon which the use of wood, its technology, is based should be well-known to the forest manager if he wishes to produce a crop of given quality useful for definite purposes."

Education, both popular and professional, was also close to Fernow's heart. During his second year as Chief of the Division of Forestry he addressed a circular to educational men in which he said:

Schools of every grade, without departing at all from their proper work, can supply some practical lessons in regard to the objects and use of forests, the nature and growth of trees, and the significance of their existence or absence, awakening thereby the interest of pupils in a kind of knowledge too little fostered in the schools of the agricultural classes. At schools of the higher grade it can be united with instruction in botany and natural history in general. In colleges forestry should be presented in lectures on its various relations to arboriculture, agriculture, and political economy.

These suggestions are wholly in line with modern efforts to have some knowledge of forestry, and also of the conservation of other natural resources, acquired by students at all levels in the educational system as an integral part of their study of other subjects. Fernow's own speeches and articles were commonly educational in character and included several series of lectures at institutions of higher learning.

"Providential Functions of Government." Fernow shared the views of many other farseeing men of the day about the dangers inherent in the forest destruction that was proceeding at an accelerating rate, and particularly about the depredations on the public lands. In 1887, as a result of a trip to the Rocky Mountains, he found it to be admitted everywhere "that the present conditions of administration have become insufferable and that the practical forestry work of the Government should first of all be directed to the protection and proper administration of its timber lands." To this end he prepared the draft of a bill providing for the establishment and management of forest reservations which was introduced by Senator Hale of Maine and helped to pave the way for the forest-reserve provision in the General Revision Act of 1891.

Closely connected with Fernow's efforts to obtain Federal reservation and administration of the public timberlands was his belief that government must take the leadership in other directions in promoting the conservation of natural resources. The view that the future of the country cannot be left wholly to the operation of enlightened self-interest, at least with respect to the basic means of subsistence, was forcefully expressed in an address entitled "The Providential Functions of Government with Special Reference to Natural Resources," which he delivered before the A.A.A.S. in 1895 as vice-president of its Section on Social and Economic Science.

In this address he pointed out that "a nation may cease to exist as well by the decay of its resources as by the extinction of its patriotic spirit." With respect to forests, he stated that

the forest resource is one, that under the active competition of private enterprise is apt to deteriorate and in its deterioration to affect other conditions of material existence unfavorably; that the maintenance of continued supplies, as well as of favorable conditions, is possible only under the supervision of permanent institutions, with whom present profit is not the only motive. It calls preeminently for the exercise of the providential functions of the State to counteract the destructive tendencies of private exploitation. In some cases restriction of the latter may suffice, in others ownership by the State or some smaller part of the community is necessary.

In spite of discouragingly meager appropriations, the Division of Forestry under Fernow's leadership had a marked influence on the thought and action of the country in forestry affairs. In addition to its specific activities in the fields of education and research, it focused public attention on the main issues involved in the forest problem, helped to make Federal forest reserves an accomplished fact, and in general paved the way for the remarkable developments that were to take place after the turn of the century.

American Forestry Association. In 1873, John A. Warder, an Ohio physician, pomologist, landscape gardener, and amateur forester, was one of the United States commissioners to the International Exhibition at Vienna, where he made a special study of forests, forest products, and forestry. Although denied by the State Department the opportunity to study forest conditions and forest practices in the forests themselves, he prepared a comprehensive report on European forestry which was published in 1876 as a House document. In this report he stressed the fact that forestry, "though quite unknown as an art in our country," must receive greater attention than heretofore. "The increasing scarcity of timber within the first century of the nation's history and that in a country famous for the richness and value of its sylva, and for the extent of its woodlands, is a subject that calls for the most serious consideration of the statesman, and perhaps also for the interference and care of government."

On September 10, 1875, the American Forestry Association was organized at the Grand Pacific Hotel in Chicago under Warder's leadership. He became its first president and continued in that capacity until the association was amalgamated with the American Forestry Congress in 1882. Fernow's biographer, Andrew Denny Rodgers III, characterizes Warder as "the leading figure of the early forestry movement in America" and "the founder of the first organized effort in the cause of forestry in America."

The next meeting of the American Forestry Association was held at Philadelphia on September 15, 1876. At that time it absorbed the American Forestry Council, a small group which had been formed following the 1873 meeting of the A.A.A.S. but which had never been active. At its 1880 meeting in Washington, the association asked Congress to appoint a commission to study forestry in Europe, and the next year Hough was sent abroad for that purpose. However, the association did not thrive, and its last meeting was held at Rochester, New York, on June 29, 1882, when plans were made for the anticipated union with the American Forestry Congress.

American Forestry Congress. The first American Forestry Congress was organized as an entirely distinct entity from the association, although Warder took a prominent part in its organization and presented no less than six papers at the congress itself. The motivation for such a conference resulted from a visit by Oberförster Baron Richard von Steuben to Cincinnati, during the course of which he commented on the need for constructive action to check forest destruction and encourage forest management in the United States. Whether or not local politics also became involved, as has been alleged, there can be no doubt as to the sincerity of those who arranged the literary part of the program.

The meeting was held at Cincinnati, Ohio, April 24 to 29, 1882, with great fanfare. Some 30,000 invitations were distributed; there was a parade of 60,000 school children; and Ohio's first Arbor Day was celebrated by initiating "a movement in miniature of the great scheme of replanting our denuded hills and valleys by planting groves of trees for each of our Presidents, and to the memory of . . . many poets, orators, and statesmen." The more spectacular part of the program was supplemented by the presentation, in full or by title, of eighty-seven papers covering a wide diversity of subjects.

The congress decided to perpetuate itself as a permanent organization and adopted a constitution the first article of which read: "The object of this Congress shall be to encourage the protection and planting of forest and ornamental trees, and to promote forest culture." The initial membership comprised seventy-three persons, including nearly all of those prominent in the forestry movement with the notable exception of Carl Schurz and Charles S. Sargent, who did not attend. Although there is perhaps some question as to how much the congress contributed to real progress in forestry, it aroused widespread popular interest and established on a permanent basis an organization that was later to exercise a powerful influence in forestry affairs.

Subsequent Meetings. The next meeting of the American Forest Congress was held at Montreal in August, 1882. At this meeting the older American Forestry Association was absorbed and its members welcomed as full-fledged members of the congress. Fewer papers were presented than at the Cincinnati meeting, but on the whole they were perhaps a bit more substantial in character. Of special interest was a paper by J. K. Ward of Montreal entitled "A Few Practical Remarks from the Lumberman's Standpoint," since that standpoint seldom found expression in the discussions of the day. A resolution adopted by the congress bestowed upon James Little, a veteran lumberman who had fought for years for the preservation of Canada's pine forests, the honorary appellation of "Nestor in American forestry"—a variation on the "father" title which Fernow later applied to Warder.

Subsequent meetings of the congress were held at St. Paul, Washington, Saratoga, Boston, Denver, Springfield (Illinois), and Atlanta. The meeting at Atlanta in 1888 was held in conjunction with the Southern Forestry Congress and resulted in a consolidation of the two organizations. The next year (1889) the combined organizations assumed the name The American Forestry Association. Under that name it has since continued its efforts as a popular organization, with membership open to all interested persons, to promote public understanding and support of sound national, state, and private forest policies—on numerous occasions with marked success.

In 1883 Fernow became corresponding secretary of the American Forestry Congress, a position which he held until 1888. From then until 1898 he served as chairman of the executive committee of the congress and the association, and from 1885 to 1898 he was editor of the *Proceedings*. His simultaneous occupation of these positions and of the position as Chief of the Division of Forestry put him in a strategic position to exercise great influence in the development of forest policy during this critical period.

Other Educational Activities. Minnesota in 1876 had the honor of establishing the first state forestry association. That same year, the state legislature appropriated $2,500, and the next year $2,000, to advance the objects of the association. The funds were used chiefly to promote tree planting, which was done on an extensive scale.

During the next ten years state forestry associations were organized in Ohio, Colorado, New York, and Pennsylvania. The Pennsylvania Forestry Association, organized in 1886, immediately started publication of a periodical *Forest Leaves*, which has had an enviable record of continuous publication from that date to this—since the winter issue of 1951 under the name *Pennsylvania Forests*.

In 1882, Hough undertook as a private enterprise to serve as editor of *The American Journal of Forestry*—a periodical "devoted to the interests of forest tree planting, the formation and care of woodlands, and ornamental planting generally, and to the various economies therein concerned." This undertaking had been suggested by Hough at the Philadelphia meeting of the American Forestry Association in 1876, when he pointed out the need for "a journal that shall do for forestry what the *American Journal of Science and Art* has done for the sciences generally." The suggestion was renewed at the Cincinnati meeting of the American Forestry Congress in 1882, and Hough was doubtless encouraged by the enthusiasm with which it was received.

The American Journal of Forestry was a highly useful monthly publication, containing original articles on all phases of tree planting and forestry, notes on current events, and bibliographic notices. Unfortunately the financial support which it received did not justify its continuation. It ran for one full year—from October, 1882, to September, 1883—when the editor and publishers announced its final suspension because its slender patronage amounted to less than the cost of publication. Hough had appraised the situation correctly in 1876 when he said, "The time will surely come when such an enterprise will be demanded, and will be sustained, although perhaps not now."

In 1884, Fernow began publication of a *Forestry Bulletin,* but it lasted for only three issues. Sargent had better success with *Garden and Forest,* which he started in 1888 as a journal of horticulture, landscape gardening, and forestry, and which continued until 1897. It appealed to a wide audience and was used by Sargent, among other things, as a vehicle for expressing his support of the establishment and businesslike administration of forest reserves. Two other works by Sargent which appeared in the 1880's and 1890's deserve special mention. These are the *Report on the Forests of North America* (1884), prepared for the Tenth Census and presenting the first truly comprehensive picture of the forest resources of the country, and the fourteen-volume "Silva of North America" (1890–1902), which is by far the most complete taxonomy of the trees of the continent yet to be published.

Move to Establish a School of Forestry. In 1880 the Chamber of Commerce of St. Paul, Minnesota, petitioned Congress to grant 300 sections (192,000 acres) of public land to the state of Minnesota for the establishment of a school of forestry. The idea was doubtless suggested by General C. C. Andrews, who served as chairman of the chamber's special committee on the subject. His interest in forestry had been aroused while he was Ambassador to Sweden, and in 1872 he had submitted to the State Department an excellent report on forests and forest culture in that country.

Before submitting the memorial to Congress, the chamber solicited the views of a number of distinguished persons as to the merits of the proposal. Of the fourteen replies received, twelve were favorable. The two unfavorable replies were from President Charles W. Eliot and Dr. Charles S. Sargent, both of Harvard University.

President Eliot did not think that such a technical school should be free; did not see why the one interest of forestry should be selected for such support rather than any other considerable industrial or commercial interest; and believed that if the government were going to spend money at all for education it should be for elementary education.

Aside from the propriety of the proposed subsidy, Dr. Sargent opposed the plan because "there are no teachers to teach and no scholars who want to be taught." He believed that the need for such schools would arise in time and that in the meanwhile attention should be devoted to the establishment of forest experiment stations. On the other hand, J. D. Ludden, a Minnesota lumberman, expressed the view that there was already a large and most inviting field for men thoroughly educated in the science and practice of forestry.

A bill providing aid for a school of forestry to be established at St. Paul was introduced in Congress by Senator McMillan of Minnesota in 1880, but no action was taken on it. Bills seeking similar aid in other states during the next few years met the same fate. Congress was evidently not interested, and schools of forestry were eventually to develop without direct support from the central government.

Special lectures and courses in forestry were, however, given at a number of institutions of higher learning prior to the inauguration of professional instruction in forestry in 1898. William H. Brewer's lectures at Yale in 1873 were perhaps the first of this character. The next year A. N. Prentiss instituted instruction in forestry at the New York State College of Agriculture at Cornell University, and in 1882 Volney M. Spalding offered a course in forestry at the University of Michigan that was included among the requirements for a degree in the School of Political Science. In his annual report as Chief of the Division of Forestry for 1887, Fernow noted that some instruction in forestry was also being offered at the New Hampshire, Massachusetts, Michigan, Iowa, and Missouri agricultural colleges, at the University of Pennsylvania (under the Michaux fund), and at the University of North Carolina. Fernow himself delivered a comprehensive series of lectures on technical forestry at the Massachusetts Agricultural College in 1894.[1]

More National Parks. Nearly twenty years after the establishment of the first national park (Yellowstone), three more parks were added to the system in 1890, all of them in California. These were the Sequoia National Park, approved by Congress on September 25, and the General Grant National Park and the Yosemite National Park, both approved by Congress on October 1.

This action constituted a belated and partial response to Secretary Schurz's plea to reserve adequate samples of the big trees and the redwoods before it was too late. The areas set aside also included some magnificent scenery in the form of mountains, waterfalls, and forests, particularly in the Yosemite. The latter did not assume its present extent until 1906, when California re-ceded to the United States the Yosemite Valley and the Mariposa Bigtree Grove which had been ceded to it by Congress in 1864 for use as a state park. The next year (1892) the three new parks were placed under the protection and administration of the United States Army.

State Activities in Forestry. Forest resources, forest destruction, and "timber culture" were the subject of much discussion and considerable legislation in the various states during the latter half of the nineteenth century, with few tangible results. Commissions of inquiry, under various names, were appointed in Michigan and Wisconsin in 1867, in Maine in

---

[1] The date is often given, apparently erroneously, as 1887.

1869, in New York in 1872, in Connecticut in 1877, in New Hampshire, Vermont, and North Carolina in 1881, in Ohio in 1885, and in Pennsylvania in 1887. These commissions usually emphasized the relation of forests to climate and stream flow as well as to the timber supply, viewed the situation with alarm, and recommended remedial legislation.

By 1897 practically all of the states and territories had legislation dealing with the prevention and control of forest fires. Most of the laws limited themselves to providing penalties and damages for the willful or negligent setting of forest fires, occasionally with special provisions for the prevention of railroad fires. A few states provided for the appointment of state and town fire wardens, but New York was the only state to establish anything approaching a modern forest-fire-protection organization. In general, the laws read well on paper, but in practice were ineffective in providing adequate protection.

The widespread interest in tree planting is demonstrated by the action of many states during the period from 1868 to 1878 in offering bounties, tax exemptions, or reduced assessments for the establishment of successful plantations. The results were in general disappointing, and most of the acts soon became inoperative or were actually repealed. Considerable planting was also done by the railroads during this period, much of which was primarily experimental in character.

New York Takes Early Action. New York was the first state to take steps toward the establishment of state forests. In 1872 it appointed a commission, of which Hough was a member, to study the feasibility and advisability of establishing a state forest preserve in the Adirondacks. This commission submitted a report recommending the prohibition of further sales of state land, which was pigeonholed by the legislature. Ten years later (1883) the legislature prohibited further sales of some 600,000 to 800,000 acres of state land in certain counties and appointed a commission, of which Sargent served as chairman, to make further studies of the situation.

As a result of the commission's recommendations, the legislature in 1885 passed the forest-fire legislation already referred to and also constituted all of the lands then owned or thereafter acquired by the state in fourteen counties in the Adirondack and Catskill regions a "forest preserve." Administration of the preserve was placed in the hands of a forest commission of three men who were to "maintain and protect the forests now on the forest preserve, and to promote as far as practicable the further growth of forests thereon." Subsequent appropriations for the purchase of lands within the preserve jumped from $25,000 in 1890 to $1,000,000 in 1897.

The constitutional convention of 1894 was responsible for the inclusion in the new constitution of the famous, some would say "infamous," Article VII, Section 7 (now Article XIV, Section 1). It provided that all lands in the preserve shall forever be kept as wild land and forbade the cutting of any timber, dead or alive, on state-owned land within its boundaries. This proviso, which of course precluded any attempt at forest management, has been and still is the subject of bitter dispute between those who do and those who do not believe that the preserve can best serve the interests of the people of New York by being kept in a state of wild nature.

Action by Other States. Pennsylvania was the only other state to embark on a program of state forests prior to 1900. In 1897 it provided for the acquisition of land for state-forest reservations, and extensive purchases were inaugurated shortly thereafter.

When Colorado became a state in 1876 it included the following section relating to forestry in its constitution: "The General Assembly shall enact laws in order to prevent the destruction of, and to keep in good preservation the forests upon the lands of the State, or upon lands of the public domain, the control of which shall be conferred by Congress upon the State." The part of this section referring to the public domain was supplemented by a memorial from the constitutional convention to the Congress of the United States asking it to turn over to the state "not only the exclusive control of all the government forests in our mountains, but also at least one-fourth of all the government lands on our plains to use in future times for forest culture." The request was based on the theory that the lands in question would be much better handled under state than under Federal ownership, and would therefore contribute more fully to the future prosperity of the state.

The first constitution of the state of Utah (1895) contained a section headed "Forestry" which was identical in substance, and almost identical in language, with the section in the Colorado constitution of 1876.

The committee appointed by the A.A.A.S. in 1873 to stimulate the inauguration of work in forestry by both Federal and state governments did not get around to making contacts with the states until 1880. In that year it addressed a memorial to all of the state and territorial legislatures, through the governors, in which it stressed the serious inconvenience and great public injury that might result from the decreasing supply and growing consumption of forest products. It suggested several steps that the legislatures might take to improve the situation, including the encouragement of highway and forest planting by premiums and preferential tax treatment; the establishment of model plantations by the state; the passage of laws to prevent the starting of forest fires; the offering of prizes for the best essays and reports on forest culture; the promotion of instruction in forestry at educational institutions; and the appointment of commissions of forestry.

Whether or not as a result of this action, California, Colorado, Ohio, and New York, in the order named, all established state forestry organizations in 1885. Of these, the only one that has had an unbroken existence to date is the New York State Forestry Commission. Other state forestry organizations were created by Maine and North Dakota in 1891, by New Hampshire in 1893, by Minnesota in 1895, and by Pennsylvania in 1897. Substantial progress in state forestry, however, did not take place until after 1900.

Forestry by Private Owners. Aside from a small amount of sporadic tree planting, no attempt at real forest management had been made by private owners prior to 1892, when Gifford Pinchot undertook the management of the Vanderbilt estate at Biltmore, North Carolina. Pinchot, a graduate of Yale University in the class of 1889, had studied forestry during the following year in France, Switzerland, and Germany and on his return had declined an offer to become assistant chief of the Division of Forestry in order to devote his energies to promoting the practice of forestry in the woods.

His employment by George W. Vanderbilt followed a year of orientation during which he had made examinations and recommendations for the management of forest lands owned by Phelps, Dodge & Company in Pennsylvania and Arizona. He himself later described his first report as "the first practical report on the application of Forestry to a particular forest ever made in America." [1]

Pinchot's work at Biltmore, and later on the much larger Pisgah Forest, included the management of the forest itself, the establishment of forest plantations, the building up of a forest nursery and an arboretum, and the preparation of a forestry exhibit for the Chicago World's Fair of 1893. All trees to be removed in either harvest cuttings or improvement cuttings were marked prior to cutting, and special attention was paid to conducting logging operations with the least possible damage to young growth. "Thus," wrote Pinchot, "Biltmore Forest became the beginning of practical Forestry in America. It was the first piece of woodland in the United States to be put under a regular system of forest management whose object was to pay the owner while improving the forest. As Dr. Fernow wrote me on July 20, 1893, 'forest management has not been put into operation on any other area in this country except Biltmore.'"

Pinchot's arrangement with Mr. Vanderbilt permitted him to take on

---

[1] This and other quotations from Gifford Pinchot, "Breaking New Ground," 1947, are used with the generous permission of the publishers, Harcourt, Brace and Company, Inc., New York.

other work, and in December, 1893, he opened an office as "consulting forester" in New York City. In this enterprise he was soon joined by Henry Solon Graves (Yale, 1892), the second American to become a professionally trained forester. Among Pinchot's accomplishments as a consulting forester was the preparation of a plan for the practical management of Ne-Ha-Sa-Ne Park, a 40,000-acre tract of northern hardwoods, pine, hemlock, and spruce in Hamilton County, New York, owned by W. Seward Webb, a brother-in-law of George W. Vanderbilt. The cutting rules embodied in the plan provided that all logging operations must be carried out under the direct supervision of the forester in charge and included the marking of all trees to be cut, avoidance of damage to young growth, and the lopping of all tops as a protection against fire.

This study, the results of which were published in a little volume by Pinchot entitled "The Adirondack Spruce," was followed almost immediately by the preparation by the Division of Forestry of more detailed plans of management for Ne-Ha-Sa-Ne Park and the adjoining Whitney Preserve of 68,000 acres. The work was handled by Henry S. Graves, Superintendent of Working Plans, under cooperative agreements with the owners, and led to publication of Bulletin 26, *Practical Forestry in the Adirondacks.*

Attitude of Industry to Forestry. The view of business as to the relation between forestry and the lumbering and woodworking industries was expressed by two speakers at a special meeting of the American Forestry Association held at Chicago in 1893 in connection with the World's Fair Congress. J. E. Defebaugh, editor of *The Timberman,* stated bluntly that "the individual lumbermen and wood-workers as such have no interest in the subject of forest preservation and culture." The reasons for this, he hastened to add, were economic and financial.

This being the case, it was worse than useless to accuse the lumberman of

the ruthless destruction of our forests. . . . So violent and unreasonable, in many cases, have been the charges against him that the lumberman has often been forced to an attitude of apparent hostility, that misrepresents his real feeling, which is one of entire indifference. The invective is misdirected and wasted, because the lumberman, as such, is not any more responsible for the bad results of his business than is the banker, the grocer, or the farmer. He is but part of a commercial and industrial system the blame for which must be divided among the fifteen millions of our voting population, and comes even to them as an inheritance.

Defebaugh reminded his hearers of the economic law that controls the proportioning of the factors of production: "Where men and not materials are the chief resource of a country, there men are cheap; where natural resources are abundant in proportion to population, there these resources are cheap and men are dear." Forest culture in this country, in his judgment, would become feasible when, and only when, "the forest area is small in proportion to population."

The editor of the *North Western Lumberman* (Met. L. Saley) agreed that "between the great bulk of lumbering business and forestry there is at present no actual relation," but was somewhat more optimistic than Defebaugh about the future.

Possibly, were lumbermen to know more about the advantages and possibilities of forestry, they would at times profit by it to a greater extent than they do. In the future there will be a decidedly intimate relation between forestry and the lumbering industry, but it will be when the hum and clatter of the great commercial mills will have nearly died away, as then there will be but few great bodies of timber from which such mills may be fed. As men now plant corn and wheat for food, so by and by will they plant trees for the needs of their children and their grand- and great-grand-children. How to cultivate trees will be taught in our colleges and universities, and the true relation between forestry and the lumbering industry—the one providing the crop which the other utilizes—will be established.

Prophetic words!

The relation of the Division of Forestry to the practices of private owners was summed up by Fernow in his last report covering the activities of the Division from 1877 to 1898:

Finally, it should be stated that a small number of timberland owners have ventured to place their woodlands under management. While in most of these cases the Division had no direct relation to the undertaking, its long-continued educational campaign, which made it apparent that decreasing supplies can only be met by intelligent recuperative methods, must have had its effect in inducing such beginnings.

Summary. The first century of American independence was marked by the steadily accelerating utilization of its forest resources, with little effort on the part of either public or private owners to check the destruction of the forests or to adopt methods of forest management that would assure its perpetuation. Except for a few men with professional training such as Fernow, Pinchot, Graves, and their associates, even the meaning and the practical possibilities of forestry were unknown to most of the population, including timberland owners and lumbermen.

In general, the period was one of education and preparation. Through the efforts of a relatively few devoted men, public interest was aroused and the way was paved for the adoption of a constructive Federal forest policy. Particularly noteworthy were the establishment and activities of the American Forestry Association and of the Division of Forestry in the Department of Agriculture.

Small but significant starts were also made in the fields of state and private forestry.

REFERENCES

Cameron, Jenks: "The Development of Governmental Forest Control in the United States," chap. 8, Johns Hopkins Press, Baltimore, 1928.

Fernow, Bernhard E.: "Report upon Forestry Investigations, 1877–1898," Government Printing Office, Washington, 1899.

Greeley, William B.: "Forests and Men," chap. 4, Doubleday & Company, Inc., New York, 1951.

Hough, Franklin B.: "Report upon Forestry," vol. I, Government Printing Office, Washington, 1878.

Ise, John: "The United States Forest Policy," chap. 2, Yale University Press, New Haven, Conn., 1920.

Rodgers, Andrew Denny, III: "Bernhard Eduard Fernow: A Story of North American Forestry," chaps. 1–4, Princeton University Press, Princeton N.J., 1951.

CHAPTER 5

Forest Reserves Arrive

The increasing interest and activity in forestry indicated in the previous chapter was reflected in efforts to improve the situation with respect to the public timberlands. From the latter part of the 1870's on, proposals to retain these in Federal ownership were almost continually before Congress.

Early Proposals for Forest Reserves. The first specific bill on the subject was introduced in 1876 by Representative Fort of Illinois "for the preservation of the forests of the national domain adjacent to the sources of the navigable rivers and other streams of the United States." Two years later Senator Plumb of Kansas introduced a more comprehensive bill which was undoubtedly inspired by Secretary Schurz and Commissioner Williamson. It provided not only for the reservation of all timberlands and the sale of timber under regulations that would assure the perpetuation of the forest, but also for the protection of these lands from fire and trespass. Nothing came of either of these measures, or of others with similar intent introduced during the next ten years, although a bill by Senator Edmunds of Vermont "to establish a reservation at the Headwaters of the Missouri River" did succeed in passing the Senate twice with little opposition.

A detailed plan for the reservation and administration of the public timberlands was prepared in 1887 by Edward A. Bowers, a lawyer who had joined the General Land Office the previous year as an inspector and who later became assistant commissioner. In these positions he became intimate with Fernow and other leaders in the forestry movement and served for a while as secretary of the American Forestry Association. After referring to the "thieveries and wanton destructions," the unchecked fires, and the "reckless wastefulness" which were devastating the public forests, he recommended the withdrawal from entry of "all the public lands valuable in any degree for timber or their forest growth." The reservations would then be administered by a bureau of forestry and public timberlands in the Department of the Interior, with power to make all necessary rules and regulations for their management on a sustained-yield basis.

Fire and timber trespass would be punishable by both fine and imprisonment. An interesting harking back to the colonial philosophy that an accused must prove his innocence was contained in the proposal "that the burden of proof should be upon the person charged with the depredation to show that the timber was lawfully cut." Because of the possible difficulty of obtaining congressional approval of a new bureau, the suggestion was made that it might at first be advisable to create a timberlands division in the General Land Office by uniting the Division of Forestry in the Department of Agriculture and the special agents employed to stop timber depredations.

Stalemate in Congress. Bowers' plan was transmitted to Congress at its request in April, 1888, and as usual was pigeonholed. Meanwhile Fernow in collaboration with Bowers and others had prepared a bill for the protection and administration of forests on the public domain which was formally approved by the American Forestry Congress at its Springfield meeting in 1887. This bill was known as the Hale bill because of its introduction in the Senate by Hale of Maine, although it was the companion bill in the House which actually received the most attention.

The bill closely resembled the proposal prepared by Bowers with respect to general principles, but differed in certain details. It also provided for the classification of public timberlands into those primarily valuable for agriculture which should be opened for homesteading; those of more value for forest purposes than for cultivation and suitable for commercial utilization; and those primarily valuable as protection forests which should be reserved from all utilization. It is interesting to note that Fernow, who was at the time Chief of the Division of Forestry in the Department of Agriculture, recommended that administration of the proposed reserves be placed in the Department of the Interior for reasons of expediency, although he stated later that their assignment to the Department of Agriculture would have been more logical.

In the House, the Committee on Public Lands proposed to embody certain portions of the Hale bill in highly emasculated form in a much more comprehensive bill repealing the timber-culture and preemption acts and otherwise amending the general land laws. The Committee on Legislation of the American Forestry Congress vigorously opposed this action, which, in its judgment, "instead of promoting the interests of forestry in our country, was calculated, by its mere semblance of doing something, to defeat our object and to prevent, for an indefinitely long period of time, further and most desirable legislation."

Nathaniel Egleston, chairman of the committee, was somewhat cynical in reporting on the matter at the annual meeting of the American Forestry Congress at Atlanta in December, 1888:

Had our bill shown a political aspect or a tendency to promote the pecuniary interests of any considerable number of persons; if it had been a scheme to take money out of the public treasury instead of being a measure for husbanding and increasing the public wealth, doubtless we should have received more consideration; but having no partisan or pecuniary advantage with which to appeal for support and aiming only at the general welfare, our bill was not allowed a hearing outside of the Committee-room. In the Senate it did not even get the consideration of the Committee to whom it was referred.

On April 23, 1889, representatives of the American Forestry Congress waited upon President Harrison, to whom they presented a memorial from the congress urging the adoption of a national forest policy. An equally important contact was made on November 14, 1889, when a joint delegation from the American Forestry Association (the new name of the congress) and the American Association for the Advancement of Science met with Secretary of the Interior John W. Noble.

Early the next year (1890) President Harrison forwarded to Congress a memorial adopted by the latter association urging the reservation of the public timberlands and the appointment of a commission to study the forest situation and recommend legislation. This memorial undoubtedly resulted from a speech by Fernow at the Toronto meeting of the association in 1889 on "Need of a Forest Administration for the United States." It constituted the first action with respect to forestry taken by the A.A.A.S. subsequent to its efforts between 1873 and 1880 to bring about the creation of Federal and state commissions of forestry.

Forest Reserve Act of 1891. The efforts exerted by the A.A.A.S. and others finally brought results in the bill for the general revision of the land laws which Congress had had under consideration for many years. When this bill went to conference it made no reference to forest reservations. When it emerged from the conference committee, it contained the following new provision as its final section:

Section 24. That the President of the United States may, from time to time, set apart and reserve, in any State or Territory having public lands wholly or in part covered with timber or undergrowth, whether of commercial value or not, as public reservations, and the President shall, by public proclamation, declare the establishment of such reservations and the limits thereof.

The report of the conference committee reached Congress on February 28, 1891, four days before the date for final adjournment. In the

Senate, which had ordinarily been hostile to measures of this sort, Senator Plumb of Kansas insisted on speedy consideration, and the bill passed with little comment. In the House, Representative McRae of Arkansas objected to Section 24 on the ground that it gave the President too much power and that no one could foresee its consequences. Representative Dunnell of Minnesota insisted that the bill should go over until it could be studied in print. Representative Payson of Illinois assured the House that there was no danger of the President's going too far, and that if he did, Congress could easily pass a joint resolution or a bill opening to settlement any lands which the President had reserved. The bill did not come before the House again until the closing hours of the session on March 2, when it was passed without debate. It was signed by the President on March 3.

Among its other provisions, the act repealed the Timber Culture Act and the Preemption Act; put a stop to auction sales of public lands except isolated tracts; forbade commutation under the Homestead Act until fourteen months after filing; tightened up the requirements for obtaining land under the Desert Land Act; and liberalized the provisions of the Free Timber Act.

The section authorizing the President to set aside forest reserves, which was attached to the bill in violation of the rule forbidding the inclusion of any new material in a conference-committee report, appears to have been added at the insistence of Secretary Noble. He had been well educated by Bowers, Fernow, and others and is reported to have told the committee that he would get the President to veto the bill unless this particular section were included. That it was prepared in considerable haste is indicated by the fact that it is not even a complete sentence, the transitive verbs "set apart and reserve" not being followed by any object. This grammatical slip, however, has not led to any question as to the intent or the validity of the act.

Circumstances Favoring Passage of the Act. The second major step in the development of a national forest policy was thus taken by Congress as casually, and as unintentionally on the part of most of the membership, as the first step in 1876 appropriating $2,000 for the preparation of a report on forestry. One must agree with the statement by John Ise that at that time "it is fairly certain that no general forest reservation measure, *plainly understood to be such,* and unconnected with other measures, would ever have had the slightest chance of passing Congress." [1]

---

[1] The quotations in this and the following paragraph are from John Ise's "The United States Forest Policy," 1920, and are used with the generous permission of the publisher, Yale University Press, New Haven, Conn.

He sums up effectively the

long chain of peculiar circumstances which made it impossible for Congress to act directly on the question. If the conference committee, like most public land committees, had included a majority of men hostile to conservation; if the forest reserve provision had been attached to anything but a conference bill; if the question had come up at the beginning instead of the close of the session; if there had been less of a public demand for revision of the land laws; if the bill had been a short one, with only a few clauses; if Congress had been a little less familiar with the general provisions of the omnibus bill under discussion and so more careful to scrutinize them, or if members had realized what important results were to follow; if any one of a score of possible contingencies had prevailed, the passage of a general forest reserve measure at this time would probably have been impossible.

Within less than a month, on March 30, 1891, President Harrison issued a proclamation setting aside the Yellowstone Park Forest Reservation. Thus the Yellowstone region had the distinction of being the birthplace of both the national park system and the national forest system. During the next two years Harrison proclaimed fourteen additional reservations in various parts of the West, bringing the total gross area to more than 13,000,000 acres. Before the end of 1893 President Cleveland set aside another 4,500,000 acres, but took no further action because of the lack of any provision for the protection and administration of the reserves.

Toward Forest Administration. In the absence of any specific authorization for their management, the Secretary of the Interior believed that creation of the reservations withdrew them not only from sale and entry but from any form of utilization; and Congress made no special provision for their protection from fire and trespass. Establishment of forest reserves therefore did not afford the protection or inaugurate the conservative management which their sponsors had anticipated. On the other hand, their withdrawal from all use was vigorously opposed by settlers, miners, stockmen, and lumbermen. The situation was satisfactory to no one.

Repeated attempts to obtain legislation that would meet the legitimate complaints of both supporters and opponents of the reserves by opening them to use under proper management met with failure. Between 1891 and 1897 no less than twenty-seven bills dealing in one way or another with forest reserves were brought before Congress. Of these, the bills that received most attention were the ones introduced by Senator Paddock of Nebraska and Representative McRae of Arkansas. The Paddock bill, which was a modification of the original Hale bill and a previous Paddock bill, was introduced on June 1, 1892. It was acted on favorably by the Senate Committee on Agriculture and Forestry but got no further. The McRae bill was first introduced on September 6, 1893, and was before Congress in one form or another until 1897. The original draft stated the purposes for which forest reserves might be created; authorized the Secretary of the Interior to make rules and regulations for their protection and administration, including the sale of timber of commercial value; authorized the Secretary of War to detail troops for the protection of the reserves; and provided that lands more suitable for agriculture than for forestry might be restored to the public domain. At one time a revised version of the bill passed both houses of Congress, but finally died in the House Committee on Public Lands, to which the many amendments made by the Senate were referred for consideration in McRae's absence from the floor. Had he been present, the bill would probably have been sent to conference and might have become a law at that time.

Forest Commission of National Academy of Sciences. In order to break the deadlock, Secretary of the Interior Hoke Smith was persuaded to request recommendations looking to "the inauguration of a rational forest policy" from the National Academy of Sciences, an organization of outstanding scientists created by Congress in 1862 for the purpose of assisting the government in any matters in the field of science. Secretary Smith, in a letter to Wolcott Gibbs, president of the academy, dated February 15, 1896, stated that he particularly desired an official expression from the academy on the following points:

(1) Is it desirable and practicable to preserve from fire and to maintain permanently as forested lands those portions of the public domain now bearing wood growth, for the supply of timber?

(2) How far does the influence of forest upon climate, soil, and water conditions make desirable a policy of forest conservation in regions where the public domain is principally situated?

(3) What specific legislation should be enacted to remedy the evils now confessedly existing?

President Gibbs felt that the academy was not in a position to make an immediate reply. He consequently appointed a committee, commonly known as the Forest Commission, to study and report on the questions asked by Secretary Smith. The commission was a distinguished one, consisting of Charles S. Sargent, Harvard University, chairman; Henry L. Abbott, formerly chief engineer of the United States Army; Arnold Hague, U.S. Geological Survey; William H. Brewer, Yale University; Alexander Agassiz, Harvard University; Gifford Pinchot, consulting forester; and Gibbs himself as a member ex officio. President Gibbs also asked Congress for an appropriation of $25,000 to defray the expenses of the commission, and this was granted in the Sundry Civil Appropriations Act for 1897. The commission later made an extensive Western trip in the course of which it visited most of the existing forest reserves and also large areas of unreserved forest.

Commission's Recommendations. Questions of strategy soon arose within the commission. Pinchot and Hague were in favor of a prompt report making recommendations for the protection and administration of forest reserves, both present and future. This course had been suggested by President Cleveland, who advised:

Take up the organization of a forest service first and then the question of more reserves. Let the plan be one that looks small, and at first costs little, and yet has in it the elements of growth; let it avoid points liable to attack by reaching its object, if possible, along other lines.

Sargent, on the other hand, felt that it was wise "to go slow and feel our way." At a meeting of the full commission on October 24, 1896, Sargent's views prevailed. It was then agreed to recommend the establishment of a number of new reserves but to make no try for legislation at the coming session of Congress. Pinchot later wrote:

There remained the chance to make a strong public statement, at the time when the new Reserves were created, that they were not to be taken out of circulation and locked up. Congress and the Western people were entitled to that. I did my best for it, failed, and came very near making a minority report, but unfortunately decided against it.

On January 29, 1897, Sargent wrote President Gibbs recommending the creation of the proposed reserves prior to the perfecting of the scheme of forest management on which the commission was working.

It believes that the solution of this difficult problem will be made easier if reserved areas are now increased, as the greater the number of persons engaged in drawing supplies from the reserved territory or in mining in them the greater will be the pressure on Congress to enact laws permitting their proper administration. . . . For this reason it is the unanimous opinion of the Commission that the establishment of the reserves described above is now a matter of the utmost importance to the development and welfare of the whole country.

A few days later (February 1) Secretary of the Interior David R. Francis (Hoke Smith's successor) wrote the President:

I respectfully suggest that the 165th anniversary of the birth of the Father of our Country could be no more appropriately commemorated than by the promulgation by yourself of proclamations establishing these grand forest reservations.

Cleveland Celebrates Washington's Birthday. In accordance with these recommendations President Cleveland celebrated February 22, 1897, by issuing proclamations creating thirteen new forest reserves with a gross area of 21,279,840 acres. This action much more than doubled the existing area of reserves, without warning and with no proposal for the utilization of their resources. The pressure on Congress which Sargent had prophesied materialized immediately, and perhaps more violently than he had anticipated. The Wyoming Legislature and the Nebraska State Senate adopted memorials praying for the abolition of the new reserves. The Seattle Chamber of Commerce, various other commercial organizations, and many individuals presented petitions and remonstrances. The American Forestry Association thanked Cleveland for his bold action, but Fernow expressed doubts about its wisdom.

On February 28 Senator Clark of Wyoming offered an amendment to the sundry civil appropriations bill then under consideration to restore the February 22 reservations to the unreserved public domain. This amendment was approved by the Senate, but the House failed to concur and requested a conference. During the debate on the report of the conference committee Representative Lacey of Iowa offered an amendment which included the more important items in the McRae bill plus a provision giving the President power to modify or revoke any executive order establishing any forest reserve. This amendment was approved by the House, and the bill returned to conference.

The second report of the conference committee omitted all reference to the subject except for the inclusion of the clause authorizing the President to modify or abolish forest reserves. In this form it passed both houses of Congress, only to be pocket-vetoed by Cleveland, who evidently feared the use that might be made of the power by a President unfriendly to the forest-reserve policy.

The temper of Congress can be illustrated by a couple of extracts from the *Congressional Record.* Senator Clark of Wyoming had this to say:

An eminent scientist and a member of that commission acknowledged to me under the press of interrogatories that not a member of the commission had ever been upon this reservation [the Big Horn] or within miles and miles of it. . . . All the reservations are made at the behest of these scientific gentlemen. I honor them for their knowledge; they are an ornament to our country; I read their reports with admiration; but they belong to that class of scientific gentlemen who think more of the forest tree than they do of the roof tree, and we have a whole lot of people in the West who think as much of their roof tree as the people of any other part of this nation.

He professed to be humiliated that the Senate should be "overridden by that power or by the threat of that power, which, for the last four years, has failed to protect anything American and has never failed to cater to anything foreign."

Representative Mondell, also of Wyoming, felt that "the action of the

President in setting aside these reservations was so outrageous and its disastrous effects so far-reaching that the attempt to remedy the conditions" by the proposed amendment would perhaps make things worse than ever, and that a thorough study of the situation was needed. He regarded the President's action as "in line with that contemptuous disregard of the interests of the people and the wishes of their Representatives which has been characteristic of the present Administration, which in the providence of God and the act of an intelligent electorate, in a little more than twelve hours shall pass beyond the power to harass and annoy the American people." There is nothing new about invective in American public life.

Bills for Administration of the Reserves. Cleveland's pocket veto of the sundry civil appropriations bill left the Washington's Birthday forest reserves intact, but it also left most of the government without funds for the next fiscal year. Consequently President McKinley soon after his inauguration called a special session of Congress to provide funds for the operation of the government and also to pass the new tariff legislation which the Republicans had promised. At this special session the legislation which those seeking sound administration of the public timberlands had been urging for twenty years was finally passed.

Early in the session Senator Pettigrew of South Dakota introduced an amendment to the sundry civil appropriations bill incorporating most of the features of the perennial McRae bill. Pettigrew's action was taken at the instigation of Charles D. Walcott, Director of the Geological Survey, who supplied the first draft of the amendment. Senator Gorman of Maryland challenged the inclusion of legislation of this sort in an appropriation bill on a point of order, but the Senate failed to uphold him by the close vote of 25 to 23. Interestingly enough, retention of the amendment was almost unanimously favored by Western senators, who were eager to have the resources of the reserves made available for use, and was generally opposed by Eastern senators, who feared that any use would result in misuse. For the first time the West was for, and the East was against, a measure which proved to be a highly constructive piece of conservation legislation.

An item in the Pettigrew amendment providing for temporary suspension of the Cleveland proclamations touched off a renewed attack on the Commission and the former President. Senator Clark took advantage of the occasion to amplify his previous remarks: "There has never been such a deathly blow struck at the theory of forest preservation in this country as was struck by the President on the 22d of February, when he issued this sweeping order of his, contrary to the facts known to exist, and that there was no necessity of the reservations being made. . . . It has almost aroused the people of the West against the whole theory of forest reservation. They have been enthusiasts in favor of it heretofore."

Senator Wilson of Washington wanted to know, "Why should we be eternally harassed and annoyed and bedeviled by these scientific gentlemen from Harvard College?" Pettigrew himself stated: "Now we come and say that this ignorant act of a corrupt Executive shall be overturned." Several senators, however, came to the defense of the reservation policy. Senator Gray of Delaware, for instance, emphasized that "the Senate should not consider that we have abandoned this great question of forest reservation in the interest of the whole people of the United States to the selfish interest of speculators—and I say this in no invidious sense— who have rushed into that country, and of course will naturally sacrifice larger interests to the particular interests they have in hand."

In the House, Representative Hartman of Montana referred to Cleveland's action as "iniquitous and unfounded." On being told by Lacey of Iowa that the trouble with him was that he assumed that the present President would make every mistake that his predecessor did, he replied: "Oh, no. I want the gentleman to understand that I do not think there is a man on this earth who is such a blunderhead that he can make even a thousandth part of the mistakes that President Cleveland made. I did not support President McKinley, but President McKinley is an American President, thank God, and he is the first one we have had for four years." Hartman of Montana characterized the proclamation as "a parting shot of the worst enemy the American people ever had."

Lacey, on the other hand, looked for the facts behind the invectives:

It is somewhat of a surprise to settlers from such states as Iowa to hear it said that a man or a great corporation should be allowed to cut timber on four sections of land free of charge, to be burned as fuel in the way of business, or to be applied to any other use that may be desired. This accounts for some of the bitterness of the hostility which this order of President Cleveland has met. . . . No wonder gentlemen complain of the loss or curtailment of such a privilege as this. . . . Nothing is so sacred as an abuse.

Several members called attention to the fact that McKinley had declined to revoke Cleveland's proclamations in spite of the fact that the Attorney General had ruled that he had a right to do so.

Forest Reserve Act of 1897. After disagreement between the two houses, their differences were ironed out by a conference committee. The bill was approved by the President and became a law on June 4, 1897. In its final form, its main provisions were as follows:

1. It appropriated $150,000 for survey by the U.S. Geological Survey of forest reserves already created or to be created.

2. It suspended Cleveland's proclamations of February 22, 1897, until March 1, 1898, when they were again to become effective.

3. It authorized the President to revoke, modify, or suspend any executive order or proclamation pertaining to forest reserves.

4. It provided that "no public forest reservation shall be established except to improve and protect the forest within the reservation for the purpose of securing favorable conditions of water flows, and to furnish a continuous supply of timber for the use and necessities of citizens of the United States."

5. It stated that it was not intended to authorize the inclusion within forest reserves "of lands more valuable for the mineral therein, or for agricultural purposes, than for forest purposes" and specifically authorized the restoration to the unreserved public domain of lands found after due examination to be better adapted for mining or agriculture than for forest usage.

6. It instructed the Secretary of the Interior to protect the forest reserves against destruction by fire and depredations and authorized him to "make such rules and regulations and establish such service as will insure the objects of such reservations, namely, to regulate their occupancy and use and to preserve the forests thereon from destruction."

7. It made any violation of the provisions of the act or of the Secretary's rules and regulations punishable by a fine of not more than $500 and imprisonment for not more than twelve months.

8. It authorized the Secretary of the Interior to sell so much of the dead, matured, or large growth of trees as may be compatible with the utilization of the forest, after advertisement, at not less than its appraised value, for use in the state or territory where the reserve is located, but not for export therefrom.

9. It authorized the Secretary of the Interior to permit the free use of timber and stone for firewood, fencing, buildings, mining, and other domestic purposes, within the state or territory where the reserve is located.

10. It provided for free egress and ingress of actual settlers, including the construction of such wagon roads and other improvements as may be necessary to reach their homes and utilize their property, under rules and regulations prescribed by the Secretary of the Interior.

11. It authorized the prospecting, locating, and developing of the mineral resources of the reserves, and the entry of any mineral lands under the laws of the United States and the rules and regulations applying thereto.

12. It authorized the settler or owner of a tract covered by an unperfected bona fide claim, or by a patent, to relinquish it to the government and in lieu thereof to select a tract of vacant land open to settlement

not exceeding in area the tract relinquished. With unperfected claims the requirements respecting settlement, residence, improvements, and so forth, must be complied with on the new claims, credit being allowed for the time spent on the relinquished claims.

13. It authorized settlers within the reserves or in the vicinity thereof to occupy not more than 2 acres of reserve land for each school and 1 acre for a church.

14. It provided that civil and criminal jurisdiction over persons within forest reserves should remain unchanged except so far as the punishment of offenses against the United States is concerned, the intent being that states should not lose jurisdiction, nor inhabitants their rights, privileges, and duties because of the establishment of the reserve.

15. It provided that all waters on the reserves may be used for domestic, mining, milling, or irrigation purposes under the laws of the state or under the laws of the United States and rules and regulations established thereunder.

For the third time a major forward step in the development of a national forest policy had been taken by attaching a last-minute rider to a bill with an entirely different primary purpose. This time, however, the action was not taken without debate, but after careful consideration and under the spotlight of widespread publicity. Although the Forest Reserve Act represented a compromise, it has proved in practice to be a statesmanlike piece of legislation. Its two really undesirable provisions—those permitting lieu selections and prohibiting the export of timber from the state in which cut—were shortly repealed. Otherwise the act still remains virtually intact and has proved a workable and effective instrument.

Report of the Forest Commission. The report of the Forest Commission of the National Academy of Sciences was submitted to the Secretary of the Interior on May 1, 1897, only six days before the Pettigrew amendment was acted on by the Senate. No reference to the report was made during the discussion of the bill in Congress, and it is doubtful whether it had much, if any, influence on the action taken. Pinchot many years later commented: "So far as I can tell there is not a word in the Pettigrew Amendment of June 4, 1897, that priceless piece of legislation, that would have been left out, not a word left out that would have been put in, if the Report of the National Forest Commission . . . had never been made at all."

The delay in the appearance of the report does not, however, mean that the activities of the commission were without value. Its vigorous espousal of "the retention of the fee of forest lands, and the sale of forest products from them at reasonable prices, under regulations looking to the perpetual reproduction of the forest," put the great influence of the National

Academy of Sciences squarely behind the whole forest-reserve movement. This was an item of considerable importance, since passage of the act of June 7 did not mean that the reserves were safe from further attack.

The commission's recommendations for specific legislation did not differ radically from that actually adopted except in one important respect. This was the proposal, fortunately never approved, that Army officers should be used in the administration of the reserves, since the duties of forest officers "are essentially military in character and should be regulated for the present on military principles." Both Sargent and Fernow had previously suggested that the Military Academy at West Point be used for the training of forest officers.

Pinchot summed up the accomplishments of the Forest Commission by saying that it "deserved well of its country. It was the means of bringing the public timberlands to national attention as they had never been brought before. It had set a marker that would remain. Never again would the forest sink to the negligible position it had occupied in the public estimation of days gone by. And for that Sargent was mainly responsible. His perversity had added one more, and the greatest of all, to the brilliant list of his services to Forestry in America."

Early Administration of the Forest Reserves. Within less than a month after the passage of the act of June 4, 1897, the Department of the Interior issued rules and regulations for the administration of the forest reserves. No funds for their enforcement were, however, available, since the act of June 4, while carrying appropriations of $90,000 for the prevention of timber trespass on the public lands and of $150,000 for surveys of the forest reserves by the Geological Survey, had not provided one cent specifically for their protection and administration. A year later Congress appropriated $75,000 for this purpose for the fiscal year 1899, and by 1904 the amount had increased to $375,000.

Each forest reserve was placed in charge of a forest supervisor, under whom the field work was handled by forest rangers. The reserves were grouped into eleven districts, each in charge of a superintendent through whom the supervisors reported to the General Land Office in Washington, and there were one or two roving inspectors. Forest officers were promptly given specific instructions to prepare detailed reports on fires, sheep grazing, timber sales, timber trespass, and areas proposed for reservation.

On November 8, 1901, Secretary Hitchcock sent the Commissioner of the General Land Office a detailed outline of the principles and practice which he had concluded should govern the administration of national forest reserves. The outline covered in constructive fashion every item of importance in the handling of the reserves and was at times sharply critical of current practice. For example:

The chief weight should hereafter be laid on field work, in contradistinction to the present plan, which administers the reserves purely on the basis of papers and reports from the office point of view, with little reference to actual work in the field. . . .

Every effort should be made to create an esprit de corps among the forest officers. . . .

[Grazing] permits should run for five years. Residents should have precedence in all cases over tramp owners and owners from other States. Local questions should be decided on local grounds and on their own merits in each separate case. . . .

The sale of mature live timber whose removal will benefit the forest should be encouraged. . . . Applications for timber cutting should reach the General Land Office accompanied by all papers necessary for a decision. . . . The cutting of unmarked timber should be absolutely prohibited. . . .

The present policy of appointing as forest officers men totally unacquainted with the conditions or requirements of their work should be discontinued. . . . The first duty of forest officers is to protect the forest against fires. Rangers should be ordered, as they are now forbidden, to leave their own beats when necessary to assist in extinguishing fires on adjoining beats. . . . The system of daily reports by forest rangers should be abolished at once. It serves no useful purpose whatever and fails wholly to secure faithful work. On the contrary, it is a constant provocation to falsehood and encourages a feeling that the report, and not the work, is the matter of first importance.

Secretary Hitchcock instructed Commissioner Hermann: "You will note carefully the plan submitted and take such steps as may be necessary to carry the same into effect." If that plan bears a remarkable resemblance to the principles and practices adopted by the Forest Service a few years later, the explanation lies in the fact that Pinchot wrote it.

Weaknesses in Administration. The intentions of the Department of the Interior, as expressed in official statements by the Secretary and the Commissioner and in such instructions as those issued by Hitchcock in 1901, were of the best. Its failure to provide effective administration of the forest reserves was due to excessive red tape and emphasis on paper work, to overcentralization of authority in Washington, to incompetent personnel, and to the susceptibility of the General Land Office to political influence. The extent and seriousness of these handicaps was made crystal clear by Edward T. Allen, a forester with actual experience in the Forestry Division of the Land Office, in an address to the Society of American Foresters in 1903. After listing the innumerable duties that fell to the lot of a forest officer, he continued:

It is evident that to do all this successfully a forest officer should not only be intelligent, but have considerable experience if not technical training. Unfortunately the present system of recruiting the force does not secure such men. The superintendents have always been appointed entirely through political influence, without regard to fitness. Some have been fairly good men, but the majority were worse than incompetent. Until within the past year it has been the same with the supervisors. They have included lawyers, doctors, editors, real estate dealers, postmasters, and even a professional cornet player. Some have been so dishonest and depraved that they disgraced the service; all have been technically disqualified, yet many are excellent men, and by intelligence, honesty and executive ability make up for their lack of timber knowledge. . . .

This state of affairs has now improved to this extent: That bad men without "pull" are removed when found and good men are sought for appointment unless the place is desired by a man with "influence." In other words, merit has become a secondary consideration, whereas before it was no consideration at all. . . .

In the first four years there was little administrative system. The regulations for grazing, patrol, and timber management were drawn by men in Washington without any knowledge of field conditions and little of the subject in general. These rules were often impractical, and always hard to execute because lacking in detail. . . . General instructions to subordinates were few and brief, and no one was able to advise his subordinates. . . .

These instructions again, both as to details and general policy, emanate from office men with small knowledge of forestry, lumbering, grazing, or even field conditions. There you have the system: An office force without knowledge of the reserves, managing them with no information other than what comes from the very men admitted to be too unreliable to decide anything themselves! . . .

The institution will never be successful as long as it is maintained without the sympathy of the people, and they will have neither sympathy nor confidence until they see a strong field force, not too closely restricted in details, and an undivided central management having knowledge of western conditions and backed by statutes enforcing its authority.

Forestry Division in General Land Office. Secretary Hitchcock, thanks to Pinchot's coaching, was well aware of the defects in the personnel and administration of the forest reserves. In an attempt to remedy the situation and thereby implement his belief that "the time for the introduction of practical forestry on the forest reserves has fully come," he decided in 1901 to create a Forestry Division in the General Land Office to handle all activities relating to the reserves under the general direction of the Commissioner. The work had previously been assigned to the Special Service Division (Division P), the primary function of which was enforcement of the land laws.

Filibert Roth, who had recently returned to the Bureau of Forestry in the Department of Agriculture after a couple of years at the New York

State College of Forestry at Cornell University, was persuaded to take charge of the new division (Division R). With the help of four foresters transferred from the Bureau of Forestry, one of whom was E. T. Allen, he made a valiant attempt to improve the situation. Early results were so encouraging as to lead Allen to remark later that "the improvement was greater than would be thought possible in so short a time."

All was not smooth sailing, however, and the heavy hand of the past, politics, and inertia made thoroughgoing and permanent reform unlikely. Consequently in 1903 Roth returned to the Bureau of Forestry and shortly afterward accepted an invitation to become chairman of the Department of Forestry at the University of Michigan, where professional instruction in forestry was being started that fall. Although the Forestry Division continued its existence, neither he nor his assistants, all of whom shortly returned to the Bureau of Forestry, were replaced by trained men. In 1905, the Department of the Interior ended its administration of the forest reserves, as it started that administration, without a single professional forester in its employ.

Lieu-land and Other Legislation. Several pieces of legislation dealing with the forest reserves, all of which were recommended by the Department of the Interior, were enacted during the period from 1898 to 1905. The most important of these dealt with the lieu-land provision of the act of June 4, 1897, the repeal or drastic modification of which was repeatedly urged by both the Secretary and the Commissioner from 1898 on. In 1901 Commissioner Hermann stated

with deep regret that many representations made to this office in advocacy of the creation of certain reserves are prompted by the desires of interested parties in possession of valueless holdings which will become a part of such reserves, thereby enabling such owners to exchange their worthless properties for valuable portions of the public domain, which under existing laws they cannot otherwise obtain. Many valueless lands belonging to the states, such as school lands and other grants, and large grants which inure to corporations which are found included within reserves are also permitted to be exchanged for large quantities of valuable lands upon the unreserved domain. Their valuable lands are not relinquished—they remain.

Both that year and the next he recommended against creating new reserves or making considerable additions to existing reserves until the law was amended to "require lands selected to be of like character or equal value with the lands relinquished."

Congress as usual was slow to act. In 1900 and 1901, it limited lieu selections to vacant, nonmineral, surveyed public lands subject to homestead entry. These acts improved the situation but fell far short of stopping abuses under the privilege. Outright repeal of the provision did not come until March 3, 1905, about a month after the transfer of the reserves to the Department of Agriculture.

An act of February 28, 1899, authorized the Secretary of the Interior to rent or lease forest-reserve lands adjacent to mineral, medicinal, or other springs for the erection of sanitariums and hotels and provided that all receipts from such leases should be covered into a special fund to be expended "in the care of public forest reservations." Enactment of this law, which had been recommended by the Department of the Interior, led Commissioner Richards to call attention to the fact that "under the present law the only resources on forest reserves which may be utilized by the Government for a money consideration are the timber thereon and mineral and other springs."

This was a straightforward statement of the Department's belief that charges for the use of the reserves could be made only when specifically authorized by Congress. It therefore requested that it be given such authority with respect to all resources, including the use of the range for grazing purposes and the use of land for sawmills, hotels, summer resorts, stores, and other establishments. It also requested that the precedent set by the mineral-springs act of covering receipts from that source into a special fund, to be used for the improvement of the reserves, be followed with respect to all other receipts. These two measures, it was predicted, would soon make the reserves self-supporting, so that "yearly appropriations for their care would cease to be necessary."

A provision included in the Sundry Civil Appropriations Act of March 3, 1898, and repeated in subsequent appropriation acts, authorized forest officers so far as practicable to aid in the enforcement of laws relating to the protection of fish and game of the states and territories in which the forest reserves are situated. The Department felt that this provision did not go far enough in promoting unity of action between Federal and state authorities in this field. It therefore, in 1903, recommended passage of "a Federal statute which would tend to harmonize State legislation upon this subject without creating a divided jurisdiction over forest reserves and which would not encroach upon the proprietary rights of the States to control the game and fish within their respective boundaries." The proposal is of interest as one of the earliest attempts to solve the problem of Federal-state relations in this touchy field.

Arrest without Warrant. A major deficiency in the act of 1897 was its failure to provide any efficient means for the arrest of persons violating the laws, rules, or regulations for the protection of the reserves. E. T. Allen summarized this weakness succinctly: "Forest officers cannot make arrests. If they find a man stealing timber or setting fires they can only report him to the supervisor, who goes to a United States Commissioner, perhaps 100 miles away, and gets a warrant. This is given to a marshal who in time goes out to look for the offender."

From 1899 on, the Department of the Interior repeatedly urged Congress to remedy this situation, which was seriously hampering effective protection of the reserves. Bill after bill failed to receive approval because opponents saw constitutional objections or feared that the power might give rangers "additional means of annoyance and irritation." Finally, on February 6, 1905, Congress passed an act authorizing the arrest by any officer of the United States, without process, of any person taken in the act of violating the laws or regulations relating to the forest reserves and the national parks. This act constituted the first authorization for a civil officer to make an arrest without a court warrant and greatly facilitated protection of the reserves.

Grazing on the Reserves. Grazing on the forest reserves was one of the most difficult problems with which the Department of the Interior had to deal and one on which it was unsuccessful in obtaining legislation. At first the grazing of sheep ("hoofed locusts") was forbidden in all reserves except those in Washington and Oregon. This action raised such a storm of protest from the sheepmen that Pinchot and Frederick V. Coville, a botanist in the Department of Agriculture who had previously studied the effect of sheep grazing on the forest in Oregon, were asked by Secretary Hitchcock to study the problem and advise him what to do about it.

Their recommendation that grazing be permitted under strict control in areas where it would not endanger either the timber or the water supply was accepted by Hitchcock, who stated in his annual report for 1901:

I have adopted the policy of permitting sheep to graze in that portion of certain reserves where it is shown, after careful examination, that such grazing is in no way injurious to or preventive of the conservation of the water supply, and that policy it is my purpose to continue. Such a policy, in my judgment, will afford all the encouragement to the wool-growing industry that it can reasonably ask in this connection from this Department, and is not inconsistent with those vast interests dependent upon irrigation, which demand consideration at my hands.

The grazing of cattle and horses, being regarded as much less dangerous to the forest, was at first subject to virtually no control. In 1900, however, permits were required for the grazing of all classes of livestock. The Secretary determined the number of stock that would be allowed to graze in each reserve and issued the necessary permits, except that supervisors were authorized to issue permits for not more than one hundred head of cattle and horses owned by actual residents on the reserve. Preference in the issuance of permits was given first to residents within the reserve; second, to nonresidents owning permanent ranches within the reserve; third, to persons living in the immediate vicinity of the reserve; and finally, to outsiders having some equitable claim.

Difficulties in the Way of Control. In February, 1902, woolgrowers' associations were allowed to recommend the allotment of permits for sheep grazing when such associations represented the majority of sheep owners or of the interests involved in woolgrowing. The arrangement did not work out well, however, and was discontinued the following October. In his report for 1903 Commissioner Richards admitted frankly that

the grazing question is the most perplexing one with which this office has to deal in connection with forest-reserve administration. Those persons who have been in the habit of ranging their stock upon the lands included within a forest reservation are insistent on continuing the practice after the reserve is established, some of them going to the extent of openly defying all rules and orders from the Department prohibiting grazing therein.

Attempts were made to stop this defiance by criminal prosecutions of the trespassers, which the Department of Justice had ruled on November 17, 1898, could be maintained under the act of June 4, 1897. Two years later, on November 14, 1900, the United States District Court for southern California decided that the provision of that act making any violation of the rules and regulations established by the Secretary of the Interior a crime was unconstitutional because it constituted a delegation of legislative power to an administrative officer. This decision did not change the opinion of the Department of Justice, which advised bringing further prosecutions in other districts. Similar suits were therefore brought in the United States District Courts in northern California, Arizona, Utah, and Washington, all of which agreed with the southern California court that the provision was unconstitutional. Since under existing law the government had no right of appeal, it was impossible to bring any of these cases before a higher court.

This situation left the government without protection against trespassers except by the cumbersome process of obtaining a court injunction —a wholly inadequate remedy, since most or all of a grazing season might have elapsed before an injunction could be obtained and served. The Department consequently recommended that Congress pass a specific act making it a criminal offense to pasture livestock in a forest reserve without a permit. Congress ignored this request, as it did the Department's other request for authority to charge a fee for use of the range.

Other Legislative Proposals. Another important recommendation made by the Department of the Interior was that the President be authorized to set apart as national parks public lands which, "for their scenic beauty, natural wonders or curiosities, ancient ruins or relics, or other objects of scientific or historic interest, or springs of medicinal or other properties, it is desirable to protect and utilize in the interest of the public," and that the Secretary of the Interior be authorized to make rules and regulations for their care and management. Much of this proposal was incorporated in the American Antiquities Act of 1906.

The Department also recommended legislation permitting the export of timber from the state in which it was cut, opening agricultural lands within the forest reserves to entry under suitable restrictions, and contributing 25 per cent of the net income from each reserve to the state or territory in which it is located. The substance of all three of these recommendations was enacted into law after the transfer of the reserves to the Department of Agriculture. The Department's weakness in the legislative field did not lie in its failure to make constructive suggestions, but in its inability to get favorable action on them; just as its weakness in the administrative field did not lie in its failure to put on paper sound ideals for the management of the reserves, but in its inability to translate those ideals into practice.

Pinchot's efforts to have the administration of the forest reserves transferred to the Department of Agriculture, within which they would of course be managed by the Forest Service, were greatly strengthened by Secretary Hitchcock's strong endorsement of the proposal. In 1900 Hitchcock expressed the view that future experience might indicate the advisability of transferring the supervision of all public forests to the Secretary of Agriculture. In 1901, he was much more sure of the wisdom of such a move: "Forestry, dealing as it does with a source of wealth produced by the soil, is properly an agricultural subject. The presence of properly trained foresters in the Agricultural Department, as well as the nature of the subject itself, makes the ultimate transfer, if found to be practicable, of the administration of the forest reserves to the Department essential to the best interests, both of the reserves and of the people who use them." With the Secretary's continued support of the proposal, reinforced of course by support from many other quarters, the "ultimate transfer" actually took place in less than four years.

Summary. In 1891 the last-minute addition to a bill revising the general land laws of a section authorizing the President to establish forest reserves inaugurated a radical change in the Federal policy of handling the public timberlands. President Harrison and President Cleveland promptly set aside a large area as forest reserves, but the failure of Congress to provide for their protection and administration created a situation that satisfied no one. After repeated attempts to obtain remedial legislation, in which the Forest Commission of the National Academy of Sciences played a prominent part, Congress passed the highly constructive Forest Reserve Act of June 4, 1897.

Early administration of the reserves by the Department proved ineffective because of excessive red tape and emphasis on paper work, over-centralization of authority in Washington, incompetent personnel, and political pressure. The system, however, remained intact, and the stage was set for subsequent progress.

REFERENCES

Cameron, Jenks: "The Development of Governmental Forest Control in the United States," chap. 8, Johns Hopkins Press, Baltimore, 1928.

Greeley, William B.: "Forests and Men," chap. 4, Doubleday & Company, Inc., New York, 1951.

Ise, John: "The United States Forest Policy," chap. 3, Yale University Press, New Haven, Conn., 1920.

Pinchot, Gifford: "Breaking New Ground," chaps. 12–23, 29–31, 34, Harcourt, Brace and Company, Inc., New York, 1947.

Robbins, Roy M.: "Our Landed Heritage," Peter Smith, New York, 1950.

Rodgers, Andrew Denny, III: "Bernhard Eduard Fernow: A Story of North American Forestry," chaps. 4–5, Princeton University Press, Princeton, N. J., 1951.

CHAPTER 6

Prelude to Progress

The brief period between 1897, when provision was made for the protection and administration of the forest reserves, and 1905, when they were placed under the jurisdiction of the Department of Agriculture, paved the way for the spurt of Federal activity in forest and range management that followed the transfer. It was dominated by the forceful leadership of Gifford Pinchot and Theodore Roosevelt and was noteworthy for the emergence of forestry as a distinct profession.

Pinchot Heads Division of Forestry. On July 1, 1898, following Fernow's resignation to become director of the newly established New York State College of Forestry at Cornell University, Pinchot took charge of the Division of Forestry with the title of Forester, not Chief. "In Washington chiefs of division were as thick as leaves in Vallombrosa," according to his view, but there was only one Forester spelled with a capital letter.

He found an organization that was desperately poor in terms of money, equipment, and manpower. The original appropriation of $2,000 in 1876 had gradually increased to $33,520 and had then been reduced to $28,520. Three years later, under Pinchot's aggressive leadership, it had risen to $185,440 for the fiscal year 1902.

During the same period the number of employees in the Division grew from eleven, of whom only two aside from Pinchot himself had any technical training, to 179 persons. Of this total, 81 were student assistants and about 25 were collaborators. The student assistants were young men studying to become foresters who were employed at $25 a month plus expenses to assist in the field work of the Division; the collaborators were scientists with permanent positions elsewhere who were employed at a salary of $300 a year for the conduct of special studies. These two groups helped materially to bolster the Division's manpower at relatively low cost.

On July 1, 1901, the Division of Forestry was promoted by Congress to become the Bureau of Forestry. The name made no change in the activities of the organization but did increase its prestige and strengthen its standing in the governmental hierarchy at Washington. Pinchot himself regarded the change in status as of "capital value" from the standpoint of increased recognition and support.

As head first of the Division of Forestry and later of the Bureau of Forestry, Pinchot had two major objectives: (1) to get forestry (always referred to as "practical" forestry) actually practiced in the woods, particularly on private lands, and (2) to get the administration of the forest reserves transferred from the Department of the Interior to the Department of Agriculture.

Assistance to Forest Owners. The first objective was publicized through Circular 21, *Practical Assistance to Farmers, Lumbermen, and Other Owners of Forest Lands*, first issued on October 15, 1898, and reissued in many revised editions. It offered to advise private owners on the management of their forest lands, with the government paying the salaries of the examining officers and the owners most of their expenses, and also to prepare complete working plans for the management of large tracts of timberlands, with all costs usually paid by the owner.

During the first year applications for assistance were received from 123 owners of 1,500,000 acres in thirty-five states. The program elicited such a favorable response that the Division and the Bureau were unable to keep abreast of the many requests for help. By 1903 applications had been received for assistance in the management of lands totaling 5,656,171 acres, of which 15,592 acres were in farm wood lots and the remainder in larger tracts. Three years later these figures had doubled. Several elaborate working plans were prepared for tracts of considerable size in many parts of the Eastern United States. A few of these plans were published as government bulletins as a means of arousing interest in other owners and of indicating the requirements and promises of "practical forestry."

Studies and recommendations were made for states as well as for private owners. The Forest, Fish, and Game Commission of New York State requested the Division to prepare plans showing how the forests in the Adirondack Preserve might be handled if the constitutional provision prohibiting the cutting of any timber in the preserve were removed. Studies of four townships were financed by the state legislature. A working plan for Township 40 in Hamilton County, said to have been "the result of the most careful study on the ground which has ever served as the basis for a forest working plan in this country," was prepared by Ralph S. Hosmer and Eugene S. Bruce and published in 1901 as Bulletin 30 of the Division of Forestry. Although the plan showed the township to be "fully adapted for practical forestry," it did not convince the people of the state that such management would be wise.

Help was also rendered to other units of the Federal government, notably the War Department. Working plans were prepared for, and cutting supervised on, several military reservations, including West Point.

Results Disappointing. In his annual report for 1902, Pinchot called attention to the rapidly increasing demands for assistance and expressed the belief that "the time for the general introduction of practical forestry in the United States is evidently at hand, provided only the necessary information and assistance can be supplied." Doubt as to the justification of this optimism is raised by his further comment: "The total area of private forests under conservative management, however, reached only the comparatively insignificant total of 372,463 acres, or 7.9 per cent of the total applications."

A year later he reported that "practical forest work in the woods was better in quality and greater in amount than ever before. But great though the progress was in comparison with other years, actually it was small. The saving of the forests by wise use is but little nearer than it was a year ago, except for the wider spread of knowledge of the nature and objects of forestry."

The obvious need was for adequate financial support. "If this Bureau can be equipped to meet the demands before destruction has gone too far, the extensive protection of woodlands by the practice of forestry will certainly be attained. The only obstacle is present inability to handle the work." Some fifteen years later Pinchot was to revise rather vehemently his opinion as to the readiness of forest owners to practice forestry if they only had the necessary knowledge.

Circular 22, *Practical Assistance to Tree Planters,* issued July 8, 1899, offered help in another direction, particularly to farmers. Planting plans would be prepared by the Division without charge for tracts of fewer than 5 acres, and at a charge covering expenses only for larger tracts. The service was wholly advisory, the actual establishment and maintenance of the plantation being the responsibility of the owner. By 1902, 262 applications for assistance had been received; 197,439 acres had been examined; and plans for the planting of 6,474 acres had been prepared. Although much interest was shown in the program, the results in terms of acreage successfully planted were not impressive.

Pinchot and the Forest Reserves. Progress toward the attainment of Pinchot's second major objective was sure but slow. Transfer of the forest reserves to the Department of Agriculture was no new idea and had been strongly urged by Fernow in his final report on the activities of the Division of Forestry. What was new was the vigorous and widespread support now accorded the proposal. In 1898 the American Forestry Association and the National Board of Trade passed resolutions recommending the transfer. During the next few years many other national and local associations followed suit. In official circles it had the backing of Secretary of Agriculture Wilson, Secretary of the Interior Hitchcock, and Presidents McKinley and Roosevelt. Opposition in Congress, however, delayed action until 1905.

Meanwhile, Pinchot found ways to get the camel's head into the tent. On December 7, 1899, Secretary Hitchcock transmitted to Secretary Wilson a recommendation from Acting Commissioner Richards of the General Land Office that the Division of Forestry "be requested to favor this office with comprehensive reports upon expert investigations of all related questions; which shall extend to including a suggested working plan for the harvesting of timber in each of the existing reserves." This arrangement had first been suggested by Pinchot to an acquaintance in the General Land Office and was of course referred to him for comment by Secretary Wilson.

In other words, Pinchot in effect both originated and approved the proposal; as he himself put it, "The play had been Pinchot to Jones to Richards to Hitchcock to Wilson to Pinchot." The tedious process of going through "channels" does not prevent skillful navigators from reaching their destination.

In due time Richards' proposal received formal approval, and the Black Hills Forest Reserve in South Dakota was selected for first attention. During 1900 and 1901 field studies were conducted under the direction of Edward M. Griffith, later for many years state forester of Wisconsin, and a detailed working plan was prepared. Field work was also conducted on five other reserves and some progress made on the preparation of working plans. None of these plans were actually put into effect until after the transfer. Their preparation nevertheless had two thoroughly worthwhile results. The work yielded much basic information concerning the resources of the reserves, and it gave the Division (and Bureau) of Forestry experience that proved invaluable when it shortly assumed responsibility for their administration.

"The Best Laid Plans." Control over the technical aspects of forest administration nearly passed to the Division several years before action by Congress. Since all of the foresters in government employ were in the Department of Agriculture, there was general agreement that some way should be found to take advantage of their expert knowledge in the management of the reserves. Accordingly, Pinchot, early in 1901, worked out a plan under which he would be appointed as Special Agent and Forester in the Department of the Interior, without pay and without relinquishing his position as Forester in the Department of Agriculture. The Department of the Interior would continue to handle the policing and patrol of the reserves, but all technical questions relating to their management would be decided by him, whether the orders putting those decisions into effect were signed by him or by someone in the Interior Department. The plan was approved by President Roosevelt and the two secretaries concerned but was never put into effect, presumably because of opposition in the General Land Office.

Failure of this particular device did not prevent Pinchot from being the power behind the throne. He served as the President's adviser in all matters relating to the forest reserves, studied the areas proposed for reservations and made recommendations with respect to their boundaries, and had charge of the preparation of all working plans. Striking evidence of the President's reliance on Pinchot is contained in Forest Service Bulletin 67, *Forest Reserves in Idaho*. This bulletin, issued in 1905, consisted of correspondence with Senator Heyburn of Idaho in which Pinchot, not the Secretary of the Interior or the Commissioner of the General Land Office, is the leading figure.

In the course of the spicy exchange of compliments, Roosevelt declined to give Heyburn or any other senator advance information as to temporary withdrawals because of the danger that entries might be made within the proposed reserves for speculative purposes. He also reiterated his strong support of the whole forest-reserve system:

When I can properly pay heed to political interests, I will do so; but I cannot for one moment consent to sacrifice the interests of the people as a whole to the real or fancied interests of any individual or of any political faction. The Government policy in the establishment of national forest reserves has been in effect for some time; its good results are already evident; it is a policy emphatically in the interest of the people as a whole, and especially the people of the West; I believe they cordially approve it, and I do not intend to abandon it.

Another Approach. The Committee on the Organization of Government Scientific Work, appointed by President Roosevelt on March 13, 1903, deserves mention as illustrating one of the many means used by Pinchot to obtain control over the forest reserves. Four months later it submitted a report in which it recommended transfer to the Department of Agriculture of all activities relating to the use of land, including those handled by the General Land Office, the Geological Survey, and the Office of Indian Affairs.

The custody and care of the National Forest Reserves and of the National Parks, now in the Department of the Interior [the report read], should be transferred to the Department of Agriculture, and examination of Forest Reserves now in the Geological Survey should be transferred to the Bureau of Forestry in the Department of Agriculture. . . . The care and administration of the National Parks is work so closely allied to that of the care of the Forest Reserves as to make it evident that they should be subject to the same control.

The recommendation is not surprising in view of the fact that the Committee was suggested by Pinchot, included him as one of its members, and held most of its meetings at his home. Although the report was not published, it probably had some influence in slanting official opinion in Washington in the desired direction.

Eastern Forest Reserves Proposed. A movement for the establishment, by purchase, of forest reserves in the Eastern United States was started by the organization at Asheville, North Carolina, on November 22, 1899, of the Appalachian National Park Association. On December 19 the association addressed a memorial to Congress proposing the establishment of a national park somewhere in the southern Appalachian region. Arguments advanced in favor of the "park" were the central location of the region, its rare natural beauty, its superb forests, the influence of these forests in preventing floods and protecting the water supply, and the large revenue that could be expected from scientific forest management.

A similar memorial was presented to Congress simultaneously by the Appalachian Mountain Club. During the next two years the proposal was endorsed by the American Association for the Advancement of Science, the American Forestry Association, the National Board of Trade, many commercial organizations, and a large number of newspapers in the Eastern United States. The Legislatures of Virginia, North Carolina, and Georgia all passed enabling legislation consenting to the acquisition by the United States of such lands as might be needed for the establishment of a national forest reserve.

Congress responded to the initial memorials by appropriating $5,000 for the fiscal year 1901 to be used by the Secretary of Agriculture "to investigate the conditions of the Southern Appalachian mountain region of Western North Carolina and adjacent States." The study was made by the Division of Forestry in cooperation with the Geological Survey in the Department of the Interior. On January 3, 1901, Secretary Wilson submitted a preliminary report recommending the establishment of a forest reserve instead of a park. It was transmitted to Congress by President McKinley with his endorsement.

The final report was submitted on December 16, 1901. Its major conclusion, with which President Roosevelt expressed full agreement, was as follows:

The preservation of the forests, of the streams, and of the agricultural interests here described can be successfully accomplished only by the purchase and creation of a national forest reserve. The States of the Southern Appalachian region own little or no land, and their revenues are inadequate to carry out this plan. Federal action is obviously necessary, is fully justified by reasons of public necessity, and may be expected to have most fortunate results.

Meanwhile, in January, 1901, Senator Pritchard of North Carolina introduced a bill appropriating 5 million dollars for the purchase of not less than 2 million acres of mountain land in the region. The bill received a favorable report from the Committee on Forest Reservations and the Protection of Game, partly because of Secretary Wilson's assurance that the proposed reserve would soon be self-supporting from the sale of timber, but made no further progress. From then on, bills dealing with the subject, one of which passed the Senate, were almost constantly before Congress until final passage of the Weeks Act in 1911.

Forestry on Chippewa Indian Reservation. The Chippewa Indian Reservation in Minnesota gave the Bureau of Forestry its first opportunity actually to practice the forestry which it was so vigorously preaching. The Nelson Act of 1889 providing for the sale of timber on the reservation at its appraised value had not worked out satisfactorily. Strong protests by many individuals and organizations led Secretary Hitchcock in March, 1899, to suspend an advertised auction and brought the entire situation to the attention of Congress. Much heated debate finally led to the passage of the Morris Act of June 27, 1902.

The Morris Act limited future cutting to merchantable pine timber, with the exception of such other trees as had to be cut in the economical conduct of logging operations, and required the burning or removal of slash in order to minimize the fire danger. It further instructed the Forester of the Department of Agriculture, with the approval of the Secretary of the Interior, to select 200,000 acres of pine land and 25,000 acres of agricultural land (defined as land containing not more than 1,500 board feet of pine to the acre), on which cutting would be done under rules and regulations prescribed by the Forester. The regulations must, however, provide for leaving 5 per cent of the merchantable pine to serve as seed trees and for the disposal of slash.

The proceeds from the sale of timber were of course to be credited to the Indians, but the government was to retain title to the lands, which were to become a forest reserve.[1] In addition to these 225,000 acres, the Forester was to select ten sections, together with certain islands and points in Cass Lake and Leech Lake, which were also to become a part of the reserve but were to be completely reserved from sale and settlement.

Eugene S. Bruce, a lumberman in the Bureau of Forestry who was

[1] The Minnesota National Forest was actually established by act of May 23, 1908.

particularly familiar with the forests of the Adirondacks, was placed in charge of the selection of the lands and the sale of the timber on them. The cutting regulations prescribed by the Bureau, in addition to reserving the required 5 per cent of the stand for seed trees, forbade the cutting of any white pine or Norway pine less than 10 inches in diameter at 3 feet from the ground, required the utilization of all merchantable pine logs 6 inches or over in diameter at the small end, and insisted upon the burning of all tops and litter.

The Morris Act constituted an important milestone in the development of Federal forest policy. It gave the Bureau of Forestry its first experience in practical forest management and thereby put government foresters to work on government land. It paved the way for the establishment of the first forest reserve to be created by act of Congress instead of by Presidential proclamation. And for the first time it wrote into law the method by which cutting should be done. Fortunately, conditions happened to be such that the dangerous practice of "legislative silviculture" worked out well in this particular case.

Oregon Land Frauds. During the early 1900's the country was startled by the exposure of widespread frauds affecting the public lands in Oregon. Of the many cases that came to trial, it may be sufficient to cite one as illustrative of the devious methods commonly used and the complications that sometimes occurred. The details of this particular case are taken from "Looters of the Public Domain," written while in jail by Stephen A. Douglas Puter, "king of the Oregon land fraud ring." All quotations are from that book.

In 1900 the official survey of Township 11 South, Range 7 East, Willamette Meridian in the Cascade Forest Reserve was approved. Claims in the township therefore became eligible for patenting and exchange under the lieu-land provision of the Forest Reserve Act of 1897. The township is located near the top of the Cascade Range, is very rough and inaccessible, and at that time there was not a habitation within 30 miles. Puter and his partner Horace Greeley McKinley decided that these facts made the area an excellent one in which to locate bogus homestead claims, since its inaccessibility minimized the chances of any field investigation of the claims.

They accordingly got ten persons to file claims to twelve homesteads of 160 acres each. Two of the group each filed two claims, of course under different names, and only three of the ten used their real names. Although none of the claimants had ever been anywhere near the township, they all signed affidavits swearing that they were living on their claims prior to the creation of the reserve and had resided there continuously up to that date; that the improvements consisted of a good house, with outbuildings, such as barn and woodshed; that they had cultivated and fenced an acre or so of ground; and that the value of their improvements amounted to several hundred dollars. The requirement that affidavits and proofs must be corroborated by two witnesses was easily met by having the claimants serve as witnesses for each other.

Complications in the "11-7" Case. A few weeks after final proofs were made, J. A. W. Heidecke who lived in Detroit, the nearest town to Township "11-7," informed McKinley that he knew that none of the entrymen had ever been on their claims and "hinted that unless he could get something out of it he would report the matter to the Commissioner of the General Land Office. The upshot of this conversation was that McKinley settled with Heidecke by paying him $50, for which amount he agreed to keep his mouth shut."

Some months later Puter learned that Special Agent C. E. Loomis had been detailed by the Land Office to make a thorough study of the claims because of charges of fraud. He promptly got in touch with Loomis; informed him that he had bought the claims in good faith and was anxious to have patents issued promptly; stated that the inspection trip would be arduous and expensive; and offered to contribute to the expense if Loomis would defer all other business and proceed to the township. He then handed Loomis a draft for $500, with a promise to give him another $500 after patents were issued. Loomis agreed to go right ahead and expressed confidence in his ability to make a favorable report.

Puter's next step was to engage Heidecke to serve as a guide to Loomis. Heidecke professed to be well acquainted with all of the entrymen and their homesteads. He promised to show these to Loomis and to introduce him to several residents of Detroit who were personally acquainted with the claimants. Puter gave Heidecke $110 to cover "expenses," offered him an additional $250 for his "extra trouble" if he succeeded in showing Loomis all the cabins and improvements, and promised to get him appointed as a forest ranger on the Cascade Reserve.

A few weeks later Loomis informed Puter that he had made a careful examination of the claims with Heidecke, whom he pronounced very much of a gentleman; that he had found all of the improvements on the twelve claims in question, although some were in a dilapidated condition on account of the heavy snowfall of the previous winter; that he was satisfied that the homesteaders had acted in good faith and had complied with the law to the best of their ability; and that he would recommend the entries to patent.

It developed afterwards that Heidecke had merely taken Loomis along some well-defined trails, that led past cabins belonging to other settlers in that part of the country, and had not been on any portion of the suspended claims with him, because it would have been a give away on both sides to have done so, and for the further reason, that they would necessarily have had to possess the agility of a goat to reach any of my claims, as they were practically inaccessible . . . after showing Loomis a certain cabin, belonging to a legitimate settler, in another township, he circled around for about half an hour, bringing up at the same cabin, but viewing it from the rear, instead of the front, as in the first instance, and later in the day, finding that he was running short of cabins, he halted Loomis, for yet a third time, at the same identical cabin, taking the precaution, on this occasion, to view it from the side.

Loomis may have overdone the matter, since it was not long before he informed Puter that he had received fresh instructions to obtain personal affidavits from the homesteaders regarding their improvements, cultivation, and residence on their claims. Puter succeeded in rounding up six of the entrymen, all of whom signed the desired affidavits. These had previously been prepared by Loomis and were in his handwriting.

The Web Grows More Tangled. Some months later, Captain Salmon B. Ormsby, superintendent of the Cascade Forest Reserve, received instructions to make a thorough examination of the claims and to obtain affidavits from disinterested parties. Puter thereupon offered Ormsby's son $500, payable on the issuance of patents, if he could persuade his father to make an immediate examination of the claims and return a favorable report thereon. Puter next arranged for Heidecke to help Ormsby in the same way that he had helped Loomis.

Shortly afterward young Ormsby told Puter that his father had sent his report to Washington, after talking with Loomis, and that he thought it was favorable. This later proved to be the case, though the reason for the elder Ormsby's action remains somewhat mysterious, since Heidecke had told him frankly that there were no settlements in the township and they had made no attempt to visit the claims. Puter, however, became nervous because at the time he was unable to learn the exact character of the report. He therefore decided to go to Washington to enlist the help of Senator John H. Mitchell, who introduced him to Binger Hermann, Commissioner of the General Land Office.

Hermann stated that the reports were all favorable but that it would be several months before patents could be issued. Shortly afterward, however, he informed Puter that he had just received a report which knocked out all twelve claims and suggested that he return to Oregon to get some additional affidavits which might strengthen the claims. Instead, Puter insisted that action be deferred until he could consult further with Senator Mitchell, to whom he offered two $1,000 bills in payment for his services. Mitchell, after "gentle protest," pocketed the money and promised to see Hermann that very evening.

The next morning he reported that he had "fixed" things up with Hermann and he was sure there would be no further trouble. A day later he stated that he had seen Hermann again and that the latter had found everything all right and had decided to issue the patents. He was well informed. Before Puter left Washington he saw the patents actually being written, and when he reached Portland patents for all twelve claims were awaiting him there. Although his troubles were now apparently at an end, the story was to have an unhappy sequel.

In the spring of 1901 Puter and McKinley had decided to repeat the procedure followed in Township 11-7 in another township which had just become subject to entry as a result of approval of the government survey. This was Township 24 South, Range 1 East, Willamette Meridian, and was also located in the Cascade Forest Reserve in a rough and inaccessible part of the Cascade Range. Claims were filed in the names of wholly fictitious persons before Marie Ware, United States Commissioner at Eugene and a sweetheart of McKinley's, who was to receive adequate remuneration for her participation in the fraud.

This case became even more complicated than the 11-7 one and led to a disagreement between two of the thieves (McKinley and George Lloyd). Lloyd complained to a special inspector of the General Land Office at Eugene and thus started a series of investigations which exposed the whole land-fraud mess. Indictments were brought against Puter, McKinley, and others in the spring of 1903 for conspiracy to defraud the government in both the 11-7 and 24-1 cases.

Trials End in Convictions. The 11-7 case came to trial on November 21, 1904, and ended on December 6 with the conviction of all of the defendants except Marie Ware, who had not been involved in this particular deal, and one of the entrymen who pleaded guilty during the trial. The failure of Puter's business friends to provide bail after his conviction, and Senator Mitchell's highly uncomplimentary remarks concerning him in the press aroused Puter's ire and decided him to put them in the same boat with himself. He accordingly supplied a grist of incriminating information to Francis J. Heney, who had been assigned by President Roosevelt as a special assistant in the Oregon land-fraud trials.

The most spectacular of the trials was that of Senator Mitchell. He was indicted for violation of Section 1782 of the Revised Statutes of 1878, which forbids any member of Congress "to receive any compensation whatever, directly or indirectly, for any services rendered . . . in relation to any . . . matter or thing in which the United States is a party, or directly or indirectly interested, before any Department, court martial, Bureau, officer, or any civil, military, or naval commission whatever." Mitchell's trial began on June 20, 1905, and ended on July 3 with his conviction. He was sentenced to imprisonment for six months and fined $1,000, but died on December 8, 1905, from a dental operation, while the case was still under appeal.

Puter managed to stay out on appeal until July 6, 1906, after playing hide-and-seek with the Secret Service for some months, when he was finally sentenced to two years in jail and a fine of $7,500. He was pardoned by President Roosevelt on December 31, 1907, after having served seventeen months of his sentence, and was actually released on January 6, 1908, his fifty-first birthday.

Extent of Land Frauds. The extent of the land frauds uncovered at this time is indicated by the fact that a Federal grand jury was in session for some time in the spring of 1904, from December 17, 1904, to April 8, 1905, and during parts of August and September, 1905. It brought in twenty-six indictments affecting more than 100 persons. Among those convicted were Congressman John N. Williamson; former United States District Attorney Franklin P. Mays; Horace G. McKinley, one of Puter's associates; and Willard N. Jones, another of Puter's associates and a member of the Oregon Legislature. Senator Charles W. Fulton escaped indictment because of the statute of limitations.

The trials attracted national attention and vigorous denunciation of the time-honored practice of stealing public lands and timber. It was rapidly becoming neither safe nor respectable to flout the law; acquittal by local juries could no longer be taken for granted. Heney clearly recognized that change in the moral climate when he congratulated the jury that convicted Senator Mitchell for demonstrating to the world "that Oregon believes in the enforcement of the laws of our country, and that in Oregon no man is above the law."

Exposure of the land frauds also did much to strengthen public opposition to the lieu-land provision of the act of 1897 and public support of more effective administration of the forest reserves. It helped to fan the moral indignation generated by the Roosevelt administration against "big business" in general and "timber barons" in particular.

Second Public Lands Commission. At about the time that the Oregon land frauds were coming to a head, President Roosevelt on October 2, 1903, at Pinchot's suggestion, appointed a Public Lands Commission "to report upon the condition, operation, and effect of the present land laws and to recommend such changes as are needed to effect the largest practicable disposition of the public lands to actual settlers who will build permanent homes upon them, and to secure in permanence the fullest and most effective use of the resources of the public lands." Its members were W. A. Richards, Commissioner of the General Land Office, chairman; F. H. Newell, Chief Engineer of the Reclamation Service; and Gifford Pinchot, Forester of the Bureau of Forestry. The Commission held formal hearings in various parts of the country, talked informally with governors, public officials, and many other individuals, and spent much time actually on the public lands.

The Commission submitted two partial reports (on March 7, 1904, and February 13, 1905), together with an appendix containing a large amount of factual information. It stated bluntly that "the present laws are not suited to meet the conditions of the remaining public domain." With respect to the theory that the major purpose in the disposal of the public lands was to place them in the hands of actual settlers, it stressed the fundamental fact that "the number of patents issued is increasing out of all proportion to the number of new homes."

Commission's Recommendations on Agricultural Lands. The many recommendations which the Commission made to improve the situation may be summarized as follows:

The remaining public lands should be studied with a view to ascertaining those chiefly valuable for agriculture. Pending such ascertainment, provision should be made "to hold under Government control and in trust for such use the lands likely to be developed by actual settlers." A homestead should be large enough to support a family, which meant that its size should vary in different localities.

The commutation clause of the Homestead Act, if retained at all, should be amended to require three years of actual (not constructive) residence on the lands before commutation is permitted. Commutation is now commonly used to obtain title to lands primarily valuable for timber or minerals rather than for agriculture. For example, in several land districts in Minnesota during the four years from 1899 to 1903, commuted entries totaled 192,189 acres in the timber belt and 26,660 acres in the prospective mineral belt as compared with 22,528 acres in the agricultural belt.

In the timber belt, the timber alone, leaving out the land, was estimated to be worth about $650,000 more than the amount received by the government, and 89.4 per cent of the lands were soon transferred to others than the original entrymen. The 1,485 entries involved are now held by fifty persons or corporations. Throughout most of the area where commutation has been heaviest there is no sign of habitation save for an occasional trapper or hunter.

The area that can be obtained under the Desert Land Acts should be reduced to 160 acres; actual residence of not less than two years should be required, with the production of a valuable crop on one-fourth of the area and real proof of adequate water supply; and assignments should be prohibited.

Recommendations on Forest and Range Lands. The Timber and Stone Act should be repealed. It has been made the vehicle of innumerable frauds and has been the cause of heavy losses to the government. It is estimated that sales of timber on 175,883 acres on the Chippewa Indian Reservation made under the Morris Act of 1902 will yield at least $2,-200,000 more than would have been received if the land had been sold under the Timber and Stone Act, and in addition the government still owns the land. A special study made in selected localities in Oregon and Washington showed that 10 per cent of the claims patented under the Timber and Stone Act in 1903 were transferred to timber and mill companies shortly after the entrymen received patent. Collusion is extremely difficult to prove under the Supreme Court ruling that a person may take up a timber claim with the object of making an immediate sale, provided there is no prior agreement to that effect.

Provision should be made for the sale of timber on the unreserved public domain, for which there is now no legal authority. Free use of timber should be limited to the actual needs of miners and settlers in place of the wholesale free use now common under the act of 1897.

The President should be authorized to set aside grazing districts by proclamation. The Commission estimated that there were more than 300 million acres of public land primarily valuable for grazing which were virtually open commons. This situation led to strife between users of the range, often accompanied by violence, and to its serious deterioration or even ruin. The proposed grazing districts should be administered and managed by the Secretary of Agriculture under rules and regulations adapted to each district, including the charging of a moderate fee for grazing permits, "with the special object of bringing about the largest permanent occupation of the country by actual settlers and home seekers." A canvass of the livestockmen made by Albert F. Potter, himself a former sheepman who was currently in charge of the Branch of Grazing in the Bureau of Forestry and later associate forester, showed them to be in favor of the proposed regulation of grazing on the public range.

Only two recommendations were made with respect to forest reserves: (1) that lands primarily valuable for agriculture be opened to homestead entry by the secretary in charge with no opportunity for commutation and (2) that the lieu-land provision of the act of 1897 be repealed. The government always lost under the latter provision, which had given rise to much speculation, scandal, and consolidation of timber holdings in the hands of large holders. Efforts were even made to have worthless or denuded lands proclaimed as forest reserves so that their owners might exchange them for valuable lands elsewhere.

Representative Fordney of Michigan had denounced the lieu-land provision as "an absolute fraud. The Government has lost millions of acres of the most valuable land in the country, and got practically nothing in return." Its operation was one of the chief causes of Western hostility to the forest reserves. In its place the Commission recommended that the government be authorized to buy needed private lands within the reserves or that owners of such lands be permitted to exchange them for specified tracts of equal area *and value* outside of the reserves.

Action on Recommendations. Three of the Commission's recommendations were later enacted into law, one of them after a lapse of fifty years. The lieu-land provision was repealed in 1905, forest homesteads were authorized in 1906, grazing districts were established in 1934, and the sections of the act of 1878 authorizing the sale of unreserved lands primarily valuable for timber or stone were repealed in 1955. Although immediate results were not spectacular, the Commission performed a useful service in focusing public attention on some of the more glaring defects in the handling of the public lands.

Pinchot's participation in its activities strengthened his preparation for the administration of the forest reserves which was so soon to be placed in his hands. The appearance of the reports at a time when the public was stirred by the Oregon land-fraud disclosures undoubtedly assured them attention that they might not otherwise have received.

State Forestry. The period between 1897 and 1905 constituted a prelude to progress in state forestry as well as in Federal forestry. Almost universal handicaps were lack of funds and trained personnel. With few exceptions, appropriations were small, and forestry agencies not uncommonly had no foresters. The situation was, nevertheless, improving slowly but surely.

New York in 1900 employed Ralph C. Bryant, the first graduate of the New York State College of Forestry at Cornell, and Pennsylvania in 1901 employed George H. Wirt, an early graduate of the Biltmore Forest School. Connecticut in 1901 was the first state to use the term "state forester" to designate the head of its forestry agency. Walter Mulford, another Cornell graduate, was appointed to the position and thus became America's first official state forester. The activities of these men and others in state employment were chiefly educational in character, and their influence on the practice of forestry was inevitably slight.

Research in Forestry. Scientific investigations of course continued to be an important activity of the Federal Division of Forestry and Bureau of Forestry under Pinchot's administration as they had been under Fernow's. Increased attention to promoting the practice of forestry on both private and public lands gave research relatively a less prominent position than it had previously occupied but did not reduce the amount of effort devoted to it. Studies dealing primarily with dendrology, silviculture, and forest management were conducted in many parts of the country, and in 1904 the first permanent sample plots were established—on private land in New Hampshire.

Research in the broad field of wood utilization, which had been discontinued in 1896 because of the withdrawal of congressional support, was resumed in 1902. Emphasis was placed on timber testing and wood preservation, with the work largely centered at cooperating institutions, among which were Yale, Purdue, Washington University (at St. Louis), and the University of Washington.

Schools of forestry have always recognized research as a major responsibility, but lack of staff and funds precluded much activity in that field during the early years of their existence.

Professional Education in Forestry Starts at Cornell. Truly professional education in forestry started in 1898 with the establishment of the New York State College of Forestry at Cornell University. Fernow left the position as Chief of the Division of Forestry in the U. S. Department of Agriculture which he had held since 1886 to become its director. Except for an interlude of four years as consulting forester, the rest of his active professional life was devoted to teaching—first at the New York State College of Forestry at Cornell University, then at Pennsylvania State College, and lastly at the University of Toronto.

The instruction at Cornell was organized as a four-year undergraduate program leading to the degree of bachelor in the science of forestry (later forest engineer). The arrangement of courses followed the pattern subsequently adopted by undergraduate schools throughout the country, with the first two years devoted almost entirely to basic subjects and the last two years chiefly to technical subjects. In addition to Fernow, the original faculty consisted of Filibert Roth, who moved with him from the Division of Forestry where he had been in charge of the investigations in timber physics, and John Gifford, an American with European training in forestry who at the time was engaged in forestry work with the New Jersey Geological Survey.

In establishing the New York State College of Forestry, the state Legislature also arranged for the administration by the college of a 30,000-acre demonstration and experimental forest in the Adirondack Mountains.

About half of it consisted of cutover spruce and pine lands, and the other half of uncut northern hardwoods with an admixture of spruce and pine. Fernow's plan of management provided for increasing the proportion of spruce. Although he expected to try out many different methods of cutting, the first large sale to the Brooklyn Cooperage Company called for clear cutting of the hardwoods, with subsequent planting where necessary to obtain satisfactory stocking of spruce. This method of cutting, which was far from the average layman's conception of "forestry," led to much criticism and to an unfavorable report by a legislative committee.

None of this criticism was aimed at the instructional work at Ithaca, but in spite of this fact, Governor Odell in 1903 vetoed the appropriation for the College of Forestry. Cornell University did not see its way clear to provide funds for the continuation of the college, which in June of that year consequently closed its doors for good. Instruction in forestry was, however, resumed in 1910 in a Department of Forestry organized as an integral part of the New York State College of Agriculture at Cornell University.

Biltmore, Yale, and Others Enter the Field. Almost simultaneously with the establishment of the original New York State College of Forestry in 1898 came the opening of the Biltmore Forest School under Carl A. Schenck at Biltmore, North Carolina. For a couple of years, Schenck had been accepting special students, but there was no formal organization of the instructional work until the fall of 1898, when the first catalogue of the school was issued.

The Biltmore Forest School was unique in the history of forestry education in America. It was essentially, though not completely, a one-man "master" school in which Schenck was always the dominating figure. Instruction, which covered one year, was both theoretical and practical, but with strong emphasis on actual field preparation and participation in all the many activities involved in the management of a large forest property. Although never recognized as having full professional standing, the school turned out many men who were leaders in the profession. After Schenck's retirement as manager of the Biltmore Forest in 1909, the school became peripatetic, spending part of the year in Europe and the remainder of the time in different regions in the United States. It was discontinued in 1913.

Still another approach to professional education in forestry was tried in 1900 with the establishment of a School of Forestry at Yale University. Only men with a bachelor's degree were admitted to the program of studies, which covered a period of two years and led to the degree of master of forestry. Funds for the support of the school were provided by an initial endowment of $150,000 from the Pinchot family, which was later increased to $300,000. Its purpose, according to Gifford Pinchot, who "had small confidence in the leadership of Dr. Fernow or Dr. Schenck," was to provide a supply of "American foresters trained by Americans in American ways for the work ahead in American forests."

Henry S. Graves left his position as assistant chief of the Division of Forestry to become director of the new school. He was assisted by James W. Toumey, formerly a professor of botany at the University of Arizona, who was at the time in charge of the Section of Economic Tree Planting in the Division. Today the school is the oldest in the country with a continuous existence from the date of its founding.

Within three years, professional education in forestry, hitherto conspicuous by its absence, had become represented by three different types of institutions. The New York State College of Forestry at Cornell University was supported by public funds and offered a four-year program leading to a bachelor's degree. This is the pattern that has been most generally followed at other institutions, most of which, however, now offer graduate as well as undergraduate work. The School of Forestry at Yale University was supported by private endowment. It offered a two-year program leading to a master's degree open only to men already possessing a bachelor's degree in arts or science. This pattern has been followed only at Harvard University (temporarily) and Duke University. The Biltmore Forest School was privately operated on a pay-as-you-go basis and offered a one-year, highly practical program. It has had no followers.

Other schools soon entered the field, several in 1903. By 1910 no less than seventeen schools, scattered from Maine south to Georgia and west to Washington, were offering programs of study. Professional education in forestry had arrived suddenly, and to stay. Perhaps in no other field has there been such a systematic effort to anticipate the need for trained men before that need became dangerously acute.

Forestry Literature. For nearly twenty years following the demise of the *American Journal of Forestry* in 1883, forestry in the United States was without a professional periodical. Considerable material of popular, and occasionally scientific, interest appeared in *Forest Leaves* (now *Pennsylvania Forests*), published by the Pennsylvania Forestry Association; in *Garden and Forest*, published by Charles S. Sargent; and in *The Forester*, first published in 1895 by the New Jersey Forestry Association and after 1898 by the American Forestry Association, with subsequent changes in name to *Forestry and Irrigation* (1902), *Conservation* (1908), *American Forestry* (1910), *American Forests and Forest Life* (1924), and *American Forests* (1931).

None of these publications, however, met the need of the growing body of foresters for a technical journal. That need was met in 1902 by the establishment by the students of the New York State College of Forestry of the *Forestry Quarterly*, of which Fernow soon became editor. The *Quarterly* started as an official publication of the college, but after the discontinuance of the college was handled by Fernow as a personal enterprise. In addition to original articles it contained book reviews and news of current happenings.

Three years later, in May, 1905, the first issue of the *Proceedings of the Society of American Foresters* made its appearance. This publication was issued at irregular intervals until 1914, when it became a quarterly. It was confined chiefly to papers presented at meetings of the society. Efforts to put it on a more substantial basis, and also to combine it with the *Forestry Quarterly*, proved unsuccessful until 1917 when the union was effected by establishment of the *Journal of Forestry*.

Other literature of a technical nature consisted mostly of the series of bulletins and circulars emanating from the Division of Forestry (later Bureau of Forestry and Forest Service) in the Department of Agriculture. These publications, issued with increasing frequency, presented the results of its investigations and other activities and contributed substantially to the growing body of forestry literature.

Birth of a Profession. Among the characteristics of a profession are the existence of a technical literature; the conduct of research and the application of the findings of research in actual practice; the development of specialized educational institutions; and the establishment of an organized association of technically trained men, with the maintenance of high standards of professional ability and ethical conduct as one of its objectives. Judged by these criteria, forestry emerged as a new profession in this country at about the beginning of the present century. The conduct of research and publication of the results had been, and continued to be, a major activity of the Division of Forestry. Sound forest practice based on scientific principles had been undertaken on a small area of private and public lands and in 1905 was extended on a large scale to the forest reserves. Ample facilities for professional education developed rapidly from 1898 on.

If a singe date were to be selected for the birth of the profession, it might well be November 30, 1900, when a group of seven men met in Pinchot's office to organize the Society of American Foresters. These men were Gifford Pinchot, Henry S. Graves, Edward T. Allen, William L. Hall, Ralph S. Hosmer, Overton W. Price, and Thomas H. Sherrard. The formal organization was completed on December 13, and two days later eight more men were elected to the society—Bernhard E. Fernow, Carl A. Schenck, James W. Toumey, Filibert Roth, Horace B. Ayers, Edward M.

Griffith, Frederick E. Olmsted, and George B. Sudworth. The names of such distinguished men as William H. Brewer, Grover Cleveland, Wolcott Gibbs, Benjamin Harrison, John W. Noble, James W. Pinchot, Theodore Roosevelt, Charles S. Sargent, and James Wilson were soon added to the list of associate members. The object of the society, as stated in the constitution, was to "further the cause of forestry in America by fostering a spirit of comradeship among foresters; by creating opportunities for a free interchange of views upon forestry and allied subjects; and by disseminating a knowledge of the purpose and achievements of forestry."

The Washington members of the society held frequent meetings at Pinchot's home, at which the standard refreshments were baked apples, gingerbread, and milk; hence the name "Baked Apple Club" by which the society was commonly known. At one of these meetings, on March 13, 1903, President Roosevelt broke through his custom of not making speeches at private houses by addressing the society on "Forestry and Foresters." In this address, he gave some sound advice to his hearers:

In the last analysis, the attitude of the lumberman toward your work will be the chief factor in the success or failure of that work. In other words, gentlemen, I cannot too often say to you, as, indeed, it cannot be too often said to any body of men of high ideals and good scientific training who are endeavoring to accomplish work of worth for the country, that you must keep your ideals high and yet seek to realize them in practical ways. . . .

You have created a new profession of the highest importance, of the highest usefulness to the State, and you are in honor bound to yourselves and to the people to make that profession stand as high as any other profession, however intimately connected with our highest and finest development as a nation. You are engaged in pioneer work in a calling whose opportunities for public service are very great. Treat that calling seriously; remember how much it means to the country as a whole. . . .

We have reached a point where American foresters trained in American forest schools are attacking American forest problems with success. That is the way to meet the larger work you have before you. You must instill your own ideals into the mass of your fellow-men and at the same time show your ability to work with them in practical and business fashion. This is the condition precedent to your being of use to the body politic.

Summary. Pinchot's appointment as head of the Division of Forestry in 1898 resulted in vigorous attempts to promote the practice of forestry by private owners and to bring about the transfer of the forest reserves from the Department of the Interior to the Department of Agriculture. Many cooperative agreements were signed and numerous working plans were prepared for both large and small forest properties, but few were put into effect. There was still too much timber beyond the horizon, and the economic advantages of forestry were still too dubious to encourage its practice by private owners on any considerable scale.

More progress was made toward attaining the second objective. The Bureau of Forestry was retained by the Department of the Interior to prepare working plans for the Black Hills and other forest reserves; Pinchot became the President's most trusted adviser in all matters pertaining to the reserves; and the way was paved for the transfer which he and many others strongly advocated.

Special significance attaches to the inauguration of technical instruction in forestry at Cornell and Biltmore in 1898, at Yale in 1900, and at several other institutions in 1903. These schools assumed leadership in preparing the trained personnel essential to direct the practice of forestry even before the need for such training was generally recognized.

Equally significant was the emergence of forestry as a distinct profession through the organization in 1900 of the Society of American Foresters. The striking advances in forest policy and forest practice that have occurred in the last half century are in large measure due to the insistence by that society on high standards of professional competence and ethical conduct.

These various developments set the stage for the great expansion of the forestry activities of the Federal government that was to start in 1905.

REFERENCES

Allen, Edward T.: The Application and Possibilities of the Federal Forest Reserve Policy, *Soc. Amer. Foresters Proc.*, 1:41–52, 1905.

Cameron, Jenks: "The Development of Governmental Forest Control in the United States," chap. 9, Johns Hopkins Press, Baltimore, 1928.

Ise, John: "The United States Forest Policy," chap. 4, Yale University Press, New Haven, Conn., 1920.

Pinchot, Gifford: "Breaking New Ground," chaps. 24–28, 33, 35, Harcourt, Brace and Company, Inc., New York, 1947.

Puter, S. A. D.: "Looters of the Public Domain," Portland Printing House, Portland, Ore., 1908.

Rodgers, Andrew Denny, III: "Bernhard Eduard Fernow: A Story of North American Forestry," chaps. 6–7, Princeton University Press, Princeton, N.J., 1951.

Roosevelt, Theodore: Forestry and Foresters, *Soc. Amer. Foresters Proc.*, 1:3–9, 1905.

CHAPTER 7

Forestry in Practice and Politics

From 1905 until Pinchot's dismissal from government service in 1910, forest policy was concerned primarily, though not exclusively, with the administration of the forest reserves (national forests). Conservation of other natural resources also became a major issue which was dramatized by the Conference of Governors in 1908. Alleged infidelity to conservation ideals on the part of the Taft administration led to the Ballinger-Pinchot controversy and to the split in the Republican party that resulted in Wilson's election as President in 1912.

American Forest Congress. The period was ushered in by an impressive American Forest Congress held at Washington, D.C., January 2 to 6, 1905. The conference was called by the American Forestry Association "to establish a broader understanding of the forest in its relation to the great industries depending upon it; to advance the conservative use of forest resources for both the present and future need of these industries; to stimulate and unite all efforts to perpetuate the forest as a permanent resource of the nation."

President Roosevelt, as honorary president of the congress, in the opening address, emphasized its two main objectives—to get the practical forester and the practical businessman working together for a common end and to get all the forest work of the government concentrated in the Department of Agriculture. Congressman Lacey of Iowa, in endorsing the latter objective, paid tribute to "the great head of forestry . . . not my dear young friend, Mr. Pinchot, but the old man, who comes from the prairie State of Iowa," Secretary James Wilson. Approval of the transfer was also voiced by Fred P. Johnson, Secretary of the National Live Stock Association, which a few days later adopted a formal resolution to that effect.

Resolutions of the Congress. Speeches from men in all walks of life dealt with all aspects of the relation of forestry to the life of the nation.

Of the eighteen resolutions adopted by the congress, the more important may be summarized as follows:

Enactment and enforcement by the states of laws for the protection of forests from fire and for reducing the burden of taxation on lands held for forest reproduction

Repeal of the Timber and Stone Act and substitution therefor of an act authorizing the sale of timber on the public lands when such sale is for the public welfare

Amendment of the lieu-land provision of the act of 1897 to limit lieu selections to lands of equivalent value or similar conditions of forest growth

Repeal of the provision prohibiting the export of forest-reserve timber from the state in which it is cut in those states where the export of such timber is in the public interest and in no others

Enactment of legislation authorizing the sale of all nonmineral products of the forest reserves and the use of receipts from such sales in the management and protection of the reserves and the construction of roads and trails therein

Immediate unification of all the forest work of the government, including the administration of the forest reserves, in the Department of Agriculture

Expansion of opportunities for general forest education in schools and colleges and for professional training in postgraduate schools

Appropriation of adequate funds for the promotion of forest education and research in the agricultural colleges and experiment stations, to be used by state forestry departments in such ways as may seem best for forestry educational purposes

Immediate passage of legislation providing for the establishment of national forest reserves in the southern Appalachian Mountains and the White Mountains

Purchase of the Calaveras Grove of Big Trees by the Federal government and reconveyance of Yosemite Valley to the government by the state of California

Encouragement of tree planting and the preservation of shade trees along public highways throughout America

Enactment of a requirement that at least 5 per cent of a homestead be planted to trees under the supervision of the Bureau of Forestry before final title is acquired

The congress was noteworthy for the high caliber of its participants, who included many of the leading figures in the fields of lumbering, railroading, grazing, irrigation, mining, education, and government; for the enthusiasm and spirit of cooperation which marked its deliberations; and for the influence which it exerted in arousing interest in the whole problem of forest conservation throughout the nation. It is generally credited with giving the final push to the legislation transferring the forest reserves to the Department of Agriculture which had been before Congress in one form or another for many years.

Transfer of the Reserves. The long-sought transfer was accomplished within less than a month after the adjournment of the American Forest Congress. The act of February 1, 1905, effecting the transfer, covered considerable ground:

1. It transferred the administration of the forest reserves from the Secretary of the Interior to the Secretary of Agriculture, effective immediately.

2. It covered all receipts from the forest reserves for a period of five years into a special fund to be expended as the Secretary of Agriculture might direct for the protection, administration, improvement, and extension of the reserves.

3. It provided that forest supervisors and rangers should be appointed, when practicable, from the states or territories in which the reserves were located.

4. It granted rights of way for dams, ditches, flumes, etc., across the reserves for various purposes under regulations prescribed by the Secretary of the Interior and subject to state laws.

The act constituted the third big step in the reservation and management of the forests on the public lands of the United States. Unlike its predecessors of 1891 and 1897, which were passed as riders attached to bills dealing chiefly with entirely different subjects, it stood on its own feet. At last a forestry measure of major importance which had been under thorough scrutiny for years was approved as an independent piece of legislation. The foresters and the forest land of the government were at last brought together under one administrative head. Roosevelt, Wilson, and Pinchot were in a position to demonstrate the advantages of the combination which they had advocated so long and so vigorously.

Wilson's Letter of Instructions. On the very day that the act was signed, Secretary Wilson addressed to Pinchot a letter of instructions, written by the recipient, stating succinctly but comprehensively the principles to be followed in the management of the reserves. Since these principles have been followed from that date to this, the letter constitutes such an important milestone in the development of Federal forest policy as to be worth quoting at length:

In the administration of the forest reserves it must be clearly borne in mind that all land is to be devoted to its most productive use for the permanent good of the whole people, and not for the temporary benefit of individuals or companies. All the resources of the reserves are for *use*, and this use must be brought about in a thoroughly prompt and businesslike manner, under such restrictions only as will insure the permanence of these resources. The vital importance of forest reserves to the great industries of the Western States will be largely increased in the near future by the continued steady increase in settlement and development. The permanence of the resources of the reserves is therefore indispensable to continued prosperity, and the policy of this department for their protection and use will invariably be guided by this fact, always bearing in mind that the *conservative use* of these resources in no way conflicts with their permanent value.

You will see to it that the water, wood, and forage of the reserves are conserved and wisely used for the benefit of the home builder first of all, upon whom depends the best permanent use of lands and resources alike. The continued prosperity of the agricultural, lumbering, mining, and livestock interests is directly dependent upon a permanent and accessible supply of water, wood, and forage, as well as upon the present and future use of their resources under businesslike regulations, enforced with promptness, effectiveness, and common sense. In the management of each reserve local questions will be decided upon local grounds; the dominant industry will be considered first, but with as little restriction to minor industries as may be possible; sudden changes in industrial conditions will be avoided by gradual adjustment after due notice; and where conflicting interests must be reconciled the question will always be decided from the standpoint of the greatest good of the greatest number in the long run.

These general principles will govern in the protection and use of the water supply, in the disposal of timber and wood, in the use of the range, and in all other matters connected with the management of the reserves. They can be successfully applied only when the administration of each reserve is left very largely in the hands of the local officers, under the eye of thoroughly trained and competent inspectors.

Revolution in Administration. Under the new regime there was a veritable revolution in the handling of the forest reserves. Administration was largely decentralized; conservative use of all resources was encouraged; local questions were decided on local grounds; business was speeded up; technical activities were directed by men with technical training; appointments and promotions were made on the basis of merit, not of pull. A lively demand suddenly developed for the young foresters who were beginning to issue in goodly numbers from the recently established schools of forestry. The Forest Service soon became recognized as one of the most efficient bureaus in Washington, with an esprit de corps that was unmatched.

Pinchot summarized the attitude of the entire organization under his inspiring leadership thus: "Every member of the Service realized that it was engaged in a great and necessary undertaking in which the whole future of their country was at stake. The Service had a clear understanding of where it was going, it was determined to get there, and it was never afraid to fight for what was right."

Fighting for the right makes both friends and enemies. In this case it made friends of the legitimate users of the forest reserves and of the great body of disinterested citizens who quickly recognized the great value of the reserves under "conservative use" in the permanent development not only of the West but of the entire nation. It made enemies of those whose unfair privileges were curtailed, whether in business or in politics.

The latter group, though much smaller in numbers, was more vociferous and temporarily had much influence in Congress. The Public Lands Convention at Denver in June, 1907, gave the malcontents, particularly among the livestock interests, an opportunity to denounce the forest reserves, the President, and the Forester in scathing terms. Although the strength of this group gradually waned, it was for a while sufficiently powerful so that, to quote Pinchot, "The Forest Service quite literally had to fight for its life." It was fully equal to the encounter.

The Forest Service was assisted in the discharge of its new responsibilities by the act of February 6, 1905, giving forest officers the authority to arrest without warrant, and the act of March 3, 1905, repealing the lieu-land clause, to which reference has already been made. Two other helpful provisions were also approved on March 3, 1905. One of these permitted timber on forest reserves to be exported from the state in which cut except in the Black Hills of South Dakota and in Idaho, from both of which the limitation was removed in 1913. The other provision, in the Agricultural Appropriations Act, changed the name of the Bureau of Forestry to the Forest Service, effective July 1, 1905. The new name was a great improvement in emphasizing "service" as the objective of the organization and in removing the taint of "bureaucracy" which existed fifty years ago as well as today.

Grazing on the Reserves. Management of the large area suitable for grazing within the forest reserves continued to be one of the most important and most difficult questions involved in their administration. Both cattle and sheep grazing were permitted under carefully drawn plans of management and under regulations that would afford reasonable protection to the water supply, to tree growth, and to the range itself. Overgrazing was reduced as rapidly as possible but was too deeply entrenched to be eliminated completely. Permits were issued to applicants in practically the same order of priority as that which had been established by the Department of the Interior. Early in 1906 official recognition was given to advisory boards from the livestock associations, and these boards did much to reduce friction between the stockmen and the Forest Service.

Fees for grazing permits constituted a particularly thorny problem. Livestockmen were naturally not enthusiastic about paying for a privilege that had previously been free. Their influence in Congress was sufficient to prevent approval of the repeated requests of the Department of the Interior for legislative authority to charge a fee and also to block the attempted inclusion of such authorization in the transfer act.

Pinchot was so thoroughly convinced that a fee was desirable both on general principles and as a means of swelling forest-reserve receipts, over the expenditure of which the Forest Service then had control, that he decided to approach the matter from a different angle. Could not the desired authority be inferred to exist already under the power granted the Secretary in the act of June 4, 1897, to make rules and regulations for the occupancy and use of the reserves?

Consequently Secretary Wilson on May 29, 1905, signed a letter to Attorney General Moody which made no reference to grazing but instead asked certain leading questions with respect to an application for the use of forest-reserve land in Alaska for a fish saltery, oil, and fertilizer plant. Did he have legal authority to grant a permit or lease for such use; could the permit be for a period longer than one year; and could he require reasonable compensation or rental for such permit?

Mr. Moody, having been properly educated not only by Pinchot but by the President himself, replied in the affirmative to all three questions. With respect to fees, the key sentence in his reply was as follows: "I have to advise you that, in my opinion, you are authorized to make a reasonable charge in connection with the use and occupation of these forest reserves, whenever, in your judgment, such a course seems consistent with insuring the objects of the reservation and the protection of the forest thereon from destruction."

This opinion was obviously couched in sufficiently broad terms to authorize the charging of a fee not only for a fish saltery but for grazing or other use of the reserves. Beginning with January 1, 1906, a charge was made for all livestock grazed on the reserves except for a limited amount of free use by actual settlers. This action caused a storm of protest, but it stuck; and in 1911 the soundness of the Attorney General's opinion was confirmed by the Supreme Court.

Grazing Regulations Upheld. Up to that time, both the Department of the Interior and the Department of Agriculture had been severely handicapped in their efforts to control grazing on the national forests by a uniform line of decisions by the Federal district courts that the provision in the Forest Reserve Act of 1897 authorizing the Secretary of the Interior to establish rules and regulations was invalid because it constituted an unconstitutional delegation of legislative power to an administrative officer. Prior to 1907 no appeal from these decisions to the United States Supreme Court was possible.

In November, 1907, criminal indictments were brought against Pierre Grimaud and J. P. Carajous for grazing sheep on the Sierra National Forest in California without having obtained the permission to do so required by the regulations promulgated by the Secretary of Agriculture. They demurred on the ground that the law making violation of such a regulation a penal offense was unconstitutional. The demurrers were sustained by the United States District Court for southern California.

The government then brought the case to the Supreme Court under that clause of the Criminal Appeals Act of March 2, 1907, which allowed a writ of error where the decision complained of was based upon the invalidity of the statute. The demurrers were confirmed by a divided court on March 14, 1910. About a month later, at the request of the government, the case was restored to the docket for reargument, and on May 3, 1911, the previous affirmance by a divided court was set aside and the judgment of the district court was unanimously reversed in a decision delivered by Justice Lamar, a former Secretary of the Interior.

A somewhat similar but not identical case was decided by the Supreme Court almost simultaneously with the Grimaud case. On April 7, 1908, the United States filed a bill in the Circuit Court for the District of Colorado alleging that Fred Light had turned out about 500 head of cattle to graze on the Holy Cross National Forest without obtaining any permit therefor as required by the Secretary of Agriculture. The court enjoined Light from further pasturing of his cattle on the forest. He thereupon appealed to the Supreme Court on the ground that the decree was erroneous and should be voided. On May 1, 1911, the Court unanimously upheld the injunction in a decision also written by Justice Lamar.

These two cases settled conclusively the following points:

1. Congress has complete power over the public lands. The acts of 1891 and 1897 providing for the creation and administration of national forests are therefore constitutional.

2. Any previous implied license to graze stock on public lands did not confer any vested right on the users, nor did it deprive the United States of the power of recalling such license.

3. The authority given the Secretary of Agriculture in the act of 1897 to make rules and regulations for the occupancy and use of the national forests is a constitutional delegation of administrative authority, not a delegation of legislative power.

4. The penalty for violation of the rules and regulations is fixed by Congress, not the Secretary, and is therefore constitutional.

5. Authority to charge for a grazing permit is implied both in the 1897 act and in other acts disposing of revenues from national forests.

6. State fencing laws do not authorize wanton or willful trespass nor afford immunity to those willfully turning stock loose under circumstances showing that they were intended to graze upon the lands of another, whether that other is an individual or the United States.

Definite settlement of these important points established beyond question the constitutionality of the national forests and greatly strengthened the hands of the Secretary of Agriculture in administering not only grazing but all other activities conducted on them. It was clear that the courts would uphold whatever action Congress might take with respect to the handling of the public lands.

Forest Homesteads. Another forward step in the administration of the forest reserves which had been recommended by the Department of the Interior came with passage of the Forest Homestead Act of June 11, 1906, commonly known as the Act of June 11. This act authorized the Secretary of Agriculture to open for homestead entry, through the Secretary of the Interior, forest-reserve lands chiefly valuable for agriculture "which, in his opinion, may be occupied for agricultural purposes without injury to the forest reserves, and which are not needed for public purposes." Each tract was to be surveyed by metes and bounds and was not to exceed 160 acres in area or 1 mile in length. Commutation, which had caused so much trouble elsewhere, was not allowed.

For several years after the passing of the act the Forest Service recommended the opening of lands to entry only in response to specific applications, which poured in at the rate of several thousand a year. Lands needed for administrative purposes or suitable only for grazing were not listed for entry. Lands suitable for cultivation but having heavy stands of timber were not listed until after sale of the timber. That this was a wise precaution to prevent speculation was shown by previous experience with homestead entries on timbered land in general and on lands eliminated from forest reserves in particular.

In 1900 and 1901, for example, 705,000 acres were eliminated from the Olympic Forest Reserve on the ground that the land was chiefly valuable for agriculture and that the settlement of the country was being retarded by its inclusion in a forest reserve. Less than ten years later, 523,720 acres had passed into the hands of owners who were holding it purely as a timber speculation, with 178,000 acres in the hands of three companies and two individuals. Of the timbered homestead claims on the eliminated area, only 570 acres were under actual cultivation, or an average of 5.7 acres per claim.

Another instance of the nondevelopment of timberlands for agricultural purposes was reported by the Forest Service in 1909. One locality in Idaho and eastern Washington had 116 perfected homestead claims on lands about equally suitable for agriculture, some of which were heavily timbered while others were nontimbered. "Of the homesteads on nontimbered land everyone was occupied and over 30 per cent of their area was under cultivation, while of those on timbered land one-half of 1 per cent was under cultivation and a large majority of the claims had been sold to lumber companies."

By June 30, 1910, the total area listed for entry was 632,412 acres. A year later the area had increased by 50 per cent to 943,718 acres, involving about 8,000 settlers. At that time the Forest Service believed that the listings undoubtedly included the cream of the land within the forests which was suitable for agriculture. In addition to the opening of individual tracts to homestead entry, large areas containing agricultural lands were completely eliminated from the forests by modification of their boundaries.

Congress Speeds Up Classification. Congress, however, was not satisfied that the work was progressing fast enough. It therefore included in the Agricultural Appropriations Act of August 10, 1912, a rider directing the Secretary of Agriculture "to select, classify and segregate, as soon as practicable, all lands within the boundaries of national forests that may be opened to settlement and entry under the homestead laws applicable to the national forests." The act carried an appropriation of $25,000 for the work, and $100,000 was made available the next year. Fortunately an attempt to instruct the Secretary of Agriculture to open for entry all lands "suitable for agriculture," regardless of their value for other purposes, did not succeed.

The Forest Service started immediately on the tremendous task of identifying and segregating the lands in national forests which were truly agricultural in character and which could pass into private ownership without jeopardizing the purposes for which the forests were established— watershed protection and timber production. This constituted the first thoroughgoing attempt at land classification ever undertaken by the Federal government. By 1919, the work was practically completed except in Alaska. In addition to the individual tracts which were opened to entry, some 12,000,000 acres were eliminated from national forests by changes in boundaries. Many of these eliminations contained large areas of agricultural land, and their removal from the forests greatly reduced the area that would otherwise have been opened to entry.

In 1921 the provisions of the Enlarged Homestead Act of 1909 were extended to the national forests, making it possible thereafter to list tracts up to 320 acres in area for entry under the Forest Homestead Act of 1906.

Altogether, more than 20,000 separate tracts with an area of about 2,500,000 acres have been listed for entry. Some of these were later withdrawn, a few were never entered, and still others were abandoned some time after entry. The total area patented amounted to about 1,800,000 acres.

In general, the Forest Service was too liberal in its estimate of the agricultural possibilities of much of the land opened to entry. This is shown by the large area entered and later abandoned and by the very small proportion of the patented area actually used for farming. For example, a detailed study of forest homesteads in Arizona and New Mexico made in 1934 showed that of 354,155 acres listed and entered, only 43 per cent went to patent. Of the area patented, 26 per cent was cultivated and 56 per cent was pastured. There can be no doubt that practically all of the land in national forests both suitable and available for agriculture under present conditions is being used for that purpose.

Contributions to Counties. Another early recommendation of the Department of the Interior—that the government make contributions in lieu of taxes to communities in which forest reserves are located—received congressional approval in the Agricultural Appropriations Act of June 30, 1906. That act provided that 10 per cent of the gross receipts from forest reserves during any fiscal year, including 1906, should be turned over to the states or territories for the benefit of the public schools and public roads of the counties in which the reserves were located, but not to the extent of more than 40 per cent of their income from other sources. The same provision, authorizing payment of the 10 per cent on a year-to-year basis, was repeated the next year.

The act of May 23, 1908, increased the payment to the states for the benefit of county schools and roads to 25 per cent of the gross receipts, eliminated the 40 per cent limitation, and made the legislation permanent. This arrangement has continued to date in spite of many suggestions that contributions should be based on the value of national forest lands rather than on receipts.

Agricultural Appropriations Act of 1907. The hostility to the forest reserves which existed in parts of the West and among certain members of Congress found expression in the Agricultural Appropriations Act of March 4, 1907. This act forbade the further creation of forest reserves except by act of Congress in the states of Washington, Oregon, Idaho, Montana, Wyoming, and Colorado. The prohibition was particularly serious because these six states contained by far the heaviest stands of timber in the West.

Pinchot and Roosevelt, however, were equal to the occasion. It happened that the Forest Service already had plans for the addition of a large area to the reserve system. These plans were hastily put in shape for action and rushed to the President, who signed proclamations adding 16 million acres to the existing area of forest reserves just before he approved the act forbidding him to do so. Irate senators complained vigorously that he could not take such action. The President gleefully replied that he had already taken it, and that was that. The limitation was later extended to California (1912), and to Arizona and New Mexico (1926), and was removed from Montana (1939).

The act also abolished the special fund, consisting of receipts from the forest reserves and controlled by the Secretary of Agriculture, which had been created by the transfer act of 1905. Congress in 1906 had started to restrict the use of the fund. Now it did away with it entirely, and in addition required the Forest Service to submit annually a classified and detailed report of receipts and estimate of expenditures. To offset the loss of funds, Congress increased the regular appropriation for protection and administration of the reserves from $900,000 to $1,900,000.

Although at the time abolition of the special fund was regarded by many as highly unfortunate, it is by no means certain that this was so. Temporarily, at least, the fund increased materially the resources at the disposal of the Secretary of Agriculture, which of course meant the Forest Service, and gave him complete freedom in determining the purposes for which they should be used. Flexibility of this sort is always attractive to an administrative officer.

On the other hand, earmarking of receipts partially removed from Congress the basic responsibility for controlling expenditures by executive agencies. It also decreased the likelihood of supplementary appropriations by Congress, however much they might be justified; and it put the Department under constant temptation to permit overutilization of the resources of the reserves in order to obtain funds clearly needed for their efficient administration. In retrospect the abolition of the special fund, whatever the motivation behind the action, appears to have been a wise move from the standpoint both of political theory and the long-run good of the reserves themselves.

A highly desirable feature of the act was its change of the name "forest reserves" to "national forests." The new name had the great merit of removing the implication that the forests were reserved from constructive utilization and of adding the implication that they had significance not only for the West but for the people of the entire nation. Perhaps to ease the shock of the provisions limiting the President's power to create reserves and abolishing the special fund, the salary of the Forester was increased from $3,500 to $5,000.

Administrative Reorganization. A major policy of the Forest Service was to place responsibility for the administration of the national forests as far as practicable in the hands of local and regional officers. Progress in this direction was made first by the establishment of three, and later (in 1907) of six, inspection districts. This step led logically, on December 1, 1908, to the transformation of the inspection districts into administrative districts. Each district was placed in charge of a district (now regional) forester, with experts of technical training and executive experience in charge of the various lines of work.

From then on, the bulk of the national-forest business previously referred to Washington was handled on the forests and in the districts. Only questions of large importance and matters involving the general administration and policy of the Forest Service were submitted to the Forester for decision. The arrangement was intended to avoid the dual evils of bureaucratic centralized administration and irresponsible decentralized administration. It was the first, and one of the most successful, moves in this direction by the Federal government.

In the same year (1908) decentralization of the research work of the Service was initiated by the establishment at Fort Valley, near Flagstaff, Arizona, of the first of a series of forest experiment stations. It was soon followed by others in Colorado, Idaho, California, Washington, and Utah. The directors of the stations reported directly to the Washington office, just as did the district foresters. The investigative and administrative organizations were thus administratively independent of each other but worked in close cooperation. In fact, most of the early work of the stations was intended to provide a sound technical foundation for the management of the national forests.

Two features of the forest (and range) experiment stations distinguished them from the start from the agricultural experiment stations. They were completely financed and controlled by the Federal government and they were organized by regions rather than by states. Since forest types do not recognize political boundaries, organization of Federal activities in the field of forest research along regional lines seemed clearly desirable.

One exception to this generalization is offered by the Forest Products Laboratory, established at Madison, Wisconsin, in 1910, which has demonstrated the advantages of centralized research in the field of wood technology. Forests themselves must be studied where they occur, but their products can be brought to a central laboratory.

Price at the Helm. In connection with the subject of administration, it is appropriate to mention the outstanding service rendered by Overton

W. Price as associate forester from 1905 to 1910. A forester with European training, a wide background of practical experience in the United States, and an exceptionally engaging personality, he possessed organizing and executive ability of the highest order.

During the last two or three years of his term in public office particularly [according to Dean Graves], he carried the main burden of internal administration of the Forest Service, doing his utmost to leave his chief, Mr. Pinchot, free to deal with the larger questions of policy and to wage his fight for national conservation . . . largely to him belongs the credit for the work which established national forestry in the United States on a sound and permanent basis.

This judgment is confirmed by Pinchot:

To say that Price was my right hand is a feeble understatement. He had more to do with the good organization and high efficiency of the Government forest work than ever I had. Most of the credit for it that came to me rightly belongs to him.

Forestry on Indian Reservations. Following its administration of cutting operations on the Chippewa Indian Reservation in Minnesota under the Morris Act of 1902, the Bureau of Forestry, at the request of the Secretary of the Interior, made recommendations for the management of the timber on several Indian reservations in Wisconsin and later supervised the logging on them. A more general agreement concerning the handling of the 6 million acres of forest lands on Indian reservations was entered into by Secretary Garfield of the Department of the Interior and Secretary Wilson of the Department of Agriculture on January 22, 1908. It provided for the management of the timber on these lands by the Forest Service at the expense of the Office of Indian Affairs.

This agreement remained in effect until July, 1909. It was then abrogated by the Department of the Interior following the passage on March 3, 1909, of an act authorizing the Commissioner of Indian Affairs to manage the timber on Indian reservations and appropriating $100,000 for the purpose, because of some question on the part of the new Secretary of the Interior (Richard A. Ballinger) as to the legality of the cooperative agreement. While it was in effect, the Forest Service supervised the cutting of some 190 million feet of timber and made management plans for a number of other reservations. The largest operation was on the Menominee Reservation in Wisconsin, where a sawmill with a daily capacity of 100,000 board feet was constructed under Forest Service supervision.

The Indian Omnibus Act of June 25, 1910, authorized the sale of mature living and dead and down timber on allotted Indian lands and all tribal lands except in Wisconsin and Minnesota, under regulations prescribed by the Secretary of the Interior. A Forestry Branch in the Office of Indian

Affairs was organized under the direction of Jay P Kinney and has subsequently been in charge of all forest activities on Indian reservations.

National Parks. The establishment of six national parks by Congress between 1898 and 1910 constituted an important step in the preservation of areas of outstanding scenic and scientific value. The first of these was Mount Rainier National Park in Washington, which had been recommended by the Forestry Commission of the National Academy of Sciences in 1897 and was actually created by the act of March 2, 1899.

That act has been severely criticized because it permitted the Northern Pacific Railroad Company to exchange its extensive, and largely worthless, holdings in the park and in the Pacific Forest Reserve for unoccupied, nonmineral lands, whether surveyed or unsurveyed, within any state into or through which the railroad ran. This provision opened the rich timberlands of Oregon to lieu selection by the railroad, although it had only a few miles of line in that state. The Northern Pacific took advantage of the opportunity to make lieu selections of some 450,000 acres, the bulk of which presently found its way into the hands of large lumber companies. The mineral-land laws were at first extended to all lands within the park, but this provision was repealed in 1908.

Crater Lake National Park in Oregon was established in 1902. Settlement, lumbering, and the conduct of any other enterprise or business occupation were prohibited, but the location and working of mining claims were permitted. Scientists, excursionists, and pleasure seekers were made welcome, and restaurants and hotels were permitted to establish places of entertainment under regulations fixed by the Secretary of the Interior, who was also authorized to make rules and regulations for the general administration of the park.

In 1906, the Sulphur Springs Reservation, which had been set aside in 1902 in the Chickasaw Indian Reservation in Oklahoma, was converted into the Platt National Park. Wind Cave National Park in South Dakota was created in 1903. An unusual feature of this act was a provision that all receipts should be covered into a special fund to be expended for the care and improvement of the park. Question has frequently been raised as to whether these two small parks are sufficiently outstanding to justify their inclusion in the national-park system.

Mesa Verde National Park in Colorado was established in 1906. The Secretary of the Interior was specifically directed to promulgate regulations for the preservation of the ruins and other works and relics of prehistoric or primitive man within the park. No reference was made to minerals until 1931, when the prospecting, development, or utilization of the mineral resources of the park was forbidden.

The act of May 11, 1910, creating Glacier National Park in Montana, specifically prohibited lieu selections by any railroad or other corporation having holdings in the park. Seven years later (1917) Congress authorized the exchange of private lands within the park for dead, decadent, or matured timber of equal value either in the park or in national forests in Montana; and in 1923 it authorized the exchange of private lands for public lands of equal value in the state of Montana. In 1911 Congress authorized the use of all receipts for the administration and improvement of the park.

Lack of adequate administrative organization, appropriations, and personnel prevented full development of these and other national parks until the creation of the National Park Service in 1916.

National Monuments. A new class of Federal reservations was authorized by the American Antiquities Act of June 8, 1906. That act authorized the President of the United States "to declare by public proclamation historic landmarks, historic and prehistoric structures, and other objects of historic or scientific interest that are situated upon the lands owned or controlled by the Government of the United States to be national monuments." Such reservations were to be limited to the smallest area compatible with the proper care and management of the objects to be protected. Administration of the national monuments was delegated to the Secretary of the Department already having jurisdiction over the lands in question. In effect, this meant the Secretaries of the Interior, Agriculture, and War.

National monuments resemble national parks in that they are reserved primarily for scientific, educational, and recreational purposes, and not for commercial utilization, as are the national forests. They differ in that they are created by proclamation by the President instead of by act of Congress; that they are reserved more because of their historical or scientific interest than because of the grandeur of their scenery; and that they are usually, though not always, smaller in area. The term "scientific interest" has been interpreted broadly enough to justify the creation of national monuments for the preservation of the Olympic elk and the geological features of Jackson Hole.

No less than twenty-three national monuments were created in eleven different states and in Alaska through the year 1910. Two of these—the Grand Canyon National Monument and the Mount Olympus National Monument—later (in 1919 and 1938) became national parks. The history of these two areas illustrates the normal difference in speed between executive action and legislative action.

Unreserved Public Lands. Not until 1897 did Congress take any action to protect the public lands from the frequent fires which were destroying their resources. An act of February 24 of that year provided a fine of $5,000 or imprisonment for two years, or both, for the willful or malicious setting of fire or for carelessly or negligently leaving any fire unattended near timber or other inflammable material on the public domain; and a fine of $1,000 or imprisonment for one year for breaking camp or leaving a fire near timber without totally extinguishing it. In 1900 this act was amended by omitting the words "carelessly or negligently" in connection with leaving any fires unattended and also by omitting the reference to breaking camp in connection with leaving a fire before extinguishing it.

For some time the General Land Office had been stressing the damage caused by forest fires and urging the passage of legislation to help control them. As an illustration of the seriousness of the situation, Commissioner Hermann in his report for 1897 cited a fire on the San Gabriel Forest Reserve in California which covered several townships and burned for nearly three months. "Depredators denude the public domain of much of its timber wealth, but fire is its greatest enemy."

Immediately after passage of the act of 1897, special agents and forest supervisors were instructed to consider the protection of the public forests from fire as one of their most important duties and were authorized to take such emergency action as might be required to bring them under control. Posters were widely distributed requesting the help of the general public and warning that violators of the law would be prosecuted to the full extent of the law.

Actual results, as usual, were disappointing. Inadequate appropriations, apathy, and lack of trained personnel were handicaps that it was difficult to overcome. The leading recommendation in Hermann's report for 1900 was for the passage of "immediate legislation to place all unreserved forest lands under the watchful care of a disciplined ranger and fire force." The response of Congress was not generous, and little progress was made until President Theodore Roosevelt converted the bulk of the public timberlands into reserves.

Repeal of the Timber and Stone Act and the Free Timber Act, both of which had been denounced almost immediately after their passage as inimical to the public interest, was repeatedly urged by the Department of the Interior. In their place the Secretary recommended that all lands primarily valuable for timber be withdrawn from entry, with adequate provision for the protection and rational use of the timber along lines similar to those governing the use of timber in forest reserves.

Both acts resulted in many serious abuses. One great improvement was, however, made in the administration of the Timber and Stone Act. Secretary of the Interior James R. Garfield, who incidentally was a close friend of both Roosevelt and Pinchot, decided that the word "minimum" in the proviso that lands should be sold "at the minimum price of two dollars and fifty cents per acre" meant what it said. On November 30, 1908, he ruled that thereafter lands to which the law applied should be appraised and sold at the appraised prices. Congress continued to make appropriations for the prevention of timber trespass on the unreserved public lands, but as in the case of fire control, substantial progress in this direction was made only by the inclusion of the bulk of the timberlands in forest reserves.

Range Lands. Control over grazing on range lands in the unreserved public domain was the subject of prolonged and bitter controversy. Both the Department of the Interior and the Department of Agriculture favored the leasing of the range lands to farmers and stockmen, each of course under its own jurisdiction. This proposal also had the support of the Public Lands Commission of 1903–1905 and of the National Live Stock Association. The cattlemen were much more favorable to such action than the sheepmen, most of whom advocated either no leasing at all or leasing by the states.

Congress ran true to form by taking no action, and a proposed item in the agricultural appropriations bill for 1907 providing for the control of grazing on the unreserved public lands by the Department of Agriculture aroused vigorous opposition. The lands continued to be a public commons until passage of the Taylor Grazing Act in 1934.

Considerable discussion also centered around fencing of the public domain. The campaign of fence destruction in the 1880's had not been entirely successful, and in 1901 Secretary Hitchcock reported 161 unlawful fences involving nearly 2,500,000 acres. Angered by the arrogance of the stockmen, he again attempted stringent enforcement of the antifencing law, but neither he nor his successors succeeded in completely stamping out the practice. The stockmen themselves argued strongly in favor of fencing as a useful tool in range management. For the same reason it had the support of Pinchot and the Forest Service, but of course under government control. Today fencing is commonly used both on the national forests and the grazing districts.

Public-land Withdrawals. Between 1905 and 1909 some 80 million acres of public lands thought to contain workable deposits of coal were withdrawn from entry by the Secretary of the Interior under instructions from President Roosevelt. The purpose was to protect the genuine prospector and to prevent monopolization of the coal lands by fraudulent use of the homestead and other land laws. About half of the withdrawals, shown by classification not actually to contain coal, were restored to entry by the end of the Roosevelt administration.

These withdrawals led in 1909 to passage of an act authorizing persons who in good faith had entered coal lands under the nonmineral laws to obtain title thereto subject to reservation of the coal to the United States. The next year another law authorized entry of coal lands under the agricultural land laws but with the retention of mineral rights by the government. In 1914 this authorization was extended to all withdrawn mineral lands. These enactments constituted the first steps toward a system of leasing mineral lands which Roosevelt had advocated on December 17, 1906, in a special message to Congress on public-land matters. "My own belief is that there should be provision for leasing coal, oil, and gas rights under proper restrictions."

On December 10, 1908, Secretary Garfield withdrew from entry approximately 4,700,000 acres of public lands in Wyoming, Utah, and Idaho which were suspected of containing potash valuable for fertilizer. Some 4,000,000 acres of oil lands were withdrawn from entry under President Roosevelt, and more than 3,000,000 acres additional during the first two years of President Taft's administration. Here again the trend was toward the leasing system that was finally adopted in 1920.

Between December 4, 1908, and February 27, 1909, at the very end of the Roosevelt administration, Secretary Garfield withdrew from entry nearly 3,500,000 acres of land in the public domain and in national forests because of their potentialities as sites for the production of water power. Richard A. Ballinger, who on March 4, 1909, succeeded Garfield as Secretary of the Interior, questioned the authority of the President to make such withdrawals, and as early as March 20 began restoring the withdrawn lands to entry. Following a vigorous protest by Pinchot, Taft not only stopped further withdrawals but instructed Ballinger to make re-withdrawals of certain of the areas which he had opened to entry. Although the re-withdrawals covered less than 500,000 acres, they are said to have included more actual power sites than had the Garfield withdrawals, which included considerable private land and much public land not adjacent to streams.

Congress Authorizes Withdrawals. Taft's views as to the inherent power of the President to take action in matters of this sort without specific authorization from Congress were much narrower than Roosevelt's. He therefore asked Congress to validate the withdrawals already made and to authorize further withdrawals "pending submission to Congress of recommendations as to legislation to meet conditions of emergencies as they arise." For once Congress acted promptly. On June 25, 1910, after some opposition, it passed an act authorizing the President to make temporary withdrawals of public lands for specific purposes from all forms of entry, to remain in force until revoked by him or by Congress. How long is "temporary" still remains a moot question.

The constitutionality of the act was upheld in 1915 by the United States Supreme Court, which reversed the findings of a lower court. In its decision, the Supreme Court stated that "the power of the President to withdraw public lands from entry has been so long exercised and recognized by Congress as to be equivalent to a grant of power." To make sure that the Withdrawal Act could not be interpreted as repealing the ban on Presidental establishment of national forests in six Western states, Congress reenacted that provision of the act of 1907.

The Pinchot-Roosevelt Team. The marked progress toward better management of the nation's forests and other natural resources that occurred during the first decade of the present century was due primarily to effective teamwork on the part of two remarkable men—Gifford Pinchot and Theodore Roosevelt. Seldom has there been a closer Damon-and-Pythias relation between two men high in public office. Both loved the outdoors, both preached and lived the strenuous life, both enjoyed a fight, and both were guided by high ideals of public service.

They saw the forests and other natural resources as assets to be used for the benefit of *all* the people, both present and future. To them the struggle to stop destruction and waste and to prevent monopoly was much more than an economic issue; it was a moral crusade. Their enthusiasm was so contagious as to inspire among their followers an almost fanatic devotion both to the men themselves and to the causes they espoused.

The partnership was formed in the fall of 1901, shortly after Roosevelt became President following the assassination of McKinley. Pinchot and his friend Frederick H. Newell, an engineer in the Geological Survey, called on the new President to enlist his interest in the forest and water resources of the country. So successful was their mission that the President asked them to let him have some material on these subjects for inclusion in his first message to Congress.

That message characterized the preservation of the forests as "an imperative business necessity"; urged the extension of the forest reserves and their transfer from the Department of the Interior to the Department of Agriculture; and proposed the construction by the government of irrigation works for the reclamation of arid lands in the West. How thoroughly the President had been convinced by his two visitors is indicated by his declaration that "the forest and water problems are perhaps the most vital internal questions of the United States."

From then on the intimacy between Pinchot and Roosevelt developed rapidly until "G.P." became one of "T.R.'s" closest personal friends and most trusted advisers. In these capacities and as a member of the "tennis cabinet" (the modern version of Jackson's original "kitchen cabinet"), he had more influence in shaping the policies of the administration than most of those in higher official positions. The President's reliance on the Forester was shown by his inclusion of the latter in practically all of the several commissions which he appointed. Among these were the Public Lands Commission, the Committee on the Organization of Government Scientific Work, the Country Life Commission, the Inland Waterways Commission, and the National Conservation Commission. These assignments also indicate the breadth of Pinchot's interests.

"Conservation" Acquires a New Meaning. Pinchot's influence was particularly strong in the field of "conservation," an old word to which he and his friends attached a new meaning. No one has yet written an entirely satisfactory definition of this new meaning, but its underlying philosophy was clear to Roosevelt, Pinchot, and their associates in the movement. Most emphatically it was not merely preservation. Rather, it was wise use for the benefit of both present and future generations. What constituted wise use must be determined by economic, social, aesthetic, and moral considerations. Its goal was the greatest good for the greatest number for the longest time. It dealt with all natural resources, recognized that each resource has a relation to every other resource, and aimed at developing an integrated program that would give each resource its proper place and its proper treatment in the picture as a whole.

Pinchot himself says that the conception of a unified program dealing with the wise use of all natural resources first occurred to him during a horseback ride in the winter of 1907. Roosevelt at once gave both the idea and the name the prestige of his unqualified support. "Conservation" thus became the keynote and the slogan of his administration. It gave a much-needed unity to a movement that had already made impressive advances on many different fronts; and it served as a common rallying ground for persons with diverse but related interests.

The great strength of conservation in its new meaning was, and is, that no one can disagree with its broad objectives. Its major weaknesses are the difficulty of obtaining general understanding and application of the concept, which too readily may mean all things to all men, and the still greater difficulty of agreeing upon ways and means of putting theory into practice. Even among equally well-intentioned and intelligent people there may be virtually complete agreement as to the goal and at the same time wide divergence of judgment as to how best to get there.

Conference of Governors. The new program for an integrated attack on natural-resource problems was soon brought to public attention in dramatic form. In the spring of 1907, during a trip by the Inland Waterways Commission down the Mississippi River, Newell suggested the holding of a conference of state governors and others to discuss the conservation of the natural resources of the nation. The proposal was formally presented to the President by the Commission in a letter dated October 3, 1907, while he was with it for a few days on another trip down the Mississippi, and was immediately approved by him.

The following month he invited the governors of all the states and territories to a conference at the White House on May 13 to 15, 1908, "To confer with the President and each other upon the conservation of natural resources." He concluded his letter as follows: "Facts, which I cannot gainsay, force me to believe that the conservation of our natural resources is the most weighty question now before the people of the United States. If this is so, the proposed conference, which is the first of its kind, will be among the most important gatherings in our history in its effect upon the welfare of all our people."

Roosevelt was right in his appraisal of the significance of the conference. It set a precedent for future meetings of the governors, now held annually for the consideration of important matters of common interest, and it focused national attention on the vital importance of natural resources and the urgent need for their conservation to an extent that would probably have been impossible in any other way. The gathering was one of the most distinguished ever to assemble in Washington. In addition to the governors and the three advisers by whom each was accompanied, the roster included many members of the Sixtieth Congress, the Supreme Court, and the Inland Waterways Commission, and representatives of scientific, professional, business, and other organizations, and the press. There were also nearly fifty general guests and four special guests—William Jennings Bryan, Andrew Carnegie, James J. Hill, and John Mitchell. Former President Grover Cleveland was prevented by illness from attending. Arrangements for the conference were made by the familiar trio of Gifford Pinchot, chairman, W J McGee, and Frederick W. Newell.

The conference was opened by President Roosevelt, who served as its permanent chairman. The President emphasized with characteristic vigor the importance of the conference and of the problem which it had been called to consider:

So vital is this question, that for the first time in our history the chief executive officers of the States separately, and of the States together forming the Nation, have met to consider it. It is the chief material question that confronts us, second only—and second always—to the great fundamental questions of morality. . . . These questions do not relate only to the next century or to the next generation. One distinguishing characteristic of really civilized men is foresight; we have to, as a nation, exercise foresight for this nation in the future; and if we do not exercise that foresight, dark will be the future! . . . All these various uses of our natural resources are so closely connected that they should be coordinated, and should be treated as part of one coherent plan and not in haphazard and piecemeal fashion.

Many papers dealing with all natural resources except wildlife (a curious omission), and varying in quality, were presented to the conference. During the course of the discussions, the President had this to say concerning the "twilight land" between the powers of the Federal government and the state governments:

My primary aim, in the legislation that I have advocated for the regulation of the great corporations, has been to provide some effective popular sovereign for each corporation. . . . If the matter is such that the State itself cannot act, then I wish, on behalf of the State, that the National Government should act.

Conference Resolutions. A Committee on Resolutions headed by Governor Newton O. Blanchard of Louisiana presented a declaration of views and recommendations which was unanimously adopted by the conference. The declaration asserted that natural resources "supply the material basis on which our civilization must continue to depend, and on which the perpetuity of the Nation itself rests"; that "this material basis is threatened with exhaustion"; that its conservation "is a subject of transcendent importance, which should engage unremittingly the attention of the Nation, the States, and the People in earnest cooperation"; that "monopoly thereof should not be tolerated"; and that "in the use of the natural resources our independent States are interdependent and bound together by ties of mutual benefits, responsibilities and duties."

The conference made five major recommendations which may be summarized as follows:

1. That the President call future similar conferences with a view to continued cooperation between states and nation.

2. That each state appoint a commission on the conservation of natural resources, to cooperate with each other and with any similar commission of the Federal government, to ascertain the present condition of our natural resources and to promote their conservation.

3. That forest policies to secure the husbanding and renewal of our diminishing timber supply, the prevention of soil erosion, the protection of headwaters, and the maintenance of the purity and navigability of our streams should be continued and extended, and that laws looking to the protection and replacement of privately owned forests be enacted.

4. That laws looking to the conservation of water resources for irrigation, water supply, power, and navigation be enacted.

5. That laws be enacted for the prevention of waste in mining and the protection of human life in the mines.

The objective of the conference was contained in the statement: "Let us conserve the foundations of our prosperity."

Creation of National Conservation Commission. In accordance with the recommendation of the Governors' Conference, President Roosevelt on June 8, 1908, appointed a National Conservation Commission to advise him as to the condition of the country's resources and to cooperate with similar state bodies. The Commission was organized in four sections of twelve men each. The chairmen and secretaries of these sections, who together constituted the executive committee, were as follows:

Section of Waters—Representative Theodore E. Burton of Ohio, chairman; W J McGee, Bureau of Soils, secretary.

Section of Forests—Senator Reed Smoot of Utah, chairman; Overton W. Price, Forest Service, secretary.

Section of Lands—Senator Knute Nelson of Minnesota, chairman; George W. Woodruff, Department of the Interior, secretary.

Section of Minerals—Representative John Dalzell of Pennsylvania, chairman; Joseph A. Holmes, Geological Survey, secretary.

Gifford Pinchot served as chairman of the Commission and of its executive committee and Thomas R. Shipp as their secretary. The executive committee promptly proceeded to prepare an inventory of the natural resources of the United States—the first ever to be made. In this herculean task it was assisted by many employees in the various Departments, each of which was instructed by the President to secure, compile, and furnish to the Commission such data as it might request.

Joint Conservation Conference. On December 8 to 10, 1908, the report of the Commission was presented to a Joint Conservation Conference attended by members of the Commission, twenty governors of states and territories, personal representatives of eleven governors and governors-elect, members of twenty-six state conservation commissions, and presidents and representatives of sixty national organizations. President-elect Taft presided at the first session of the conference and Chairman Pinchot at subsequent sessions. Taft paid a generous tribute to the two indiivduals primarily responsible for launching the conservation movement and committed himself to carrying on the work they had started.

President Roosevelt and Mr. Pinchot have brought about an unprecedented condition of affairs. They have gone into the States and brought the governors here, and they have created by that very fact a public opinion and a public interest in this great subject that I think could have been created in no other way. . . .

I had some notes that I was going to read, but the truth is they contained so many expert statements that I am afraid you might suspect their authorship, and so if you will excuse me from going into a civil-service examination on the subject of waterways and water and where it conceals itself and how it ought to be treated, I shall content myself only with the statement of my deep sympathy with this movement and with my purpose . . . to do everything I can to carry on the work so admirably begun and so wonderfully shown forth by President Roosevelt.

The Joint Conference unanimously adopted a resolution approving the report of the National Conservation Commission. It especially commended the principle of cooperation among the states and between the states and the Federal government; favored the maintenance of conservation commissions in every state; and urged on Congress the high desirability of maintaining a national commission on the conservation of the natural resources of the country. More specifically, it urged adoption of the policy of separate disposal of the surface rights and mineral rights on the remaining public lands; the disposal of mineral rights by lease only; treatment of all watersheds as units; approval of the broad plan recommended by the Inland Waterways Commission for waterway development under an executive board or commission appointed by the President; and the enactment by states of laws regulating the cutting and removal of timber and slash on private lands. Finally, it arranged for the appointment of a joint committee consisting of six members of state conservation commissions and three members of the National Conservation Commission to prepare and present to the state and national commissions, and through them to the governors and the President, a plan for united action by all organizations concerned with the conservation of natural resources.

Commission's Report. The exhaustive, three-volume report of the National Conservation Commission viewed with alarm the existing situation with respect to waters, forests, lands, and minerals, and made many recommendations for improvement, in the execution of which it urged harmonious cooperation and collaboration between the state and Federal governments. The report was transmitted to Congress by the President on January 22, 1909, with a message in which he summed up his views as to the meaning and importance of conservation and its place in his administration:

With the statements and conclusions of this report I heartily concur, and I commend it to the thoughtful consideration both of the Congress and of our people generally. It is one of the most fundamentally important documents ever laid before the American people. It contains the first inventory of its natural resources ever made by any nation. . . .

The policy of conservation is perhaps the most typical example of the policies which this Government has made peculiarly its own during the opening years of the present century. The function of our Government is to insure to all its citizens, now and hereafter, their rights to life, liberty, and the pursuit of happiness. If we of this generation destroy the resources from which our children would otherwise derive their livelihood, we reduce the capacity of our land to support a population, and so either degrade the standard of living or deprive the coming generations of their right to life on this continent.

The President concluded his message with a request for an appropriation of at least $50,000 to cover the expenses of the National Conservation Commission. "I know of no other way in which the appropriation of so small a sum would result in so large a benefit to the whole nation." Congress' reply was to adopt an amendment to the sundry civil appropriations bill, proposed by Representative Tawney of Minnesota, forbidding the payment of salaries or expenses of any commission not authorized by law or the use of any employee of the government on work connected with the activities of any such commission. Whether this action was occasioned by the President's request for an appropriation to continue the work of the National Conservation Commission or by his request for an appropriation of $25,000 to print the report of the Country Life Commission, it certainly evidenced a violent allergy to Presidential commissions. Roosevelt stated that if he did not believe the Tawney amendment to be unconstitutional he would have vetoed the bill which contained it and that if he were to remain in office he would have refused to obey it. Taft proved less belligerent.

The Conference of Governors and the report of the National Conservation Commission led to the creation of two important organizations, the National Conservation Congress and the National Conservation Association. The National Conservation Congress was organized by the Washington Conservation Association and held its first meeting at Seattle from August 26 to 28, 1909. Subsequent congresses were held at St. Paul in 1910, at Kansas City in 1911, at Indianapolis in 1912, and at Washington in 1913.

The National Conservation Association was formally launched in October, 1909, for the purpose of helping, through a large individual membership, to give practical effect to the declaration of principles adopted by the Conference of Governors. President Charles W. Eliot of Harvard University was elected as its first president, but resigned the following January (1910) in favor of Gifford Pinchot.

North American Conservation Conference. Roosevelt's next move was to advance conservation on the international front. Canada, Newfoundland, and Mexico were invited to send representatives to a North American Conservation Conference "to consider mutual interests involved in the conservation of natural resources, and in this great field deliberate upon the practicability of preparing a general plan adapted to promote the welfare of the Nations concerned." The conference differed radically from the Conference of Governors in that attendance was limited to a few official delegates from each country and there were no public meetings. It was held in Washington on February 18, 1909. Forester Gifford Pinchot, Secretary of State Robert Bacon, and Secretary of the Interior James R. Garfield represented the United States.

At the opening session of the conference, President Roosevelt stressed the international significance of natural resources and their wise use:

In international relations the great feature of the growth of the last century has been the gradual recognition of the fact that instead of its being normally to the interest of one nation to see another depressed, it is normally to the interest of each nation to see the others elevated. I believe that the movement you this day initiate, is one of the utmost importance to the world at large.

Declaration of Principles. The conference adopted a declaration of principles recognizing that the conservation of natural resources is indispensable for the continued prosperity of any nation.

We agree that no Nation acting alone can adequately conserve them, and we recommend the adoption of concurrent measures for conserving the material foundations of the welfare of all the Nations concerned, and for ascertaining their location and extent. . . . We agree that those resources which are necessaries of life should be regarded as public utilities, and that their ownership entails specific duties to the public, and that as far as possible, effective measures should be adopted to guard against monopoly.

The declaration dealt in specific terms with the principles and practices that should be observed in the conservation of public health, forests, waters, lands, minerals, and game. For the most part these were in full accord with the report of the National Conservation Commission. With respect to forests, the conference advocated creation of many and large forest reservations under government control; early completion of inventories of forest resources; extension of technical education and practical field instruction; separation of land and timber for taxation purposes; public acquisition of forest lands at the headwaters of streams; encouragement of reforestation by private owners; effective protection of forest lands from fire; and rigid regulation of all lumbering operations. Finally, the conference recommended the establishment in each country of permanent conservation commissions which should keep in close touch with each other and the calling of a world-wide conference on the inventory, conservation, and wise utilization of natural resources.

The government of the United States had already anticipated the latter recommendation in January by sounding out the other principal governments as to whether they would look with favor on an invitation to send delegates to such a conference. Since the replies were uniformly favorable, invitations were sent to fifty-eight nations to meet at the Peace Palace in The Hague in September, 1909. President Taft did not follow through on the proposal, and the conference never materialized. The proposal may, however, be regarded as the forerunner of the United Nations Scientific Conference on the Conservation and Utilization of Resources held at Lake Success, New York, in 1949.

Ballinger and the Alaskan Coal Claims. Richard A. Ballinger, formerly Mayor of Seattle and Commissioner of the General Land Office for a year under Garfield, became Secretary of the Interior under Taft on March 4, 1909. His opening to entry of many of the water-power sites withdrawn by Garfield, his unfriendliness to the Reclamation Service in general and to Newell in particular, his abrogation of the cooperative agreement between the Forest Service and the Office of Indian Affairs, and his actions with respect to the Cunningham coal claims in Alaska brought him into immediate conflict with Pinchot, who regarded his whole attitude on conservation matters as thoroughly unsound. The Alaskan coal claims constituted the issue which attracted the greatest public attention and eventually resulted in the so-called Ballinger-Pinchot controversy.

These claims, covering 5,280 acres in the Bering coal field, had been entered by Clarence Cunningham as agent for thirty-three claimants. The laws of 1900 and 1904, under which the claims were made, permitted a claimant to take up not more than 160 acres at $10 per acre, "for his own benefit and not directly or indirectly, in whole or in part, in behalf of any person or persons whatsoever," but allowed the claims to be filed by an agent.

The provision was unworkable since 160 acres was too small an area for economic development. Like many other public-land laws it almost invited violation. As early as 1905, complaints arose that such violation was in process and that the Cunningham claims were actually being taken up for the benefit of the Morgan-Guggenheim syndicate.

Several preliminary reports by special agents of the General Land Office indicated collusion. Early in December, 1907, Louis R. Glavis, then chief of the Portland Division of the Land Office, was ordered to make a thorough investigation of the claims. Before he could do so, Commissioner Ballinger, on December 26, after a conference with former Governor Miles C. Moore of Washington, one of the claimants, ordered that twenty-six of the claims be clear-listed for patenting. When Glavis heard of this action, he protested so vigorously that Ballinger withdrew the order.

Little progress was made on the investigation during 1908. Glavis did, however, succeed in getting hold of Cunningham's journal, which contained rather convincing evidence of the intention of the claimants to pool their claims. Ballinger, in his capacity as a private citizen, assisted the claimants in preparing an affidavit which he presented in person to Secretary Garfield. Because of this involvement, he asked Assistant Secretary Pierce to handle the Cunningham claims when he himself became Secretary of the Interior in March, 1909.

On June 20, 1909, the chief of the Field Service in the Land Office asked Glavis for an immediate report on the Cunningham claims. On July 8 Glavis submitted an unfavorable report and urged that hearings on the matter be delayed until after completion of the field inspections of the claims. A young lawyer from Chicago was then placed in charge of the cases and shortly afterward submitted a report sustaining Glavis' position that the claims were fraudulent.

Forest Service Becomes Involved. By this time Glavis had become convinced that the Department was determined to approve the claims. Since they were located in the Chugach National Forest, he next turned to the Forest Service for help. At the suggestion of A. C. Shaw, one of the legal staff in the Service, Secretary Wilson requested the Department of the Interior to postpone the hearings. Glavis also conferred with Associate Forester Overton W. Price, and on August 9 presented the matter to Pinchot, who was attending the National Irrigation Congress at Spokane. Pinchot advised Glavis to lay the case in person before President Taft, to whom he gave a letter of introduction.

Shaw helped Glavis to prepare the statement which he presented to Taft at Beverly, Massachusetts, on August 18, 1909. Taft referred the statement to Secretary Ballinger for reply, and on September 6 conferred with him and with Assistant Attorney General Oscar Lawler, assigned to the Interior Department. Shortly afterward he discussed the case with Attorney General Wickersham. On September 13 he sent Ballinger a lengthy letter, which became known as the "white-wash letter," completely exonerating him of incompetence, disloyalty, or dishonesty and authorizing him to dismiss Glavis.

Two months later, on November 13, 1909, *Collier's Weekly* published an article by Glavis entitled "The White-washing of Ballinger—Are the Guggenheims in Charge of the Department of the Interior?" This article, which had been reviewed by Shaw and Price prior to its publication, pulled no punches and forced Congress to take cognizance of the matter. On December 21 the Senate asked Taft for all papers relating to the case, and on January 6, 1910, the President transmitted to it the papers which he said he had used in preparing his letter of September 13.

Pinchot's Dismissal. On that same day Senator Dolliver of Iowa read in the Senate a letter which he had received from Pinchot in reply to one from him. Pinchot's letter discussed various aspects of the case, including Shaw's and Price's involvement in it. He stated that he had reprimanded both men for violating a rule of propriety but recommended without hesitation that no further action was required. This letter was at least a technical violation of an Executive order forbidding any subordinate to give information to a member of Congress without the consent of the head of the Department. Pinchot and Wilson disagreed as to whether he had such consent.

The Dolliver letter brought immediately the dismissal which Pinchot had been courting. On January 7, Taft wrote him that his letter was

in effect an improper appeal to Congress and the public to excuse in advance the guilt of your subordinates before I could act, and against my decision in the Glavis case before the whole evidence on which that was based could be considered . . . if I were to pass over this matter in silence, it would be most demoralizing to the discipline of the executive branch of the Government.

By your own conduct you have destroyed your usefulness as a helpful subordinate of the Government, and it therefore now becomes my duty to direct the Secretary of Agriculture to remove you from your office as the Forester.

Price and Shaw were also dismissed. The next morning Pinchot bade good-by to the members of the Forest Service in Washington in a brief talk thanking them for their loyalty and urging them to continue their efforts in behalf of conservation with renewed energy. It was an inspiring, fighting, unrepentant call to the colors.

Taft's admirable appointment of Henry S. Graves, Pinchot's close personal friend and previous assistant, as the new Forester, and of Albert F. Potter as Associate Forester, did not restore public confidence in him. Although most of the newspapers regarded Taft's dismissal of Pinchot as justified, and under the circumstances inevitable, a few of them expressed the deep concern felt by the people generally. According to the *Providence Journal*, "The verdict will be that the President has cast his lot with the enemies of conservation, and no amount of argument will bring conviction to the contrary."

Said the *Colorado Springs Gazette*:

If there is in the United States a public land or timber grabber or a plunderer of water power sites who is not wearing a broad smile of satisfaction today, it is because he has not learned the news from Washington. President Taft's summary dismissal of Gifford Pinchot is the greatest thing that has happened to these gentry since they began operations on the public domain, and it is safe to say that every one of them threw his hat in the air and hip-hurrahed when he heard of it. . . . Rightly or wrongly, his removal from office is construed as the severance by Mr. Taft of the last cord that binds the present administration to its predecessor.

Congress Investigates. On January 19 Congress appointed a Joint Committee to Investigate the Department of the Interior and the "Bureau of Forestry," which had been nonexistent for nearly five years. The Committee consisted of seven regular Republicans, one insurgent Republican (E. H. Madison of Kansas), and four Democrats. It held exhaustive hearings which ran for four months and occasionally produced some violent explosions.

Louis D. Brandeis, attorney for Glavis and later Justice of the United States Supreme Court, succeeded in bringing out two facts that were highly damaging to the reputation of the President and others. One was that the President, in spite of several requests for all documents pertaining to the case, had failed to give the Committee the draft of a letter of exoneration from Taft to Ballinger which Taft had asked Assistant Attorney General Lawler to write "as if he were President" and the very existence of which had been repeatedly denied. The other fact was that a searching analysis of the case by Attorney General Wickersham, which Taft claimed to have had and to have studied carefully prior to writing his letter of September 13 to Ballinger, had actually been completed in November and then backdated.

After some parliamentary sparring in September, during which a minority group almost succeeded in getting the upper hand, the Joint Committee finally submitted its report to Congress on December 6, 1910, in thirteen volumes totaling about 760 pages. The seven regular Republicans stated that Ballinger's critics had "wholly failed to make out a case." They did, however, express the belief that it would be the height of unwisdom to permit the great coal fields of Alaska to be monopolized and therefore recommended that they be permanently withdrawn from entry and utilized under leases granted at fair royalties and for reasonable periods.

The four Democrats were severely critical of the Secretary of the Interior both as a witness and a public official. Their final conclusion was: "That Mr. Ballinger has not been true to the trust reposed in him as Secretary of the Interior; that he is not deserving of public confidence, and that he should be requested by the proper authority to resign his office as Secretary of the Interior."

Representative Madison submitted a separate report in which he made an obvious attempt to be impartial. His final conclusions were that the charges made by Glavis and Pinchot should be sustained and that Secretary Ballinger had "not shown himself to be that character of friend to the policy of conservation of our natural resources that the man should be who occupies the important post of Secretary of the Interior."

Aftermath. After it became clear that Congress would take no action on the Committee's report, Ballinger resigned in March, 1911, against the wishes of the President, who continued to express complete confidence in his integrity. He was succeeded by Walter L. Fisher of Chicago, whose attitude with respect to conservation is illustrated by the fact that at the time he was vice-president of the National Conservation Association, of which Pinchot was president. The Cunningham claims were presently canceled; and in 1914 Congress stopped the sale of coal lands in Alaska, authorized the reservation of certain lands, and provided for the leasing of unreserved coal lands.

On December 31, 1909, shortly before his dismissal, Pinchot had written at length to Roosevelt in Africa informing him in detail of Taft's defection from the Roosevelt policies, not only in conservation but in other fields. A few months later, he was able to elaborate on this account in person when the two men met on the Italian Riviera. The ensuing split between Roosevelt and Taft resulted in the organization of the Progressive ("Bull Moose") Party and the election of Woodrow Wilson as President of the United States in the fall of 1912. The Ballinger-Pinchot controversy had its share in making political history, but it did not seriously or permanently retard the development of the conservation movement.

Forestry in the States. The first decade of the present century witnessed a striking increase of interest in forestry by state governments as well as by the Federal government. In 1900 only nine states had agencies of one sort or another dealing specifically with forestry. Ten years later the number had increased to twenty-five. Louisiana, in 1904, was the first Southern state to join this group. Sometimes the forestry agency stood on its own feet; often it was part of an organization with broader responsibilities, such as a state board of agriculture, forestry and immigration; a department of game, fish and forestry; a state geological and economic survey; or a department of conservation. The number of conservation departments, hitherto small, increased markedly following the recommendation of the Conference of Governors for the creation of such a department in each state.

The objectives of state forestry agencies at this time were largely educational. They devoted much effort to the collection and dissemination of information for the dual purpose of arousing public interest in and support of forestry activities and of helping forest owners in the actual management of their lands.

Protection of forests from fire received steadily increasing attention and in a few states became a really major activity. A notable step in this direction was the designation in 1909 of practically all of northern Maine, nearly 10 million acres, as a "forestry district," in which fire-control activities were handled by the state land agent and forest commissioner with funds obtained through a special tax levy on all landowners in the district.

Some states made a small start toward giving technical assistance to forest owners in the management of their lands and toward raising nursery stock for their use in forest planting. Modification of the general property tax to reduce its tendency to force premature cutting came in for much discussion but little action.

The most tangible progress was in the establishment of state forests, which were commonly known as forest reserves, or forest preserves. By 1909 there were nearly 3,000,000 acres of state forests in eleven states. New York led with 1,612,000 in the Adirondacks and Catskills, much of which had been acquired by purchase and in all of which cutting was forbidden by the state constitution. Pennsylvania came next with 863,000 acres, most of which had been purchased and all of which was being placed under management as rapidly as legislative appropriations permitted. Wisconsin was third with 254,000 acres, most of which was tax-delinquent land that had been retained by the state. The eight other states, all of which were in the Northeast and the Lake states, together held nearly 100,000 acres.

On the whole, the period was one of distinct progress in arousing public interest and in laying the foundations for later substantial accomplishment in state forestry activities. Some twenty state forestry associations were influential in obtaining constructive legislation and in supporting and promoting the work of state forestry departments.

Forestry on Private Lands. The early 1900's were years of tremendous industrial expansion and increasing demand for forest products. Lumber production reached an all-time high in 1906 and 1907 with an estimated cut of 46 billion board feet. The peak of production had already passed in the Lake states; more and more of the cut was coming from the virgin forests of the South and the Pacific Northwest. Taking the country as a whole, the "timber famine" which the alarmists had been predicting was not yet in sight, stumpage prices were still low, and competition was keen.

Under these circumstances it is not surprising that forest management for the production of future crops was not widely practiced by private owners. There were, however, enough exceptions to the general rule to indicate that a change in attitude and in practice was on its way. The wag who remarked that "forestry is now being practiced everywhere except in the woods" was not entirely right.

To cite only a few examples, woods forestry was actually being practiced at Pisgah Forest in North Carolina, at Ne-Ha-Sa-Ne Park and the
Whitney Preserve in New York, by the Armstrong Forest Company in
Pennsylvania, and by the Crossett Lumber Company in Arkansas. Many
other companies were obtaining basic information, both through the
Forest Service and through their own efforts, for later use in improved
management practices. In the Northwest several associations for the control of forest fires were organized by private owners and did effective
work. The Western Forestry and Conservation Association, organized in
1909, gave strong leadership both in this field and in the development of
private forestry.

These advances did not, of course, meet the high hopes with which
Pinchot in 1898 initiated his campaign to bring about the widespread
practice of forestry by private owners and which for a while seemed to
be justified by the favorable response to his offer of cooperation. Years
later he admitted that:

> In spite of early promise and later increase, the attempt to introduce the
> practice of Forestry on private lands by proving that it would pay did not
> succeed on any general scale. Even in 1910, the peak year, when it had been
> introduced on about 1,500,000 acres, still less than one-half of 1 per cent of all
> privately owned forest land at that time was under some form of forest
> management.

He attributed this disappointing outcome to the fact that after 1905 the
Forest Service had to concentrate its attention on administration of the
national forests, to the sudden fall in stumpage and lumber prices caused
by the depression of 1907, to the Ballinger-Pinchot controversy, and to
the end of Roosevelt's term as President.

Attitude of National Lumber Manufacturers Association. From the
time of its organization in December, 1902, the National Lumber Manufacturers Association took a lively interest in the subject of private
forestry. At its annual meeting in 1903 the association pledged its "earnest
cooperation in every practical plan for the better handling of our forest
properties" and urged its membership to labor for the "enactment of such
laws as will tend to the fullest encouragement of all practical reforestation effort." In 1905 it undertook to raise a fund, which was completed
several years later, for the endowment of a "Chair of Applied Forestry
and Practical Lumbering" at the Yale School of Forestry, and in
1908 it appointed a standing Committee for the Conservation of Our
Forests.

Economic conditions made it difficult to translate interest in forestry
into action. As a speaker at the 1909 convention of the association put it,

"Nothing but higher values can give much encouragement to any effort for the saving of trees." The next year Chief Forester Graves stated that "for the lumberman forestry is a business proposition pure and simple" and expressed doubt as to whether its immediate practice on a large scale was practicable. At the same time he urged timberland owners generally to test its practicability on a small part of their holdings. The enthusiasm of members of the association for any major reform in cutting methods was probably also dampened to some extent by widespread newspaper criticism of lumbermen as despoilers of the forest—a criticism which the industry resented as being unfair and malicious.

Views of Representative Lumbermen. At the American Forest Congress in 1905 J. E. Defebaugh, then editor of the *American Lumberman*, emphasized the changing attitude of timberland owners and lumbermen. Up to that time, he pointed out, lumbermen had been engaged in a pioneer enterprise involving much risk and cheap raw material. Economic factors precluded the possibility of making expenditures for the purpose of preventing waste and assuring the perpetuation of the forest. "The lumbermen did with their property only what would yield the best returns." Now timber supplies were decreasing and stumpage prices increasing to the point where conservative forest management was becoming economically practicable.

The change of heart on the part of the lumbermen was also encouraged "by the disciples of forestry when they ceased to preach the doctrine of indirect and deferred benefits and began to demonstrate that direct benefits could be made to result from forestry as a science and as a practice." The millennium would not arrive overnight, but a start had been made. "Lumbermen realize that not only is it possible to carry on their work in this manner but circumstances are so adjusting themselves that it is imperative that it shall so be done."

At the Conference of Governors in 1908 R. A. Long of Kansas City, after dwelling at length on the aesthetic and indirect values of forests, finally got down to business with the statement that "conservation and perpetuation of our forests and unremunerative prices for lumber cannot travel the same road . . . low prices of any commodity mean waste and neglect." He deplored the current vogue of denouncing a "lumber trust, which does not and never did exist" and stated that "compulsory competition is our present commercial nightmare."

Long warned that the low tariff on lumber (at that time $2 per M board feet under the Dingley Act of 1897) would bring American lumber into direct and disastrous competition with Canadian lumber and would thereby still further decrease the likelihood of improved forest management by private owners. His basic conclusion was that the problem is essentially a public problem. "Whatever plan is adopted must furnish an incentive, a substantial inducement to the timber owner, to forego a present gain for the public good, and in this matter it can only be accomplished by governmental cooperation. And what is done should be done quickly, for the time is fast approaching when our forests will be so nearly gone that it will be too late."

Many other proposals involving government cooperation, regulation, and ownership were to be made before forestry on private lands came to be practiced on any considerable scale.

Lumber Tariffs and Forestry. Although manufactures of wood of various sorts have been included in tariff acts since 1794, it was not until 1872 that a specific rate on lumber made its appearance. In the tariff act of that year a duty of $1 per M board feet was levied on hemlock, white wood, sycamore, and basswood, and a rate of $2 per M on pine and other kinds of sawed lumber. These duties aroused considerable resentment in Canada, which has always been the main source of American imports of lumber and logs. Ontario retaliated by placing an export duty of $2 per M on logs, with serious results to such Lake states sawmills as were largely dependent on Canada for their supply.

A compromise was effected under which the McKinley Tariff Act of 1890 reduced the duty on pine and other species to $1 per M in return for the abolition by Ontario of its log-export tax. A few years later the Wilson Act of 1894 eliminated the duty on pine and provided for virtually free trade in sawed lumber. Both acts provided for restoring the previous duties if any foreign country (meaning Canada) should impose a log-export duty.

Efforts to restore the tariff on lumber were started at once by the lumber industry, which was particularly concerned over Canadian competition in the lower grades. A national conference held at Cincinnati in 1896 formulated the views of the lumber manufacturers on the subject and was influential in obtaining the inclusion in the Dingley Tariff Act of 1897 of a duty of $1 per M on sawed lumber of white wood, sycamore, and basswood, and of $2 per M on all other species.

The act provided for increasing these rates by the amount of any export log duty that might be imposed by any foreign country, but made no provision for retaliation in case of a complete embargo. Ontario therefore forbade altogether the exportation of logs cut from Crown Lands after April 1, 1898, and its lead was presently followed by other provinces. The result was to encourage the investment of American capital in sawmills on the Canadian side of the line.

The Payne-Aldrich Tariff Act of 1909 reduced the basic rates on sawed lumber from $2 to $1.25 per M. It was followed in 1913 by the Underwood Tariff Act, which placed all lumber except cabinet woods on the free list, where it remained until 1930.

What effect these various tariffs had on lumber prices and consumption is difficult to determine. That they benefited some American manufacturers is reasonably certain, but that they had any major influence on a national scale is doubtful.

There is no evidence that the tariff was instrumental in promoting the practice of forestry on private lands despite the argument that it would increase the prosperity of the lumber industry and that a prosperous industry would be better able and more disposed to take satisfactory care of its forest lands. Too many other complicating factors were involved to enable the tariff alone, whatever its effect on the profits of individual manufacturers and timberland owners, to play a decisive role in promoting private forestry. The economics and psychology of the situation in the first few decades of the 1900's were not such as to favor any general advance in this field.

Summary. Thanks chiefly to Roosevelt and Pinchot, "conservation" became a well-known and generally accepted goal during the early years of the present century. It was no fault of theirs if everyone did not recognize that natural resources are the foundation of our material prosperity, that they were being dangerously wasted and depleted, and that drastic measures to remedy the situation must be taken without delay.

The period was one of ferment and of feverish activity on the part of Federal agencies. Commissions were appointed, reservations were created, and legislation was enacted to protect, to develop, and so far as practicable to perpetuate our resources in water, soil, forests, range, wildlife, and minerals. Perhaps the greatest progress was made with forest lands, about a fifth of which were permanently reserved in Federal ownership as national forests and placed under the competent administration of the Forest Service.

The Conference of Governors in 1908 and the subsequent activities of the National Conservation Commission did much to stimulate both state and private interest in natural resources. Numerous state forestry agencies were created and nearly three million acres of state forests were established, but concrete results in getting forestry more widely practiced either by the states themselves or by private owners were relatively small. Economic conditions were still not such as to encourage the widespread adoption of forest management on private lands. Nevertheless, there was a noticeable change of attitude on the part of many timberland owners and lumbermen which boded well for later progress.

REFERENCES

Cameron, Jenks: "The Development of Governmental Forest Control in the United States," chap. 9, Johns Hopkins Press, Baltimore, 1928.

Conference of the Governors of the United States, *Proc.*, 1909.

Greeley, William B.: "Forests and Men," chap. 5, Doubleday & Company, Inc., New York, 1951.

Ise, John: "The United States Forest Policy," chaps. 4–5, 7–9, Yale University Press, New Haven, Conn., 1920.

Lillard, Richard G.: "The Great Forest," part III, Alfred A. Knopf, Inc., New York, 1947.

Mason, A. T.: "Bureaucracy Convicts Itself," The Viking Press, Inc., New York, 1941.

Peffer, E. Louise: "The Closing of the Public Domain," chap. 2, Stanford University Press, Stanford, Calif., 1951.

Pinchot, Gifford: "Breaking New Ground," chaps, 45–89, Harcourt, Brace and Company, Inc., New York, 1947.

Proc. Amer. Forest Cong., 1905, American Forestry Association, Washington, 1905.

U.S. Congress, Joint Committee to Investigate the Department of the Interior and the Bureau of Forestry: "Investigation of the Department of the Interior and of the Bureau of Forestry," 13 vols., vol. 1, 1911.

CHAPTER 8

New Approaches

By the end of the Roosevelt-Pinchot regime, large areas of public land chiefly valuable for timber production or watershed protection had been included in national forests and placed under management. Much land thought to be chiefly valuable for minerals or for the development of water power had been withdrawn from entry. Congress had given the President specific authority to make temporary withdrawals, and Taft had added materially to the mineral and power-site withdrawals, but no authority existed for the utilization of these lands either by the government itself or by private owners under lease. Range lands were still an unreserved, unmanaged commons. There was strong sentiment both for and against ceding all of the public lands to the states in which they were located.

Change of Emphasis. Franklin K. Lane, a Californian who in 1913 had succeeded Walter L. Fisher as Secretary of the Interior, in his first annual report to President Wilson summed up the evolution in philosophy and practice that had led to the current situation:

Congress has always been most generous as to the disposition of the national lands. One cannot read our land laws without being struck with the fixed determination which they show that it was wisest to be quit of our lands as quickly as possible. It might almost be said that the Government regarded its lands as a burden rather than an asset. We gave generously to our railroads and to the States. There was land for all, and it was the Government's glad function to distribute it and let those profit who could. There was no thought then of creating timber barons or cattle kings, or of coal monopoly. The sooner the land got into hands other than those of the Government the better. And this generous donor was not so petty as to discriminate between kinds of lands, the uses to which they could be put, or the purposes which those might have who got them. Land is land, save when it contains minerals; this was roughly the broad principle adopted. To classify was a task too difficult or not worth while. The lands would classify themselves when they arrived in individual ownership. And so the door was opened for monopoly and for fraud.

If the Government did not appreciate the invaluable nature of its assets there

were men who did. Great fortunes were laid in the vast holdings of what had
but a short time since been the property of the people. There was danger that
the many still to pour into the West would by necessity become the servitors
of a fortunate and early few. On this discovery our indifference at once took
flight. And so out of the abuse of the Nation's generosity there came a reaction
against a policy that was so liberal as to be dangerous.

The Nation wanted home makers, but found its lands drifting into the hands
of corporations which were withdrawing them from the market, awaiting a
time when lands would be more scarce; it gave opportunity for many compet-
ing coal operators and iron manufacturers, but found the sources of raw material centering into a few large holdings; it wished its lands to be cleared of
forests to make way for farms, but it found hundreds of consecutive miles
reserved from use by the fiat of those who appreciated their worth, and many
more miles of watershed despoiled of its needed covering in places where homes
were not possible.

The West Asks Action

A reaction was inevitable [continued the Secretary]. If lands were to be withdrawn from public service, why might not the Government do the withdrawing
itself? The old philosophy that "land is land" was evidently unfitted to a country
where land is sometimes timber and sometimes coal; indeed, where land may
mean water—water for tens of thousands of needy neighboring acres. . . .

So there has slowly evolved in the public mind the conception of a new
policy—that land should be used for that purpose to which it is best fitted, and
it should be disposed of by the Government with respect to that use. To this
policy I believe the West is now reconciled. The West no longer urges a return
to the hazards of the "land is land" policy. *But it does ask action.* It is reconciled to the Government making all proper safeguards against monopoly and
against the subversion of the spirit of all our land laws, which is in essence that
all suitable lands shall go into homes, and all other lands shall be developed for
that purpose which shall make them of greatest service. But it asks that the
machinery be promptly established in the law by which the lands may be used.
And this demand is reasonable.

The machinery that the West demanded had of course been in existence so far as timberlands were concerned since passage of the Forest
Reserve Act of 1897 and had actually been operating efficiently since
passage of the Transfer Act of 1905. There was, however, need to
strengthen the administration of the national forests, to extend the national-forest system to the East, and to expand the cooperative activities
of the Forest Service with the states and with private owners. Yet the
machinery so urgently needed for the intelligent handling of mineral
lands and water-power sites was not provided by Congress until 1920,
and for range lands not until 1934. New policies in the public interest
were not to be established without a fight.

National-forest Problems. When Henry S. Graves became Forester in 1910, he found two major threats to the integrity and permanence of the national forests—wholesale opening to homestead entry and cession to the states. The seriousness with which Graves regarded these threats is shown by the vigor with which he discussed them at the Fourth National Conservation Congress at Indianapolis on October 3, 1912:

An amendment which was attached to the Agricultural Appropriation Bill last June, and which passed the Senate but was rejected by the House, would have required, had it become law, the opening to private acquisition under the homestead laws of all lands "fit and suitable for agriculture" within national forests, irrespective of their value for other purposes or of their importance for public use. The result would have been not to facilitate but to block agricultural development. It would also have been to transfer to powerful private interests timberlands, water power sites, and other areas, possession of which would tend to private monopoly of resources now under public control. . . .

Experience has amply proved that the elimination, under pressure, of national forest lands locally considered or alleged to be of agricultural value but in point of fact more valuable for other purposes has led to their early acquisition by timberland speculators, great lumber interests, water-power companies, livestock companies, and others who desire the lands for other ends than agriculture. . . .

All but an entirely insignificant part of the national forests is not susceptible of profitable cultivation. . . . The areas which form an exception to this condition are not over four per cent of the total; and such areas are now being sought out by the Forest Service and will, under the existing law, be made available for homestead entry as fast as they can be opened without defeating the purpose of the law itself. It is necessary that the country should understand the manner in which bona fide settlement is being brought about in the national forests, and also the motive of those who are trying to break down the system of forest Conservation under the guise of promoting settlement.

Cession to the States. With respect to cession, he had this to say:

There has been during the past two or three years a steadily growing movement to turn over the national forests to the individual States. During the past session of Congress a rider to the Agricultural Appropriation Bill was offered in the Senate, providing for the grant of the national forests to the several States, together with all other public lands, including "all coal, mineral, timber, grazing, agricultural, and other lands, and all water and power rights and claims, and all rights upon lands of any character whatever." While the amendment was ruled out on a point of order, it received a surprisingly large amount of support.

The proposition so far as the national forests is concerned is to turn over to the individual States property owned by the Nation covering a net area of over one hundred and sixty million acres. This property has an actual measurable value of at least two billion dollars, while from the standpoint of its indirect value to the public no estimate on a money basis could possibly be made. These are public resources which should be handled in the interests of the public. Moreover, the problems involved are such that they should definitely remain in the hands of the National rather than be turned over to the State governments. The property belongs to the Nation as a whole, and every citizen has an interest in it. . . .

The underlying purpose of the proposed transfer of the national forests to the States is really not to substitute State for Federal control, but rather to substitute individual for public control. Its most earnest advocates are the very interests which wish to secure such control. . . . The proposition is one which the people as a whole would repudiate in an instant if they understood what is proposed.

Graves rightly concluded that the most effective way to meet these and other threats to the integrity and permanence of the national forests was so to manage them as to demonstrate conclusively that all their resources were being used for the purpose for which they were best fitted and for the benefit of all the people. While by no means avoiding publicity, which he recognized as essential to maintain public support, he focused his efforts mainly on sound administration of the veritable empire in his custody. In this task he was materially helped by court decisions and legislation.

Campaign for Acquisition Gains Ground. Interestingly enough, while cession of the public lands to the states was being advocated, the campaign to increase Federal landownership by purchase was steadily gaining ground. The original proposal for the acquisition of forest lands in the southern Appalachians was soon followed by agitation for the purchase of forest reserves in the White Mountains. Several bills looking to this end were introduced in Congress but got nowhere. Then New England and the South decided to join forces, and in 1906 a bill providing for the acquisition of lands in both the southern Appalachians and the White Mountains passed the Senate but stalled in the House.

In 1907, an appropriation of $25,000 was made for a survey by the Forest Service of forest, land, and water conditions in the two regions. On the basis of this study Secretary of Agriculture Wilson the next year submitted a brief, businesslike report recommending the purchase of not more than 600,000 acres in the White Mountains and of not more than 5 million acres in the southern Appalachian Mountains, and the immediate appropriation of $1,500,000 and $3,500,000 for purchases in the respective regions. At about the same time the National Academy of Sciences adopted a resolution urging Congress to pass legislation "to acquire in the southern Appalachian Mountains and the White Mountains such forest lands as are necessary to protect the navigable streams which have their sources therein and to make permanent the timber supply of the eastern part of the United States."

Interest in the purchase of lands for national forests also developed in other regions. Between 1905 and 1908 bills were introduced in Congress providing for such purchases on the watershed of the Potomac River, in the Ozarks, at the head of the Mississippi River and the Red River, and in the highlands of the Hudson River. A bill sponsored by Senator Brandegee of Connecticut appropriating $5,000,000 for the purchasing of forest lands, without geographic limitation, passed both houses, but was killed by a filibuster of a few Western senators when the amendments adopted by the House came before the Senate for consideration in the closing days of the Sixtieth Congress (March, 1909).

Constitutionality of Acquisition. Opponents of the legislation argued that it was unconstitutional; that the real purpose was to conserve forests, not to aid navigation; that the effects of forests on stream flow are negligible; that the ultimate cost would be enormous; that the government would be fleeced by speculators; and that the whole proposal was one calling for state rather than national action. The argument that the Constitution does not give the government authority to buy forest land was countered by the argument that the Constitution does give Congress power "to regulate Commerce among the several States" and that the purchase of forests for the purpose of improving the flow of navigable streams that might be used in interstate commerce clearly came within this power.

Much of the debate in Congress centered about these opposing points of view. After careful consideration, the House Committee on the Judiciary, in May, 1908, with one dissenting vote, adopted the following resolutions on the subject:

Resolved, that the committee is of the opinion that the Federal Government has no power to acquire lands within a state solely for forest purposes; but under its constitutional power over navigation the Federal Government may appropriate for the purchase of lands and forest reserves in a state, provided it is made clearly to appear that such lands and forest reserves have a direct and substantial connection with the conservation and improvement of the navigability of a river actually navigable in whole or in part; and that any appropriation therefor is limited to that purpose.

Resolved, that the bills [in question] are not confined to such last-mentioned purpose and are therefore unconstitutional.

In other words, the committee regarded the proposed purchase program as constitutional if it would *in fact* exert a beneficial influence on the flow of navigable streams. The committee did not, however, believe that forests do have any substantial effect on stream flow, or even if they do, that change in ownership would substantially alter that effect. The situation was complicated by radical disagreement among alleged experts as to the facts.

Influence of Forests on Runoff. In general, foresters and geologists claimed that forests in hilly country have a marked influence in reducing surface runoff of water and therefore in regularizing stream flow, while meteorologists and many engineers vigorously disputed this claim. In 1908 Lieut.-Col. Hiram M. Chittenden of the Army Engineer Corps had published an exhaustive article entitled "Forests and Reservoirs in Relation to Streamflow, with Particular Reference to Navigable Rivers," which probably constitutes the most effective presentation yet made of the point of view that forests have virtually no effect on precipitation and little or no effect on stream flow, particularly in times of extremely high or low water.

Much more controversy was aroused by a report on *The Influence of Forests on Climate and Floods* prepared by Willis L. Moore, Chief of the U.S. Weather Bureau, at the request of the House Committee on Agriculture. He too expressed the belief that forests have little, if any, effect on climate or stream flow and that "the runoff of rivers is not materially affected by any other factor than the precipitation." Both Chittenden and Moore professed complete sympathy with efforts to protect and perpetuate the forests as producers of timber, but decried the attempt to bolster these efforts by claiming beneficial indirect influences which in their judgment did not exist.

Filibert Roth, head of the Department of Forestry at the University of Michigan, commented caustically on Moore's report:

The entire paper is a jumble, it deals with a lot of irrelevant stuff crudely and poorly put together. It is full of fallacy and contradiction and is an insult to thinking and observing people of our country. . . . It is evident that the thinking of the whole civilized world with respect to forest influences is based on observation, good sense, and experience, all of which seem sadly lacking in Mr. Moore's paper.

The paper did, however, have the beneficial effect of furthering cooperation between the Forest Service and the Weather Bureau in a field study, under controlled conditions, of the effect of forests on runoff. The Forest Service had begun to make plans for such a study in 1909 and solicited the participation of the Weather Bureau. On June 1, 1910, the two bureaus jointly initiated at Wagon Wheel Gap, Colorado, on a tributary of the Rio Grande River, the first investigation of its kind in the world. From June, 1911, to July, 1919, complete records of meteorological factors, runoff, and erosion were maintained on two small, adjacent watersheds as nearly alike as possible in geologic, topographic, and forest conditions. The entire forest cover on one watershed was then cut, and observations to determine the effect of the cutting were continued until October, 1926.

The results showed a slight increase in runoff from the cutover area during the entire year and a marked increase during the period of spring floods, when there was also a great increase in erosion. The main conclusion drawn from the study was that forests do have an effect on runoff, but that this effect depends on so many factors which vary widely from watershed to watershed that generalizations on the subject are impossible.

Passage of the Weeks Act. How many votes were changed by the scientific and pseudoscientific arguments as to the influence of forests on stream flow is open to question. There is no doubt, however, that the opinion of the House Judiciary Committee that purchase of forest lands was constitutional if such an influence existed materially facilitated passage of the legislation. The same argument was also used to justify financial cooperation with states in the protection from fire of forests on the watersheds of navigable streams, which was included in later bills on the subject.

Almost exactly ten years after the introduction of the first acquisition bill, Congress approved of legislation providing for the attainment of both objectives. The bill finally passed was introduced by Representative John W. Weeks of Massachusetts on July 23, 1909, passed the House by a rather close vote on June 24, 1910, passed the Senate by an overwhelming majority on February 15, 1911, and was signed by President Taft on March 1, 1911.

Cooperation and Acquisition to the Fore. The Weeks Act blazed three new trails of major importance in the development of forest policy in this country:

1. It authorized the states to enter into agreements or compacts with each other for the purpose of conserving the forests and the water supply of the states concerned. Congress thus gave its consent in advance to any interstate compact that might later be negotiated in the fields indicated. The authorization is significant because of its recognition of the interstate character of the problems involved and of the need for coordinated state as well as national action in their solution. Although no use has been made of this section of the act, it paved the way for the interstate forest-fire-protection compact negotiated by the Northeastern states in 1949, for which specific congressional approval was obtained.

2. It appropriated $200,000, to be available until expended, to enable the Secretary of Agriculture to cooperate with any state or group of states

in the protection from fire of any private or state forest lands on the watersheds of navigable streams. Cooperation was limited to states which had provided by law for a system of forest-fire protection, and the amount spent in any state during any fiscal year was not to exceed that spent by the state. This provision, with later amendments, constitutes the basis for the cooperative fire-control programs which now exist in all of the forested states. It was influential in encouraging many states to adopt effective fire-control legislation and fire-control organizations and in increasing the efficiency of fire-control activities throughout the country.

3. It appropriated 1 million dollars for the fiscal year 1910 (already past) and 2 million dollars for each subsequent fiscal year through 1915 "for use in the examination, survey, and acquirement of lands located on the headwaters of navigable streams or those which are being or which may be developed for navigable purposes." There was no geographic limitation as to where acquisition might take place, although it was generally understood that at least the first purchases would be in the southern Appalachians and the White Mountains.

Further Provisions Relating to Acquisition. The act created a National Forest Reservation Commission, consisting of the Secretary of War, the Secretary of Agriculture, the Secretary of the Interior, and two members each from the Senate and the House of Representatives, to pass upon lands recommended for purchase and to fix the price at which purchases should be made. The composition of the Commission was interesting and unusual in that it included representatives from both the executive and the legislative branches of the government, with the latter in the majority. Differences of opinion between the two groups have been surprisingly few.

The Secretary of Agriculture was authorized to recommend the acquisition of such lands as in his judgment were necessary for regulating the flow of navigable streams and to purchase such lands as might be approved by the Commission. No purchases were to be made until the Geological Survey had reported that control of the lands in question would promote or protect the navigation of streams on whose watersheds they lay and until the Legislature of the state in which the land was located should have consented to its acquisition by the United States for the purpose of promoting the navigability of navigable streams. The object of these provisions was obviously to put a check on any exaggerated ideas the Forest Service might have as to the influence of forests on stream flow and to observe the rights and wishes of the states in which purchases were contemplated.

Sellers were permitted to make reservation of any or all of the minerals or merchantable timber on purchased lands under terms definitely specified in the deed. In 1913 this provision was broadened to permit the purchase of lands having rights of way, easements, or reservations, which, in the judgment of the National Forest Reservation Commission and the Secretary of Agriculture, would not interfere with the use of the lands so encumbered for the purposes of the act. The amendment facilitated the acquisition of needed lands which might not otherwise have been available for purchase.

The Secretary of Agriculture was authorized to sell acquired lands chiefly valuable for agriculture, which in his opinion could be occupied for agricultural purposes without injury to the forests or to stream flow and which were not needed for public purposes, at their true value, in tracts not exceeding 80 acres, under rules and regulations prescribed jointly by him and the Secretary of the Interior. With this exception, no claim could be initiated or perfected to any lands acquired under the act. This provision precluded the entry of acquired land under any of the mineral laws applicable to the public lands until 1916, when the Secretary of Agriculture was authorized to permit the prospecting, development, and utilization of mineral resources on acquired lands.

The Secretary of Agriculture was further authorized to organize acquired lands as national forests to be administered under the provisions of the act of March 3, 1891, and acts supplementary to and amendatory thereof. In practice, areas under acquisition are designated as "purchase units" until the acreage in government ownership becomes sufficiently large to justify their organization as national forests for administrative purposes.

Each state was to receive 5 per cent of the gross receipts from acquired lands for the benefit of the public schools and public roads of the counties in which they were located, provided that this payment should not constitute more than 40 per cent of the counties' income from all other sources.[1] In 1914, the contribution was increased to 25 per cent of the gross receipts, in order to equalize payments to all counties having national forests, irrespective of their origin.

Jurisdiction, both civil and criminal, over persons on acquired lands was not to be affected by their acquisition and administration as national forests, except so far as the punishment of offenses against the United States was concerned. As in the case of public-land national forests, the purpose was to make sure that creation of an acquired-land national forest did not change the jurisdiction of the state or the rights, privileges, and duties of its citizens.

A later act of August 11, 1916, authorized the President to establish refuges for the protection of game animals, birds, or fish on any of the lands purchased under the Weeks Law.

[1] This restriction was removed in 1950.

Progress under the Act. Field studies by the Geological Survey soon convinced it that there was no question as to the effect of forests on stream flow in the southern Appalachian Mountains. More extensive studies in the White Mountains, where the relationship was less clear, led to the same conclusion. No tract which the Forest Service wished to buy has been turned down because of an adverse report by the Survey.

Most of the states in which purchases were contemplated promptly and enthusiastically gave their consent. Maryland first gave its consent, but some years later withdrew it. In some states in other regions, consent was limited to a specific part of the state, as in Pennsylvania, and in others to a maximum area, as in Wisconsin. In general, the requirement of state approval has not interfered with the prosecution of the acquisition program.

Funds made available by Congress for the purchase of lands varied greatly from year to year after June 30, 1915, when the original appropriations came to an end. On the whole, they were disappointingly small, and the erratic fluctuations made it difficult to maintain an effective purchase organization. Purchases were limited chiefly to cutover areas and were made at extremely reasonable prices. In the 1920's dissatisfaction with the slow progress of the program became so acute as to incite a vigorous drive for increased appropriations that resulted in the McNary-Woodruff Act of 1928, which will be discussed in Chapter 10.

It is difficult to overemphasize the importance of the Weeks Law in broadening the forestry activities of the Federal government. The act launched the government on one wholly new enterprise—the expansion of the national-forest system by purchase instead of merely by reservation. It also extended the system to the Eastern states and made the government a large landowner and administrator in regions where the absence of public lands had given the people no experience with it as landlord.

In addition, the act enlarged and greatly strengthened cooperative relations between the Federal government and the states. These relations had previously been sufficient to justify the maintenance in the Forest Service of an Office of State and Private Cooperation, but its activities had been limited chiefly to a few studies dealing with such matters as forest resources, forest and market conditions, growth and yield, and practicable methods of forest management in the various states. From now on, the government was to participate at least indirectly in the formulation and administration of state policies in the overwhelmingly important field of forest-fire control, and with hard cash it was to help out states that were willing to help themselves to at least an equal extent.

Moreover, this pattern of financial assistance, which inevitably involved a certain degree of Federal leadership, was one that could be, and presently was, extended to other activities.

Finally, the act performed a useful service by calling attention to the interstate compact as a device of possible value in the field of forest protection, even though it did not result in any immediate action along that line.

These new approaches, approved by Congress after a full decade of debate, were to have a major and cumulative influence on the development of policy in the fields of Federal, state, and private forestry.

Roads in National Forests. The Agricultural Appropriations Act of August 10, 1912, made 10 per cent of all receipts from national forests available to the Secretary of Agriculture for the construction of roads and trails within national forests in the states from which the receipts were derived. Cooperation with states was authorized in the furtherance of any system of highways of which such roads might be made a part. This legislation was made permanent the next year. It constituted a partial reversion to the 1905 policy of placing all receipts from national forests at the disposal of the Secretary of Agriculture, but differed in two major respects. Only a small part of the total receipts was involved, and this part had to be used for a specific purpose.

Further assistance in the development of transportation facilities was given by the act of July 11, 1916, which appropriated $10,000,000, available over a ten-year period, for the construction of roads and trails within or partly within national forests. The Post Office Appropriations Act of February 28, 1919, appropriated $3,000,000 for each of the fiscal years 1919, 1920, and 1921, to be available until expended, for cooperation with states in the construction and maintenance of roads and trails necessary for the use and development of resources or desirable for the proper administration, protection, and improvement of the national forests.

The Federal Highway Act of November 9, 1921, appropriated $4,-400,000 for the construction of "forest development roads" intended to promote the protection, utilization, and administration of the national forests, and $9,500,000 for the construction of "forest highways" intended to develop the general highway system of the states in which the forests were located. Cooperation with local units of government was permitted but was not mandatory.

These acts established the precedent of making special appropriations for the development of transportation facilities in the national forests. It has since been continued and has proved most helpful in making their resources more available and in improving their administration.

Other National-forest Legislation. Still another provision of the Agricultural Appropriations Act of August 10, 1912, authorized and directed the Secretary of Agriculture to sell timber in national forests at actual cost to homestead settlers and farmers for their domestic uses. The object was not to do away with free use but to give settlers an opportunity to obtain at a reasonable price more timber than the Secretary might feel justified in granting them under free-use permits. "Actual cost" was interpreted to mean all expenses involved in making the sale exclusive of stumpage. Sales of this character never ran into large figures. During the two years following the passage of the act they barely reached $10,000 a year, and in subsequent years seldom exceeded $20,000.

In the same year, but by another act, California was added to the list of states in which national forests could not be created or extended except by act of Congress. Arizona and New Mexico were added in 1926. Thereafter until 1939, when Montana was removed from the enumerated list, Utah and Nevada were the only Western states in which national forests could be established or enlarged by the President.

An act of March 4, 1913, removed the existing prohibition against the export of timber from national forests in Idaho and the Black Hills in South Dakota, thus permitting the Secretary of Agriculture in his discretion to permit the export of timber cut on any national forest in any state. This authorization was reenacted in somewhat modified form in 1926 by specifying that export was permitted if in the judgment of the Secretary it would not endanger the supply of timber for local use. The same provisions were extended to timber cut on unreserved public lands in Alaska under the jurisdiction of the Secretary of the Interior. Sale of timber from these lands had been authorized in 1898, and the export of pulpwood and wood pulp had been permitted in 1905, but the export of timber from the territory had previously been forbidden.

Recreational Use Recognized. In 1915 Congress first recognized recreation as a legitimate and important use of the national forests. It authorized the Secretary of Agriculture to grant permits to responsible persons or associations for the construction and use of summer homes, hotels, stores, or other structures "needed for recreation or public convenience," in tracts of not more than 5 acres and for periods of not more than thirty years.

Special-use permits for recreational purposes had previously been issued by the Secretary under his general authority to make rules and regulations for the occupancy and use of the national forests; but the situation was not satisfactory because the revocable nature of the permits discouraged the permittee from making the investment in permanent improvements often necessary for the most effective use of a given site.

Fr 400 Syllabus

Text: Dana, Forest and Range Policy
Highlights in the History of Forest Conservation

Assignments:

Sept. 8 — 10   Introduction

13 — 17   Dana 1 — 2

20 — 24   3

24 — Oct. 1   4

Oct. 4 — 8   Exam. Dana, 5

11 — 15   Dana, 6 — 7

18 — 22   7 — 8

25 — 29   Exam. 9 — 10

Nov. 1 — 6   11

The continuing concern of Congress for the settler was indicated by the inclusion in the act of a proviso that the authority granted the Secretary should "not be construed to interfere with the right to enter homesteads upon agricultural lands in national forests as now provided by law."

Branch of Research. The research activities of the Forest Service were greatly strengthened by the establishment on June 1, 1915, of the Branch of Research. The reorganization had two major purposes. One was to bring about more effective correlation of all investigative work by placing it under a single administrative head. With research in so many different fields (forest management, range management, forest and range economics, forest influences, and forest products), conducted in part from the Washington office, in part from the district offices, and in part at the forest experiment stations, the need for such correlation was clear.

The second purpose was to give the research work and personnel fullest recognition by developing and strengthening research as a coordinate division of the Service with the same organizational status as administrative activities. In spite of repeated suggestions that research in the field should be placed under the direction of district (now regional) foresters in order to tie it in more closely with administration, no substantial change has since been made in the organization adopted at that time.

Under the imaginative and aggressive leadership of Earle H. Clapp, rapid progress was made toward attainment of the desired objectives. Coordination, expansion, and above all the successful conduct by competent personnel of studies yielding fruitful results gradually brought the Branch of Research recognition as one of the most useful and most important parts of the Forest Service. The important role that it played in the First World War is discussed later in this chapter.

In 1921 the first regional forest experiment station in the Eastern United States was established with headquarters at Asheville, North Carolina. Others soon followed until the entire country was covered by a network of regional stations. It is significant both of their value and of the growing interest of private owners in forestry that funds for the establishment of the Eastern stations came largely as a result of pressure by timberland owners on Congress. At all the stations, both East and West, research activities are oriented to meet the needs of public and private owners alike. Many of them have the benefit of advisory committees which help in the formulation of programs and in other ways.

Forestry on Indian Reservations. Indian lands have a peculiar status in that they are the property of the Indians but are subject to control by Congress in its role as trustee. Responsibility for the management of these lands has been placed in the Indian Service in the Department of the Interior, which thus has the task of handling them in the best interests of the Indians to whom they belong. So far as forest and range lands are concerned, the task is made more difficult by the frequent desire of the Indians for a larger immediate return than is justified by their management for continuous production on a sustained-yield basis.

The forest policy evolved by the Forestry Branch of the Indian Service after Jay P Kinney took charge of the work in 1910 recognized this situation. It sought to administer all tribal lands primarily adapted to timber production or watershed protection so as to obtain the highest present economic return consistent with sound management and to administer all allotted timberlands so as to obtain the highest present economic return consistent with the probable future use of the land. In spite of the continuing handicap of inadequate funds and personnel, marked progress has been made toward the attainment of these objectives.

Kinney's efforts to have Indian lands primarily valuable for timber production given a definite status as "Indian forests" and managed for the permanent good of the tribe, just as national forests are managed for the permanent good of all the people, met with only partial success. Congress did establish the Red Lake Indian Forest of 110,000 acres in northern Minnesota in 1916, but went no further in this direction. The act of March 28, 1908, providing for harvesting the dead and mature timber on the Menominee Reservation in Wisconsin did not specifically create a Menominee Indian Forest, but did indicate clearly that the forest was to be maintained in a productive condition. Cutting operations have consistently been conducted with that end in view.

Proposals that timberlands in Indian reservations be placed under the management of the Forest Service or that they be purchased for inclusion in the national-forest system have elicited no response from Congress. A more promising proposal is for the coordinated management of the timber in national forests and in adjacent Indian reservations on a sustained-yield basis. It is, however, difficult of practical application in the absence of legislation that would provide greater assurance of stability in the handling of Indian timberlands by giving them a permanent status, as has been done in the case of the Red Lake Indian Forest.

Range Management on Indian Reservations. Progress in the adoption of sound methods of management was much slower with range lands in Indian reservations than with forest lands. For many years pressure for large current income, coupled with lack of trained personnel, resulted in overgrazing with consequent depletion of the range, damage to tree reproduction, erosion, and irregular runoff of water. Practically the only exception to the rule was on the few reservations where the superintendent assigned the supervision of grazing to foresters in the employ of the Forestry Branch.

Kinney repeatedly called attention to the situation and urged that steps be taken to remedy it. In 1914, for example, following a field trip to New Mexico, he wrote: "The great fault with grazing matters on the Jicarilla and other reservations in the past has been that practically everything has been determined by clerks in the Washington Office and in the local Agency office who had no information as to actual conditions on the range and often neither theoretical nor practical knowledge of the stock business." [1]

In spite of constant worsening of conditions in the field, no steps of consequence to improve matters were taken until 1929, when Ray Lyman Wilbur became Secretary of the Interior. A few months later a special representative of the Secretary who had been assigned to study the situation recommended that supervision of all grazing activities be assigned to the Forestry Branch. Kinney was averse to assuming responsibility for grazing on reservations in the Great Plains region but was overruled. On April 15, 1930, Secretary Wilbur transferred the administration of grazing on all Indian lands from the Administrative Division, where it had previously been located, to the Forestry Branch.

That branch immediately initiated a grazing reconnaissance on all reservations to obtain the basic information necessary for the introduction of constructive methods of range management. The next spring (1931) new regulations for the control of grazing on Indian lands were issued, and after a period of adjustment were accepted both by the stockmen and the Indians. Range management on Indian reservations had at long last arrived.

Bureau of Corporations Estimate of Standing Timber. In 1906 and 1907 the House of Representatives and the Senate in separate but similar resolutions requested the Secretary of Commerce and Labor to investigate the causes of the high prices of lumber, including specifically the possibility that these high prices might have resulted from any contract, agreement, or combination in restraint of trade. A brief preliminary report on this investigation was submitted by the Bureau of Corporations in 1911 and a voluminous final report in 1913 and 1914. The final report was in four parts dealing with standing timber, concentration of timber ownership in important selected regions, land holdings of large timber owners, and conditions in production and wholesale distribution, including wholesale prices.

One of the most important parts of the Bureau's study was an estimate

[1] Quoted from Jay P Kinney, "Indian Forest and Range," published by Forestry Enterprises, Washington, 1950, with the generous permission of the author.

of the volume of standing timber in the country. The estimate was based (1) on figures furnished by private timberland owners and obtained by field investigations by the Bureau in the "investigation area" (roughly the Pacific Northwest, the southern pine region, and the Lake states) and (2) on figures furnished by the Forest Service for timber in private ownership in the rest of the country and for all timber in public ownership. It showed a total stand of about 2,800 billion board feet, mill scale, of material large enough for utilization as lumber. Of this total, about 2,200 billion board feet was in private ownership, with nearly 80 per cent in the investigation area.

The Bureau's estimate was universally recognized as the best made up to that time. It contrasted interestingly with estimates of 856 billion feet (excluding Douglas fir and ponderosa pine) by C. S. Sargent for the Tenth Census (1880); of 1,400 billion feet by G. W. Hotchkiss in "History of the Lumber and Forest Industry of the Northwest" (1898); of 1,390 billion feet by Henry Gannett for the Twelfth Census (1900); of 2,000 billion feet by B. E. Fernow in "Economics of Forestry" (1902); of 1,970 billion feet by *The American Lumberman* (1905); of 2,500 billion feet by R. S. Kellogg in Forest Service Circular 166, *The Timber Supply of the United States* (1909). The general tendency for each estimate to be larger than its predecessor is doubtless due to more accurate information and to the inclusion of trees of size, quality, and species not previously regarded as suitable for the production of lumber.

Concentration of Timber Ownership. With respect to concentration of ownership, the Bureau of Corporations found that in the investigation area 195 holders, many interrelated, controlled half of the privately owned timber. Three holders (the Southern Pacific Company, the Weyerhaeuser Timber Company, and the Northern Pacific Railway Company) controlled 14 per cent of such timber, or nearly 11 per cent of all the privately owned timber in the country. In 1870, three-fourths of the timber included in the study was publicly owned; forty years later, as a result of the public-land policy of the government, about four-fifths of it was privately owned.

Vast speculative purchases had, in the judgment of the Bureau, led to dominating control of the standing timber by a comparatively few large holders. There had been an enormous increase in the value of stumpage which "the holder neither created nor substantially enhances" and which promised the fortunate owners still greater profits than those already realized.

The Bureau regarded any direct combination in restraint of trade by the very large number of manufacturers and distributors of lumber as unlikely, but believed that indirect control over the entire lumber industry might readily be exercised through concentration of ownership of standing timber. "Those who control the standing timber can control the manufacture of lumber; any permanently effective combination in the manufacture of lumber or in the wholesale trade must almost certainly be a combination among timber owners; and to them will go any increase of profit which such a combination may secure."

As evidence that the industry itself also held this point of view, the Bureau quoted a statement by the manager of the National Lumber Manufacturers Association: "The day of cheap lumber is passing and soon will be gone, but the men who make the money will be those who own timber and can hold it until the supply in other parts of the country is gone. Then they can ask and get their own price." On the other hand, that same man told the association "that students and promoters of forest conservation regret the form in which that report was put out as calculated to mislead the public and consequently to set back the cause of conservation."

Although the Bureau of Corporations found that the concentration in timber ownership was already sufficient to cause serious concern, with still more impressive possibilities for the future, it pointed out that the process had not yet gone as far as in some other fields. As a partial remedy for the situation it recommended strengthening the position of the government, which in spite of its land-disposal activities was still much the largest single timberland owner in the country. This was to be done by maintaining the integrity of the national forests and by placing other publicly owned timber under similar management, including any timberlands to which the government might recover title by forfeiture of railroad grants. The report was decidedly vague as to other measures that might be taken to avoid the development of "impregnable monopolistic conditions" which it foresaw as a future possibility.

Lumber Prices. The Bureau's apprehensive report was followed in 1916 by a book, "The Organization of the Lumber Industry," by Wilson Compton, an economist who later served for many years as manager of the National Lumber Manufacturers Association, which dealt in part with the same subject. Compton came to the conclusion that the more rapid rise in the price of lumber than in the price of commodities in general was due to natural rather than to artificial influences. By "natural influences" he meant the depletion of the best timber in the more accessible regions and the consequent movement of the lumber industry to regions further removed from the center of population, with resultant increases in the cost of getting the lumber into the hands of the consumer.

Although he recognized that "associations have frequently aspired to price fixation," he believed that the fundamental organization of the lumber industry was such as to prevent attainment of sufficient control over supply to achieve that purpose. Concentration of timber ownership as a cause of high lumber prices he looked upon as a future rather than a present danger. "As an influence, distinct from general relative exhaustion of supply, tending to raise lumber prices, concentration per se in timber holding in certain regions has not been effective in the past, eithei directly or indirectly. It is, however, a powerful potential factor, the true significance of which is dependent upon future exigencies of the timber supply."

Forest Service Study of Lumber Industry. Shortly after the completion of the investigation by the Bureau of Corporations, the Forest Service undertook a study of conditions in the lumber industry in cooperation with that Bureau and its successor, the Federal Trade Commission. One of the reports resulting from this study, "Some Public and Economic Aspects of the Lumber Industry," by W. B. Greeley, published in 1917, is of particular interest in this connection.

Greeley came to the conclusion that rising costs of transportation as timber shortage moved the mills farther and farther from the bulk of consumers had been an important factor in increasing the cost of lumber. He pointed out that abundant forests had been the country's best guaranty of cheap lumber and that their depletion was a national waste which bade fair to be felt sooner or later in the cost of forest products. At the same time, competition between lumber and other materials, competition between different kinds of lumber and between individual owners and manufacturers, and the steadying influence of large public holdings were likely to prevent any artificial rise in the price of lumber as a result of any attempted combination in restraint of trade.

The transfer of large bodies of timber from public to private ownership had not been an unmixed blessing so far as the private owners were concerned. In the West, speculation had resulted in the accumulation in private hands of some 700 billion board feet more timber than the existing mills could carry. Carrying charges in the form of interest and taxes on this surplus were mortgaging liberal advances in future worth, so that in the long run Western timberlands could hardly earn the profits expected of them. Contrary to the view of the Bureau of Corporations, he believed that "the large speculative gains in buying stumpage which have tided lumbermen over many tight places are mostly over. The future earnings of the industry probably will have to be made in its milling and merchandising."

Greeley believed that the situation called for a marked extension of public ownership and for public assistance to and regulation of private owners in order to replace speculative ownership of timberlands by stable ownership based on management of the forest for sustained production.

The burden of carrying a surplus of 708 billion feet is too heavy for the industry alone. . . . If tendencies toward monopoly develop . . . extensive public forests can be effective in keeping the industry competitive. . . . Public owner-ship should be backed up by reasonable requirements imposed upon the han-dling of private forest lands. . . . Public cooperation in fire protection and taxation will be an important factor.

Forfeiture of Oregon and California Railroad Land Grant. In 1866 Congress granted to the Oregon and California Railroad Company all of the odd-numbered sections of nonmineral public land to a distance of 20 miles on each side of a proposed railroad from Portland to the Cal-ifornia line. Lieu selections could be made within an indemnity strip 10 miles wide on each side of the main grant. Three years later the act was amended to provide that the lands granted must be sold to actual settlers only, in quantities not greater than 160 acres to one purchaser and at a price not exceeding $2.50 per acre. All three of these conditions were flagrantly violated, and in 1903 the Southern Pacific Company, of which the Oregon and California Railroad Company had become a subsidiary, for all practical purposes withdrew the remaining lands from the market.

Congress in 1908 authorized the Attorney General to start proceedings looking to the forfeiture of the grant. Suit was accordingly brought against the railroad company and also against some forty-five purchasers who had bought more than 1,000 acres each from the company. The total area involved was nearly 3 million acres containing some of the finest timber in Oregon.

On August 20, 1912, Congress authorized settlement of the suit against the forty-five purchasers on the basis of forfeiture of the lands to the government, with privilege of repurchase by the holders at $2.50 per acre. The same act provided that after one year no suits should be brought for the recovery of smaller tracts purchased in violation of the terms of the grant.

In the main suit against the Southern Pacific Company, the Federal District Court in 1913 declared the lands forfeited to the government on the ground that the grant was "on condition subsequent" and therefore forfeitable if the conditions were broken. Two years later the Supreme Court reversed this ruling on the ground that the terms of the grant did not constitute a condition subsequent but a covenant, violation of which could be handled by Congress as it saw fit. The Court therefore enjoined the railroad company from disposing of any of the lands or timber until Congress should have a reasonable opportunity to act.

Thus put on the spot, Congress passed the Chamberlain-Ferris Act of June 9, 1916, which Ise characterized as "a fitting climax to the long list of blunders dealing with the public lands." The act revested in the United

States title to all unsold lands to which the Oregon and California Railroad Company (the O. and C.) had received patent and to which it was entitled to receive patent and provided for their disposal. The validity of the act was upheld by the Supreme Court on April 23, 1917.

Disposal of Revested Lands. Instead of placing the revested lands in national forests, with which they were partly intermingled, the Secretary of the Interior was directed to divide them into three classes to be disposed of as follows:

Class 1. Power-site lands, to include only lands chiefly valuable for water-power sites. These lands were to be subject to withdrawal and such use and disposition as was provided by law for other public lands of like character.

Class 2. Timberlands, to include lands having not less than 300 M board feet of timber on each 40-acre subdivision. The timber on these lands was to be sold under competitive bidding as rapidly as reasonable prices could be secured in a normal market. The cutover lands were then to become subject to disposal in the same way as Class 3 lands, except that payment of the $2.50 per acre necessary to acquire title to "agricultural" lands was not required.

Class 3. Agricultural lands, to include all lands not falling into either of the other two classes, regardless of their actual character. These lands were to be opened to entry under the general provisions of the homestead laws, except that a charge of $2.50 per acre was made and commutation was not allowed.

All lands except power-site lands were to be open for exploration, entry, and disposition under the mineral laws. Unpaid taxes on the revested lands were to be paid by the government.

Provision was made to determine the railroad company's equity in the lands by subtracting from the amount it would have received if all of the land had been sold at $2.50 per acre the amount it actually did receive. Receipts by the government from the sale of land and timber were to be applied first to paying the company the amount due it and next to reimbursing the government for accrued taxes which it had paid. Of the remainder, 25 per cent was to go to the state to become part of its irreducible school fund; 25 per cent to the counties for common schools, roads, highways, bridges, and port districts; 40 per cent to the reclamation fund; and 10 per cent to the general fund in the United States Treasury.

Since receipts during the first ten years left no remainder to distribute, an act of July 13, 1926, provided that the government should pay to the eighteen counties involved amounts equal to the taxes they would have received during the period from 1916 to 1926 if the revested lands had remained on the tax rolls, and should also continue such payments in the future until a balance became available for distribution to the counties. Both past and future payments were to be based on 1915 valuations, irrespective of any changes in actual value that might have taken place since that time.

Handling of Revested Lands. The total area of revested lands was subsequently reduced, by alienation to private owners and by transfer to national-forest status, to about 2,575,000 acres containing some 50 billion board feet of timber, or about a sixth of the total sawtimber stand in the Douglas fir region of Oregon. By decision of the Commissioner of the General Land Office in 1919 and of the Assistant Secretary of the Interior in 1923, no change was made in departmental jurisdiction over some 465,000 acres in the indemnity limits which had been included in national forests between 1892 and 1907 and to which patents to the company had neither been received nor were pending. Fire protection on the entire revested area was handled partly by the Oregon Forest Fire Association and partly by the Forest Service.

The basic weakness of the 1916 act lay in its assumption that practically all of the revested lands were agricultural in character. Actually they are in rough, mountainous country where what little agricultural land there is had already been taken up. Most settlers who tried to take advantage of the homestead provisions of the act were soon starved out. Because of the erroneous assumption that the lands would eventually be cleared for agriculture, no attempt was made either by Congress or the Secretary of the Interior at conservative management of the timber, which was the real asset. Even after the original revestment act was amended in 1928 to authorize the Secretary of the Interior to make such rules and regulations for the cutting and removal of the timber as he deemed necessary, the only change in practice was to require the purchaser to file a bond to assure disposal of slash resulting from cuttings.

By 1933 the government had paid more than 4 million dollars to the railroad company and nearly 13 million to the counties, while it had taken in receipts of less than 8 million dollars. Cutting had been done on some 120,000 acres with no plan for sustained-yield management of the forest. The situation was not to change materially until an aroused public opinion led to passage of the act of 1937 providing for a new and effective method of handling the lands for permanent timber production.

Coos Bay Wagon Road Lands. Closely connected with the litigation concerning the O. and C. lands were about 93,000 acres in a grant which had been made to the state of Oregon in 1869 to aid in the construction of the Coos Bay Military Wagon Road and which had later been acquired by the Southern Oregon Company. Suit was brought for the re-vestment of these lands because of violation of the terms of the grant. While the case was still before the Supreme Court on appeal, Congress in an act of February 26, 1919, authorized dismissal of the suit under an arrangement by which the company would deed to the United States all of the land it still held in return for payment of an amount (about $232,000) that would bring its total receipts up to $2.50 per acre for the entire area included in the grant.

The lands thus reconveyed were to be classified and disposed of under the provisions of the O. and C. act of June 9, 1916. Receipts were to be used first to pay the company the amount due it and to reimburse the government for unpaid taxes which it had advanced. Of the remainder, 25 per cent was to go to the counties and 75 per cent to the United States Treasury.

Administration of the lands followed the same unsatisfactory course as that of the O. and C. lands. Exchange of lands in both grants for adjacent private lands was made possible under acts passed in 1918 and 1920, wherever such exchange will advantageously consolidate government holdings.

Stockraising Homestead Act. Unregulated use of the unreserved public lands primarily valuable for grazing was rapidly resulting in their serious deterioration and even destruction. An attempt to improve the situation by opening the way for private ownership of the range lands was made in the Stockraising Homestead Act of December 29, 1916. That act authorized the Secretary of the Interior to designate and open for entry under the homestead laws "lands the surface of which is, in his opinion, chiefly valuable for grazing and raising forage crops, do not contain merchantable timber, are not susceptible of irrigation from any known source of water supply, and are of such character that six hundred and forty acres are reasonably required for the support of a family."

Instead of cultivation, the entryman was required to make permanent improvements to the extent of $1.25 per acre. Commutation was not allowed. All of the coal and other minerals were to be subject to entry under the coal and mineral-land laws. Lands containing water holes and other bodies of water needed or used by the public for watering purposes were not to be designated, but might be reserved from entry under the withdrawal act of June 25, 1910, and held open for public use. The Secretary of the Interior could also withdraw from entry lands needed to ensure access by the public to watering places and needed for use in the movement of stock to summer and winter ranges or to shipping points. Finally, the Secretary was authorized to make all necessary rules and regulations for the carrying out of the act.

Disappointing Results. Marked interest in this new method of acquiring title to public lands is indicated by the fact that within four months after passage of the act the Interior Department had received applications for the designation of some 24,000,000 acres. By 1923, 21,523 patents had been issued covering an area of 7,141,175 acres, with a corresponding area embraced in pending entries.

In his annual report for that year the Commissioner of the General Land Office stated that land capable of supporting a family by stockraising in bodies of not more than 640 acres had already been largely appropriated and that current entries were becoming more and more speculative in character. Entrymen finding that they could not make a living on the land proceeded to lease it to some large stockman in the vicinity. The Commissioner therefore recommended that the Stockraising Homestead Act be supplemented by an act authorizing the Secretary of the Interior to reserve lands chiefly valuable for grazing purposes and to lease them in such bodies and under such regulations as would ensure the preservation of their forage values.

No action was taken by Congress on this recommendation, which was repeated in 1924 with the comment that unless some constructive policy were soon adopted "the subject matter itself will disappear." By 1925 the Commissioner had come to the conclusion that the Stockraising Homestead Act had outlived its usefulness. He therefore recommended its outright repeal and its replacement by a leasing act which would "permit the control of stock grazing on our public lands, protect and conserve the waters in connection with such use, and allow the fullest development of our mineral resources." In 1928 he proposed that the area subject to entry as a stockraising homestead be increased to 1,280 acres if Congress were unwilling, after repeated urging, to adopt the course which he still preferred of replacing the act of 1916 by a leasing system.

The complete inadequacy of the Stockraising Homestead Act to bring about satisfactory management of the public range lands soon became generally recognized. Even those who originally favored its passage saw that it was applicable to only a small part of the total area involved and that it was leading to speculation and to concentration of ownership in large holdings rather than to new homes. Recommendations similar to those emanating from the Commissioner of the General Land Office were consequently made by many individuals and organizations, including the Department of Agriculture.

Congress finally responded by passing the Taylor Grazing Act of 1934, which will be discussed in Chapter 11. The Stockraising Homestead Act is still on the statute books, but practically no suitable land is now available for designation under it.

National Park Service. Thirteen national parks had been established before specific provision for their administration was made by the act of August 25, 1916. That act created a National Park Service in the Department of the Interior to promote and regulate the use of national parks, monuments, and other reservations of like character. The purpose of these areas was "to conserve the scenery and the natural and historic objects and the wild life therein and to provide for the enjoyment of the same in such manner and by such means as will leave them unimpaired for the enjoyment of future generations."

The Secretary of the Interior was authorized to make rules and regulations for the use and management of the national parks, monuments, and reservations under the jurisdiction of the National Park Service, and violation of these rules and regulations was made a penal offense. More specifically, the Secretary was authorized to sell or dispose of timber when necessary to control the attacks of insects or diseases or otherwise to conserve the scenery or natural or historic objects; to provide for the destruction of such animals and plant life as may be detrimental to the use of the reservations; to permit the grazing of livestock in any reservation except the Yellowstone National Park when in his judgment such use will not be detrimental to the primary purpose for which the reservation was created; and to grant leases for the use of land for the accommodation of visitors for periods not exceeding twenty years.

The mineral leasing act of 1920 specified that the provisions of that act should not apply to national parks. Receipts from national parks are covered into the United States Treasury, with no contributions to local communities in lieu of taxes.

National Park Policy. The National Park Service was organized in 1917 with Stephen T. Mather as its first director. Secretary Lane's admirable letter of instructions to him (May 13, 1918) stated that the administrative policy to which the Service would adhere was based on three broad principles:

First, that the national parks should be maintained in absolutely unimpaired form for the use of future generations as well as those of our own time; second, that they are set apart for the use, observation, health, and pleasure of the people; and third, that the national interest must dictate all decisions affecting public or private enterprise in the parks.

Every activity of the Service is subordinate to the duties imposed upon it to faithfully preserve the parks for posterity in essentially their natural state. The commercial use of these reservations, except as specially authorized by law, or such as may be incidental to the accommodation and entertainment of visitors, will not be permitted under any circumstances. . . .

In studying new park projects you should seek to find "scenery of supreme and distinctive quality or some natural feature so extraordinary or unique as to be of national interest and importance." . . .

The national park system as now constituted should not be lowered in standard, dignity, and prestige by the inclusion of areas which express in less than the highest terms the particular class or kind of exhibit which they represent.

The policies promulgated in these instructions have continued to guide the administration of the national parks. The chief difficulties in their application have been to maintain standards in the face of local pressures for the inclusion of substandard areas; to prevent commercialization; and to accomplish the almost impossible task of providing for the present enjoyment of the parks by as many people as possible and at the same time leaving them "unimpaired for the enjoyment of future generations."

Decision is often difficult as to whether or not a given area actually meets acceptable park standards. When borderline cases involve heavily forested lands, a further question arises as to the relative value of the land for "museum" purposes and for economic use as a source of timber supply. Sharp difference of opinion on this point between the National Park Service and the Forest Service has not been uncommon. The possibility of removing substandard areas from the system was demonstrated in 1931, when Sullys Hill National Park, after seventeen years of joint administration as a national park and a game preserve, was changed into the Sullys Hill National Game Preserve and placed under the sole administration of the Biological Survey in the Department of Agriculture.

A new type of reservation was proposed by the Secretary of the Interior in 1925, when he recommended that he be authorized to withdraw and set aside public lands for health and recreational purposes, city and county parks, and the protection of municipal water supplies. The proposal has never been approved.

Education in Forestry. The success with which forestry is practiced in the woods obviously depends on the ability of its practitioners. Prior to 1910 technical training in forestry was offered at fifteen schools, of which only one was in the South and four in the West. The establishment of so many schools in so short a time, with widely varying curricula and quality of instruction, led Pinchot, then Chief of the Forest Service, to call a conference of forest schools for the primary purpose of considering standards of education in forestry. The conference met at Washington, D.C., December 30 to 31, 1909. It recommended the organization of an association of forest schools and appointed a committee, with Henry S. Graves, Dean of the Yale School of Forestry, as chairman, to formulate a standard of forest education.

That committee submitted a provisional report at a national conference on forest education held at Washington, December 28 to 29, 1911. It emphasized the fact that the standard of professional training in forestry must be high if it is to meet the country's needs. As an indication of what such a standard should be, the committee presented in detail the content of each technical subject that in its judgment should be included in a four-year curriculum, and the minimum number of hours that should be devoted to each subject. The totals came to 1,500 hours, of which it was recommended that 585 hours be in classwork and 915 hours in field or laboratory work.

The conference expressed general agreement with the committee's proposals and authorized it to prepare a final report, which appeared in the September, 1912, issue of the *Forestry Quarterly*. For nearly a decade, the standards thus established continued to serve as a guide to the schools of forestry, the number of which had been increased by eight during the five years between 1909 and 1914. Divergencies in the programs offered by the various schools naturally arose and gave rise to the belief that the entire situation should be reviewed.

Such a review took place at a "second" (really the third) national conference on education in forestry held at New Haven, Conn., December 17 to 18, 1920, under the chairmanship of James W. Toumey of the Yale School of Forestry. Reports were submitted by nine different committees, each of which submitted specific recommendations concerning the subject assigned to it. These reports were later published in full by the U.S. Bureau of Education (Bulletin 44, 1921).

The one resolution adopted by the conference requested the Society of American Foresters to appoint a continuing Committee on Forest Education to consider all suggestions made at the conference together with such other phases of forest education as it deemed desirable. Reports from the eleven subcommittees into which this committee was in turn divided were presented at the annual meeting of the Society at Toronto on December 27 to 28, 1921. Thereafter the committee's activities became increasingly sporadic until 1934, when it was replaced by the Division of Education.

The careful, collective consideration given to professional education in forestry during a critical period in its development helped materially to raise standards in an activity on which the competence of the forester so largely depends.

Society of American Foresters. Another major influence in any profession is the standing of its national organization. The Society of American Foresters, established in 1900 with a handful of members, grew steadily in members and in strength. Its influence was felt most strongly in raising the standards of technical competence and in establishing an exceptionally strong *esprit de corps* throughout the profession.

One weakness—the practice of holding all meetings of the society in Washington, where they could be attended by only a few members—was remedied in 1912 by the establishment of a Northern Rocky Mountain Section at Missoula, Montana. This precedent was followed in other parts of the country until there are now (1955) twenty-one regional sections, with numerous subsections. The opportunity thus afforded for practically all members of the society to participate directly in its activities has broadened their outlook, increased their efficiency, and solidified the profession.

An important move was made on January 1, 1917, when the society assumed responsibility for publishing the *Journal of Forestry,* a new periodical formed by amalgamation of the *Forestry Quarterly* and the *Proceedings of the Society of American Foresters.* B. E. Fernow, formerly editor of the *Forestry Quarterly,* served as editor-in-chief, and Raphael Zon, chief of the Office of Silvics in the Forest Service, as managing editor. At first published eight times a year, it became a monthly in 1935.[1] It remained the only professional publication in the United States devoted wholly to forestry and related subjects until the appearance in March, 1955, of *Forest Science,* "a quarterly journal of research and technical progress."

Association of Eastern Foresters. Probably the earliest organization of professional foresters next to the Society of American Foresters was the Association of Eastern Foresters. It was started on January 11, 1908, when six foresters (Alfred Gaskill, C. R. Pettis, H. H. Chapman, G. H. Wirt, F. W. Besley, and A. F. Hawes) met informally in New York City and decided to hold annual or semiannual meetings. Formal organization was effected on November 11, 1910, with twenty-one charter members, and a constitution was ratified on January 12, 1911. Membership was limited to forest officials and to instructors at universities and schools of forestry in New England, New York, New Jersey, Pennsylvania, Delaware, and Maryland, with such other professional foresters as might be elected.

The association concerned itself both with forest practice and forest policy. In 1913, for example, it vigorously supported the Forest Service in opposing the proposal to cede the national forests to the states. In 1915 it urgently advocated reappropriation of the 3 million dollars that were not used under the Weeks Act of 1911; and in 1919 it urged that the Federal appropriation for cooperation with the states in the protection of forests from fire be increased to $500,000, and that the requirement limiting such cooperation to the watersheds of navigable streams be repealed.

By the fall of 1920 the subject of a national forest policy had become so important in the eyes of the state foresters as to make its consideration

[1] Twelve issues were published in 1925, and nine issues in 1934.

on a broader scale desirable. The meeting held at Atlantic City on November 12 was accordingly attended by representatives of the Forest Service and of sixteen states. This group unanimously agreed to form a State Foresters' Association, which was duly organized at Harrisburg on December 8 to 9, 1920.

The establishment of the new organization and of strong local sections in the Society of American Foresters led to a loss of interest by the state foresters in the Association of Eastern Foresters, which did not meet again. It had performed a useful service to the profession and the nation during its relatively brief but decidedly lively existence.

War and Wood. The entrance of the United States into the First World War had immediate and far-reaching effects on both industry and government. The lumber and wood-using industries were suddenly called upon to meet the heavy and often exacting requirements of the Army and Navy on a time schedule that brooked no delay. Essential civilian needs also had to be met. The situation was not one to encourage even the more progressive owners to improve their forest practices.

So far as the Forest Service was concerned, the national forests had to be managed with a greatly reduced force, but the heaviest impact was felt in the field of research, in which all investigations not concerned with war problems were soon halted. The situation was well summarized in the Forester's annual report for 1918, when the war was at its height:

Wood products are among the most important materials of the war. Lumber is required in immense quantities for the extension of military posts, for the construction of buildings in connection with the new training camps, for ship building, for industrial housing, etc. The war has also brought an unusual demand for special wood products such as material for artillery carriages, escort wagons, and other vehicles, rifle stocks, airplanes, shipping containers, and various military materials and equipment requiring the use of by-products.

Critical problems have arisen relating to wood supplies, technical qualities of woods heretofore little used, drying processes, the development of waterproof glue, the design and construction of laminated structures and plywood, the manufacture of by-products of wood, and many other matters. . . . To secure this special information speedily, practically the entire research organization of the Forest Service has been placed on special war investigations and the organization has been increased in size more than five times to meet the situation. Information has been required by practically all the war-work branches of the Government having to do with the purchase of wood materials. . . . In addition, there has been cooperation and assistance to the allied Governments and to the industries furnishing war materials.

The Forest Products Laboratory became an invaluable part of the war machine not only through the conduct of fundamental research but also by assisting military and civilian agencies in every phase of the selection, purchase, and utilization of wood for the countless purposes for which it was used. Hardly less valuable were the field studies conducted by other parts of the Forest Service to locate adequate supplies of timber of the species, size, and quality essential for aircraft, ships, vehicles, gunstocks, tanning materials, wood distillation, cooperage, containers, and other products. Forest research, especially in the field of wood technology, suddenly assumed an urgency previously unknown.

Logs and Labor. One of the major difficulties faced by the lumber industry in meeting wartime needs arose from its relations with labor. The efforts of the Industrial Workers of the World during the preceding decade to organize lumberjacks and mill workers in a mass drive to raise starvation wages, to decrease long hours of work, and to improve disgraceful working conditions had accomplished little but to stir up occasional violence in which both parties were the aggressors. But the coming of war, with its urgent demands for wood that must be supplied without delay, gave the "Wobblies" an opportunity of which they were quick to take advantage.

The situation was particularly critical in the Pacific Northwest, where strikes seriously endangered production of adequate supplies of Sitka spruce for airplane construction and of heavy Douglas fir timbers for a wide variety of purposes. Government and industry found the answer in the establishment of the Loyal Legion of Loggers and Lumbermen, created by General Bruce E. Disque, head of the Spruce Production Corporation.

The "4 L" sought to promote understanding and cooperation in the war effort by including equal numbers of workers and employers throughout the organization, from a camp committee to the board of governors. It succeeded not only in maintaining production at the necessary level, but in permanently improving labor-management relations by effecting a substantial increase in wages, reduction in working hours, and betterment in living conditions.

Forest Engineers in France. Another war activity of major importance was the supplying to the American Expeditionary Force in France, and to a less extent our French and British allies, the enormous amounts of wood needed by armies in the field. Shortage of shipping virtually precluded the export of lumber and other forest products from the United States to the war zone, with the result that 75 per cent of the timber used by the American Army had to be cut from French forests by our own troops.

To meet this situation, steps were taken as early as May, 1917, at the

request of the War Department, to organize a regiment known as the
Tenth Engineers (Forest) to handle the necessary logging and milling.
The regiment included skilled woodsmen and millmen from all the main
lumber regions of the country, while about half of the officers were
trained foresters. The Forester, Henry S. Graves, was sent abroad by the
War Department to make preliminary arrangements for the work.

Some six months later steps were taken to form another forestry regiment (the Twentieth Engineers), battalions of which were sent to France
as rapidly as they could be organized and equipped. Early in 1918 the
two regiments were combined and thereafter operated as the Forestry
Division in the Service of Supply. By the date of the armistice the
Forestry Section, at that time in charge of Lieut.-Col. William B. Greeley,
numbered some 12,000 engineer troops and 9,000 service troops.

These troops provided the wood needed by the Army for lumber, railroad ties, large timbers for docks, barges, trestles, and bridges, piling up
to 100 feet in length, telephone and telegraph poles, fuel, and small material for such uses as wire-entanglement stakes and pickets for supporting
camouflage nets. By the end of the war there were about ninety active
operations, most of them with sawmills. All the cutting had to be done
under instructions prepared by the French Forest Service, which even
under the stress of war took pains to make sure that the American lumberjacks did not jeopardize the future of one of the country's most precious
assets. Logging was normally restricted to trees which could properly be
removed from the standpoint of maturity and regeneration of the forest
and was conducted in substantial compliance with the requirements of
the French foresters.

It is a tribute to the efficiency of French forestry that the forests were
able to meet the tremendous demands suddenly placed upon them and
to do so without serious injury to their future. So far as the American
troops are concerned, the experience was a highly educational one which
materially broadened the training of the foresters and gave the practical woodsman a different attitude toward forestry than he had had
before.

Summary. Passage of the Weeks Act of 1911 established two new
Federal policies of major importance—cooperation with the states in the
protection of forests from fire and acquisition of forest lands for the protection of the watersheds of navigable streams. Other advances included
a strengthening of national-forest administration, establishment of sound
forest management on Indian reservations, and creation of the National
Park Service. Title to nearly 3 million acres included in the O. and C.
land grant was revested in the government because of violation of the
terms of the grant. The Stockraising Homestead Act of 1916 attempted

to solve the problem of what to do with the public range lands with little success.

Forest research was greatly strengthened by organization within the Forest Service of a Branch of Research, coordinate in rank with the administrative branches. Intensive studies of the lumber industry were made by the Bureau of Corporations and the Forest Service. The Society of American Foresters played an increasingly influential part in raising standards of technical competence and in promoting solidarity among professional foresters. The First World War placed a heavy strain on industry and government to provide the wood needed for a wide variety of military purposes, diverted Forest Service research almost entirely to this end, and sent two regiments composed chiefly of foresters, woodsmen, and millmen to France to supply the wood needed by the armies of the United States and its allies in the field.

REFERENCES

Cameron, Jenks: "The Development of Governmental Forest Control in the United States," chap. 9, Johns Hopkins Press, Baltimore, 1928.

Greeley, William B.: "Some Public and Economic Aspects of the Lumber Industry," Report 114, Office of the Secretary of Agriculture, 1917.

Ise, John: "The United States Forest Policy," chaps. 6, 10–12, Yale University Press, New Haven, Conn., 1920.

Lillard, Richard G.: "The Great Forest," part III, Alfred A. Knopf, Inc., New York, 1947.

Peffer, E. Louise: "The Closing of the Public Domain," chaps. 3–7, 10, Stanford University Press, Stanford, Calif., 1951.

Robbins, Roy M.: "Our Landed Heritage," Peter Smith, New York, 1950.

U.S. Department of Commerce and Labor, Bureau of Corporations: "The Lumber Industry," 3 vols., part I, Standing Timber, 1913.

CHAPTER 9

Controversy and Cooperation

Shortly after the close of the First World War, separate but similar drives to bring about public regulation of cutting on privately owned forest lands were started by the Forest Service and the Society of American Foresters. What was new about this approach was not the idea itself but the determined effort to put the idea into practice.

Forest Service Advocates Regulation. Chief Forester Graves opened the campaign on February 24, 1919, before a forestry congress held under the auspices of the Boston Chamber of Commerce and the Massachusetts Forestry Association. He pointed out that

most timber-land owners do not intend to hold their lands after cutting the timber, and they see no reason why they should expend money or effort on the land to secure public benefits or to avoid injury to the community. It is the speculative character of ownership that explains the lack of incentive to timber-land owners to handle their lands constructively; and we may not expect that such owners will take any different view or action on their own initiative. . . .

The public in its own protection should prohibit destructive methods of cutting that injure the community and the public at large. With the co-operation of the public, constructive measures of forestry are feasible. They should be mandatory.

Two months later, in an address to the American Lumber Congress at Chicago, Graves stated that the transfer of great bodies of timber from public to private ownership had been a grave mistake which had created a perplexing dilemma.

On the one hand the public is deeply concerned that the private forests be handled in a way to provide for forest renewal and growth. We have on the other hand the timber owners struggling under a responsibility that has never been fully sensed or accepted. The result is that while considerations of public interests demand that something be done, nothing substantial is actually being accomplished. . . .

As I see it, either private owners must assume the full responsibility of properly caring for their timber-lands, including protection and forest renewal; or the public must take over the responsibility that it once had and surrendered; or the public must share with the owners both the responsibility and the burden of securing the objectives that are essential to safeguard the public welfare. My own view is that the last is the only fair and practical method from the standpoint of all concerned. . . .

I am ready to advocate a policy more far-reaching in all respects than has generally been offered. I would afford whatever public assistance is needed to make possible the conservative handling of our forests, and I would then make fire protection, conservative production of lumber, and right methods of removal a matter of requirement, with such public direction and control as is necessary to realize the aims desired by the public.

At Graves' request many organizations, such as the National Lumber Manufacturers Association, the American Paper and Pulp Association, and the Association of Wood-Using Industries, appointed committees to cooperate with the Forest Service in attempting to work out a policy for the solution of the problems which he posed. Numerous conferences were held throughout the country with representatives not only of timberland owners and wood-using industries but of other business groups, of state departments of forestry, and of public organizations such as the American Forestry Association.

Society of American Foresters Acts. Meanwhile, Frederick E. Olmsted, president of the Society of American Foresters, in the spring of 1919 appointed a Committee for the Application of Forestry, headed by Gifford Pinchot, to recommend action for the prevention of forest devastation on privately owned timberlands in the United States. Olmsted explained his reasons for the appointment of the committee in an article in the *Journal of Forestry* for March, 1919:

On this whole question of forestry for the private owner and forestry for the Government we have recently been thinking in a rut. We grant that the Government should practice forestry on its own lands and agree that it is now doing so quite successfully. Then we put the question, "Can the private owner of forest lands practice forestry?" and the pedantic answer is, "Oh dear, no; he could not afford to wait 80 or 100 years for another crop of timber which might net him 3 or 4 per cent on a somewhat hazardous investment. He might do so, perhaps, if allowed to bring his individual interest into a lumber syndicate of such pregnant power as to rival the Government itself; but existing conditions are indeed most discouraging." And there we have dropped the matter.

That is muddle-headed reasoning. Quite regardless of whether forest lands may be publicly or privately owned in the future or of intricate guesses as to future financial returns, *private forest lands now being logged must be kept productive, for otherwise they will be of no value to any future owner.* . . .

Let us see to it, first of all, that we keep trees growing in order that we may have wood, and then, being assured of wood, let us determine, with all necessary deliberation, who, ultimately, should own and manage the lands upon which this wood is produced. In plain words, this means that the public must compel the lumberman to treat his forest decently, and that the forester, without delay or quibble, must show the public how this may be done. . . .

The [lumber] industry as a whole is archaic, individually self-centered, and pennywise. . . .

For many years past the forester has endeavored to persuade the lumberman that measures for keeping his lands productive were worth adopting. Persuasion has utterly failed, resulting in little more than mild amusement on the lumberman's part. The lumberman, to be sure, has always expressed a keen desire to "co-operate" with the forester in all ways possible; but the difficulty there has been that in practice the lumberman's understanding of co-operation has been to accept everything to his immediate liking and to yield nothing in return. Precisely at the point where the forester's suggestions imply a present restraint on the lumberman's part in order to insure the future welfare of both the public and the lumberman, the forester becomes a theorist, an idealist, and what chambers of commerce are pleased to term an "insidious influence." The humor of the situation is that the lumber industry has fallen to its present level because of a total lack of theory and ideals. The time for persuasion has passed.

Public Ownership the Alternative? An article by Harry T. Kendall of the Kirby Lumber Company, Houston, Texas, in the April, 1919, issue of *Cut-Over Lands*, indicated that at least some representatives of the lumber industry regarded public ownership as the only solution of the forest problem. Said Mr. Kirby:

Lumbermen know that without reforestation the lands which now carry a stand of timber must become exhausted, to the ultimate destruction of their business and to the embarrassment and injury of posterity. The Texas lumbermen likewise know that there is little probability that the laws, organic and statutory, will be so changed that forest lands under private ownership may be carried for a sufficient time to enable those lands to produce a crop of trees. . . .

I wish to again impress upon you this fact; that, as a lumberman, my interest in forestry is nil. Nothing can now be done by the State that will bring one penny's benefit to any lumberman now engaged in the manufacture or sale of forest products. When the lumberman of today saws the trees he owns and scraps his plant, his capital will enable him to become the banker, the ranchman, or the manufacturer of some other commodity, or with his capital and experience he can continue business in other sections of the country where timber is still available. But, on the other hand, the lumberman, as a private citizen, being closely in touch with the situation and realizing, as the great body of citizens must soon realize, the necessity of some definite forestry project under the control of the State, his interest is a serious one, and as a citizen I raise my voice to appeal to you and to the other citizens of this State to face the forestry problem fairly and squarely and deal with it in the only possible way, viz., the State to acquire suitable lands for reforestation purposes and to safeguard and handle the growing forests along recognized lines.

Recommendations of Committee for the Application of Forestry. The Committee for the Application of Forestry presented its report on November 1, 1919, under the title, *Forest Devastation: A National Danger and Plan to Meet It.* Certain reservations were expressed by two members of the committee. The report emphasized the nation's requirements for wood, the danger of a serious timber shortage resulting from the devastation of privately owned forest lands, and the consequent need to keep all forest lands in continuous production.

The plan which it proposed to achieve this goal was "based upon the following *fundamental principles*":

1. Prosperity in peace and safety in war require a generous and unfailing supply of forest products.
2. The national timber supply must be made secure (*a*) by forbidding the devastation of private forest lands and by promoting the conditions necessary to keep these lands permanently productive; and (*b*) by the production of forest crops on public forests owned and operated by communities, States and the Nation.
3. The transformation of productive forests into idle wastes impoverishes the Nation, damages the individual, is wholly needless, and must be stopped.
4. Unless and until lands can be more profitably employed for other purposes they should be used to produce forest crops.
5. The ownership of forest land carries with it a special obligation not to injure the public.
6. The secure and steady operation of the lumber industry is of vital concern to the public.
7. The lumber industry being nation-wide, uniform and adequate control over it must be national.
8. National legislation to prevent forest devastation should [provide] such control over private forest lands . . . as may be necessary to insure the continuous production of forest crops . . . and to place forest industries on a stable basis in harmony with public interests . . . [with] transfer of control back to the forest industries as soon as they become able and willing to assume responsibility for respecting the public interests.
9. The National, State, and community forests should be maintained and largely increased.

The plan itself was most comprehensive. With respect to the central issue of regulation, it recommended the creation of a Federal commission with authority "to fix standards and promulgate rules to prevent the devastation and provide for the perpetuation of forest growth and the production of forest crops on privately owned timberlands operated for commercial purposes," except upon farm wood lots and other areas which might be exempted with safety to the public interests. The commission would also be authorized to control timber production whenever necessary in the public good in times of economic stress and to sanction the cooperative combination of lumber manufacturers for all purposes resulting in economies in production and marketing whenever satisfied that such action would promote the public interest.

Other proposals dealt with the acquisition of additional Federal and state forest lands; with the conduct of logging operations by the Secretary of Agriculture; with forest-fire control, forest taxation, forest insurance, and forest credit; and with the organization of regional and national councils of lumber employers and employees.

"The Lines Are Drawn." Pinchot, in a personal statement accompanying the report, expressed himself with his usual vigor:

The fight to conserve the forest resources of our public domain has been won. . . .

Another and a bigger fight has now begun, with a far greater issue at stake. I use the word fight, because I mean precisely that. Forest devastation will not be stopped through persuasion, a method which has been thoroughly tried out for the past twenty years and has failed utterly. Since otherwise they will not do so, private owners of forest land must now be compelled to manage their properties in harmony with the public good. Pressure from without, in the form of public sentiment, crystallized in compulsory nation-wide legislation, is the only method that promises adequate results. To apply this method successfully means to fight. . . .

The issue is real and immediate because forest devastation increases with appalling rapidity; because the need for governmental control on private timberlands is now self-evident; because without such control the general practice of forestry in this country will never become a reality; and because unless enough forestry is practiced to prevent forest devastation the danger to our prosperity in peace and safety in war will grow steadily worse. The field is cleared for action and the lines are plainly drawn. He who is not for forestry is against it. The choice lies between the convenience of the lumbermen and the public good.

Federal versus State Regulation. Graves' speeches and the report of the Committee for the Application of Forestry focused attention sharply on two main issues:

1. Is public regulation of cutting on privately owned forest lands essential to the public interest?

2. If so, should such regulation be exercised by the Federal government or the state governments?

While there was by no means complete acceptance of the accuracy of the alarming picture painted by the committee, there was general agreement that the situation was sufficiently disturbing so that something ought to be done about it, and done promptly. Although most timberland owners and operators would doubtless have preferred to limit either Federal or state action to cooperation, their spokesmen rather generally supported state regulation as a means of avoiding the greater evil of Federal regulation. The chief arguments against the latter were that it was un-American and unconstitutional and that it would lead to endless jurisdictional conflicts between the Federal government and the states, whose right to regulate cutting on private land under their police powers was seldom questioned.

The report of the Committee for the Application of Forestry was submitted to the entire membership of the Society of American Foresters for an itemized referendum ballot, in which 166 members participated. All of the fourteen "principles" underlying the committee's proposed plan were approved by substantial majorities, with the closest vote on the principle of national control. All but a minor one of the twenty-four proposals in the suggested legislation were also approved, although for the most part by less substantial majorities. The controversial proposals to authorize the commission to sanction cooperative combination of lumber manufacturers under certain conditions and to authorize the Secretary of Agriculture to conduct logging operations on national forests carried by large majorities, while the proposal to authorize the commission to control production in times of economic stress was approved by a much closer vote.

At the suggestion of the Forest Service, the following question was added to the ballot to determine whether there was any general agreement on the basic principle of public control: "Regardless of method or machinery, do you favor the prevention of forest devastation on privately owned commercial timber lands by the enactment and enforcement of effective and fair legislation?" The answer was overwhelmingly "Yes." Considerable dissatisfaction was, however, expressed that the ballot had not offered a clear-cut choice between Federal and state control. The society at its annual meeting in December, 1920, therefore authorized another ballot on these opposing principles as exemplified in two bills then before Congress. The result showed 195 members in favor of state control and 109 members in favor of Federal control.

Position of Forest Service. The official position of the Forest Service was that the problem is a national one in which Federal leadership is necessary but in which regulatory measures should be handled by the states under their police powers. As Graves said in an article in the *Journal of Forestry* for December, 1919:

The States have not only the function of handling the public forests owned by them, but they have also a direct responsibility in the protection and continuance of private forests. In this, the Federal Government should take part to meet interstate and national problems, to stimulate action by the States, and to bring into harmony the efforts of the different States. In the problem of private forestry, the Government would work through and in cooperation with the States. The legislation affecting the private owner in the matter of protection and continuance of forests should be by the States. The Government should help the States in formulating plans and developing methods and should give direct assistance in carrying them out. The assistance offered by the Government should be contingent upon the States taking legislative and administrative action to provide for the protection and renewal of their forests.

William B. Greeley, who became Chief Forester on April 15, 1920, following Graves' resignation, in his annual report for 1920 stated that the Forest Service program was

based on the conviction that the problem of halting forest devastation is fundamentally a national, not a local, problem, and must be faced and handled as such. At the same time it is felt that the speediest, surest, and most equitable action can be secured through dependence on the police powers of the States for the enforcement of such reasonable requirements as should be made of private owners and on the State governments for providing organized protection of private lands against fire. Because the problem itself is essentially national—that is, one affecting the public welfare of the entire country and requiring to be attacked as a whole, not piecemeal—both Federal leadership and a large measure of Federal aid are obligatory. It should be obligatory upon private owners to apply the safeguards necessary to prevent devastation. There is practical unanimity of agreement that the first and most essential step is nation-wide protection from forest fires, applicable to all classes of forest land and borne jointly by the landowner and the public.

Pinchot and Olmsted continued adamant in their insistence on direct Federal control.

National Forestry Program Committee. It should be emphasized that none of the proposals for regulation of cutting on private lands limited themselves to that subject. The urgent need for improved forest practice on these lands led to the suggestion of public regulation as one of the means of attaining this end, and the highly controversial character of the proposal kept the limelight focused on it; but Graves, Greeley, Olmsted, and Pinchot all recognized the importance of taking other steps to maintain and increase production on all forest lands, public and private alike.

Extension of publicly owned forests of all classes, strengthening of both state and Federal cooperation with private owners in fire control and forest management, improving current methods of forest taxation, providing forest insurance and forest credit on reasonable terms, and expanding research by public agencies were included in practically all proposed programs as essential items in any really adequate policy of forestry for the nation. In fact, as discussion proceeded and the prospect of obtaining any agreement among the advocates of different types of regulation became increasingly dim, more and more attention was paid to these other items.

Leadership in securing the adoption and enactment of a genuinely comprehensive Federal forestry policy was assumed by the National Forestry Program Committee. The committee was organized at meetings held on October 15 and December 8, 1920, "To coordinate the various plans and measures proposed to provide an adequate and permanent timber supply for the people of the United States, and to promote national legislation to this end." It consisted initially of representatives of the American Forestry Association, American Newspaper Publishers Association, American Paper and Pulp Association, Association of Wood-Using Industries, National Lumber Manufacturers Association, National Wholesale Lumber Dealers Association, Newsprint Service Bureau, Society for the Protection of New Hampshire Forests, and Western Forestry and Conservation Association.

Royal S. Kellogg, secretary of the Newsprint Service Bureau, who called the first meeting of the group, served as chairman of the committee during the eight years of its existence. Close cooperation was maintained with the Chamber of Commerce of the United States and with the Forest Service.

Snell Bills and Clarke Bills. One of the first activities of the committee was to prepare an all-inclusive bill covering every phase of the Federal government's interest in forestry. One section authorized Federal financial assistance to states passing legislation which required private owners to handle their forest lands so as to keep them continuously productive. The bill was introduced in Congress by Representative Bertrand H. Snell of New York on December 22, 1920.

Hearings on the bill were held during January, 1921. It was strongly endorsed by witnesses from both public and private agencies, including the provision for Federal encouragement of state regulatory legislation. As one representative of industry (E. T. Allen) said:

We believe, when all the facts are known and the maximum of voluntary participation is inspired, it will be found that our forest problem is in the main solved. But if any regulatory steps are necessary to assure equitable participation by lumbermen in this movement, then we must expect them to be taken. . . . We are for this forestry movement notwithstanding that we realize fully it may to some extent restrict individual independence for the common good.

All we ask is that it be worked out in the most practical way, with all concerned sharing equally without discrimination, and with our industry treated as a cooperating ally and not as an enemy.

The one outspoken opponent of the bill was Gifford Pinchot. He was certain that state legislation would never be effective because in the heavily timbered states the legislatures would be controlled by the timberland owners and lumbermen. He also feared that passage of this part of the bill would do more harm than good by giving the public the impression that nothing more was necessary, when actually the Federal government alone was big enough and strong enough to take effective action.

Congress adjourned on March 4, 1921, without acting on the bill, which was reintroduced by Representative Snell in the new Congress on April 11. A four-day hearing on this second Snell bill was held before the House Committee on Agriculture in January, 1922. It was virtually a repetition of the previous hearing on a somewhat larger scale, with the same line-up of forces and the same arguments pro and con.

When it became clear that the struggle between the state regulationists and the Federal regulationists was likely to result in a stalemate, Greeley recommended that that feature of the bill be omitted and attention concentrated on obtaining favorable action on the remainder of the bill, on which there was practically unanimous agreement. This was done, and in February, 1923, Representative John D. Clarke of New York introduced two bills which contained the substance of the Snell bill exclusive of the item relating to regulation. No action was taken on these bills, but they served to give the new proposal publicity both in Congress and elsewhere.

Capper Bills. Meanwhile other developments had been taking place. On May 20, 1920, Senator Arthur Capper of Kansas introduced in Congress a bill which incorporated practically all of the recommendations of the Committee for the Application of Forestry, including the creation of a Federal Forest Commission with direct administrative control over cutting on privately owned forest lands. It was of course supported by Pinchot and by the National Conservation Association of which he was president and was opposed by the Forest Service and the National Forestry Program Committee.

Since the proposed control was based on the government's power to regulate interstate commerce, serious question as to its constitutionality was raised by the Supreme Court decision of June 3, 1918, invalidating the child labor law of 1916 on the ground that it violated the Tenth Amendment by invading powers reserved to the states. In this decision, which appeared to be fully as applicable to forest products as to other commodities, the Court stated:

The making of goods and the mining of coal are not commerce, nor does the fact that these things are to be afterwards shipped or used in interstate commerce, make their production a part thereof. Over interstate transportation, or its incidents, the regulatory power of Congress is ample, but the production of articles, intended for interstate commerce, is a matter of local regulation.

Consequently the second Capper bill, introduced on May 2, 1921, sought to exercise Federal control by imposing an "excise tax on the privilege or franchise of conducting the business of harvesting forest crops." The tax was to be 5 cents per M board feet for all timber cut in accordance with the regulations of the Federal Forest Commission and $5 per M for all timber cut in violation of those regulations.

This approach also appeared to be headed for difficulty in the courts when the Supreme Court on May 15, 1922, declared unconstitutional the second child labor act of 1919, which was based on exactly the same principle. In this case, the Court said:

Grant the validity of this law, and all that Congress would need to do, hereafter, in seeking to take over to its control any one of the great number of subjects of public interest, jurisdiction of which the States have never parted with, and which are reserved to them by the Tenth Amendment, would be to enact a detailed measure of complete regulation of the subject and enforce it by a so-called tax on departures from it. To give such magic to the word "tax" would be to break down all constitutional limitation of the powers of Congress and completely wipe out the sovereignty of the States.

In an attempt to meet this difficulty the third Capper bill, introduced on February 16, 1924, imposed an excise tax of $5 per M board feet on all timber cut on private lands, but provided a bounty of $4.95 for all timber cut in accordance with the regulations of the Commission. This bill also provided for an alternative arrangement by which the states would operate under the cooperative plan. It was again introduced in the next Congress on December 18, 1925.

No hearings were held on any of the Capper bills, largely no doubt because of the evident preponderance of opinion in favor of state regulation.

Capper Report. Another move by Senator Capper was to get Senate approval in February, 1920, of a request to the Secretary of Agriculture to submit information on *Timber Depletion, Lumber Exports, and Concentration of Timber Ownership*. A report bearing this title, commonly known as the *Capper Report*, was prepared by the Forest Service and transmitted to the Senate on June 1, 1920. In brief, its findings were that:

1. Timber depletion had already gone far and was steadily progressing.

2. Lumber prices had recently risen to unprecedented heights, with timber depletion an important contributing cause.

3. Lumber exports had no serious influence on timber depletion except for high-grade timber, particularly of hardwoods.

4. Concentration of timber ownership had not yet approached monopolistic conditions, but might do so in the future.

Throughout the report emphasis was placed on the extent and seriousness of forest depletion, the remedy for which, Secretary Meredith stated, lay in "a national policy of reforestation. Increased and widely distributed production of wood is the most effective attack on excessive prices and monopolistic tendencies. Depletion has not resulted from the use of forests but from their devastation, from our failure, while drawing upon our reservoirs of virgin timber, to use also our timber-growing land."

Among the many measures recommended to stop forest devastation and to restore idle land to timber production, the Secretary laid special stress on "The immediate urgency of legislation (1) which will permit effective cooperation between the Federal Government and the several states in preventing forest fires and growing timber on cut-over lands, and (2) which will greatly extend the National Forests."

Other Forest Service Reports. The *Capper Report* was merely the first of a series of publications by the Forest Service dealing with the subject of forest depletion and what to do about it. The *Yearbook of the Department of Agriculture* for 1922 contained a lengthy article on "Timber: Mine or Crop?" while that for 1923 was devoted largely to "The Utilization of Our Lands for Crops, Pasture, and Forests," in which current and desirable utilization of land for forests came in for extended treatment. In 1924 a bulletin entitled *How the United States Can Meet Its Present and Future Pulpwood Requirements* emphasized improved forest management and consequent increase in timber production as the answer to that problem.

While the discussion of public regulation was at its height, the Forest Service decided that the proposal could be clarified and support for it obtained by explicit statements as to the "minimum requirements" likely to be imposed by such regulation. This idea resulted in the publication between 1926 and 1935 of twelve bulletins dealing with *Timber Growing and Logging Practice* in the twelve regions covered. Each bulletin presented for its region the minimum measures regarded as necessary to prevent timber-bearing lands from becoming barren and also the more intensive measures regarded as constituting desirable forestry practice. How much immediate effect on forest practice the bulletins had there is no way of knowing, but they served a useful purpose in arousing interest and in supplying important basic information.

Cause for Concern. The various analyses of the situation made by the Forest Service and the Society of American Foresters indicated clearly why it gave cause for so much concern. They showed that the process of conquering the continent—of building its houses, producing its food, and developing its industries—had made heavy drafts on its forest resources. The *Capper Report*, for example, estimated that as of 1920 the area of commercial forest land had shrunk from 822 million to 463 million acres; that the stand of sawtimber had been reduced from 5,200 billion to 2,215 billion board feet; and that the annual cut and destruction of timber of all sizes was 4.3 times the growth.

The forest resources of one region after another had been depleted until 51 per cent of the total remaining stand of sawtimber and 65 per cent of the softwood stand were on the Pacific Coast. Decreasing supplies and increasing costs of transportation had contributed materially to the rising price of lumber. Destructive methods of cutting and the fires that commonly followed logging had left some 81 million acres with practically no forest growth. The transfer of the great bulk of the more productive forest land from public to private ownership, followed by a rather general cut-out-and-get-out policy on the part of the private owners, had resulted in an instability of forest ownership that seriously interfered with planned management for sustained production.

Regardless of whether public policy, private policy, or irresistible economic, social, and psychological forces were responsible for the situation, something obviously needed to be done about it if the American people were to continue to enjoy an adequate supply of wood at reasonable prices. Realization of the problem was accompanied by acceptance of the fact that both private owners and public agencies must participate in its solution.

General Exchange Act. The first legislative achievement of the resulting campaign for an expanded program of forestry was passage of the General Exchange Act of March 20, 1922. After repeal in 1905 of the lieu-land provision of the Forest Reserve Act of 1897, the only way in which exchange of private and public lands within national forests could be effected was by specific act of Congress. In his annual report for 1920 Greeley expressed dissatisfaction with this piecemeal method of meeting a problem that was found on every forest. "The importance and necessity of general legislation whereby our National Forests may be properly completed and adequately protected through consolidation by exchanges with private owners cannot be too strongly urged."

Perhaps because Congress was already familiar with the problem through the many proposals for exchange brought to it for consideration, it acted promptly on this recommendation, which also had the support of the National Forestry Program Committee. The result was approval of an act which authorized the Secretary of the Interior, at the request of the Secretary of Agriculture, to exchange surveyed, nonmineral national forest land or national forest timber for any privately owned or state lands of equal value within the exterior boundaries of national forests in the same state. The values in each case were to be determined by the Secretary of Agriculture. All timber included in such exchanges must be cut and removed under the laws and regulations relating to national forests.

The original act was amended in 1925 to permit either party to an exchange to make reservations of timber, minerals, or easements, the value of which was to be duly considered in determining the value of the exchanged lands. In the same year the provisions of the act were extended to national forests acquired by purchase, subject to the approval of the National Forest Reservation Commission.

The law has proved of much value in permitting the consolidation of Federal holdings in national forests. Its administration has, however, been criticized with the charge that private owners are sometimes subjected to undue pressure to persuade them to exchange land for national-forest timber. In practice it has proved sufficiently flexible to permit receipts from the sale of national-forest timber to one party to be used in acquiring land from a third party. When this is done, approval of the transaction by the counties concerned is obtained, since income from such sales does not go into the United States Treasury, and the counties therefore do not get their usual 25 per cent of all receipts.

Although the law superficially resembled the old lieu-land law, it differed from it radically in three important respects. All exchanges were placed on the basis of equal value, rather than equal area; they must be approved both by the Secretary of Agriculture and the private or state owner, instead of leaving the decision entirely to the latter; and either Federal land or timber could be exchanged for private or state land, whereas formerly the exchange always had to be on a land-for-land basis. This last provision made the law a vehicle for increasing the net area of national forests, which is desirable or not depending on one's point of view.

Senate Committee on Reforestation. The next step forward was the adoption by the Senate on January 21, 1923, of a resolution providing for the appointment of a Select Committee on Reforestation "to investigate problems relating to reforestation, with a view to establishing a comprehensive national policy for lands chiefly suited for timber production in order to insure a perpetual supply of timber for the use and necessities of citizens of the United States."

The Committee, of which Senator Charles L. McNary of Oregon served as chairman, held twenty-four hearings in sixteen states and the District of Columbia. Its report made many specific recommendations, all aimed primarily at the extension of public forest ownership and the encouragement of private reforestation. Even before it was submitted to the Senate on January 10, 1924, bills incorporating its proposals had been introduced in the Senate by Senator McNary and in the House by Representative Clarke. These bills resulted in the Clarke-McNary Act of June 7, 1924, one of four measures passed between 1922 and 1928 (to be discussed later), which together comprised a well-rounded forest policy.

For this achievement much of the credit must go to the National Forestry Program Committee. That committee was also influential in 1921 in opposing a bill introduced by Senator King of Utah which would have returned the national forests to the Department of the Interior and in getting the Chamber of Commerce of the United States to appoint a Committee on National Forestry Policy. The committee of the chamber held hearings throughout the country, and the following year submitted a report which led in October, 1923, to a national referendum by the chamber. In essence, the referendum resulted in approval of the general principles embodied in the Snell bill.

Clarke-McNary Act. The Clarke-McNary Act of June 17, 1924, is a lineal descendant of the original Snell bill. It differed from that bill, however, in several important respects, notably in the omission of any reference to public regulation of timber cutting and to forest research. The latter omission was not due to any lack of recognition of the importance of research, but rather to the belief that the subject could be handled more effectively in a separate bill. The many provisions of the act, and their significance, are summarized below.

Cooperation in Fire Control. The act authorized the Secretary of Agriculture to recommend for each forest region of the United States such systems of forest-fire prevention and suppression as would adequately protect the timbered and cutover lands therein, with a view to the protection of timber on lands chiefly suitable therefor. This made possible the establishment of standards by which the effectiveness of fire-control activities under the widely varying conditions in different parts of the country could be fairly appraised.

It authorized the Secretary of Agriculture to cooperate in forest-fire control with states in which the system and practice of forest-fire prevention and suppression substantially promote the objects indicated. In no case, except for preliminary investigation, was the amount spent by the government in any fiscal year to exceed that spent by the state for the same purpose, including the expenditures of forest owners or operators which are required by state law or which are made in pursuance of the forest-protection system of the state under state supervision. Due consideration must be given to the protection of watersheds of navigable streams, but the government's cooperation could be extended to any timbered or forest-producing lands. These provisions permitted the government to limit its cooperation to states actually having reasonably effective systems of forest-fire protection and also to extend such cooperation to virtually all forest lands, regardless of their relation to navigation.

It authorized an annual appropriation of not more than $2,500,000 to enable the Secretary of Agriculture to cooperate with the states in fire-control activities, with permission to use such part of the appropriations actually made as he thought advisable for studies of forest taxation and forest insurance. The purpose of these studies was to devise "tax laws designed to encourage the conservation and growing of timber, and to investigate and promote practical methods of insuring standing timber on growing forests from losses by fire and other causes." Thoroughgoing studies in both fields resulted in publications analyzing the existing situation and making recommendations for its improvement. Of even more immediate practical importance was the sizable authorization for cooperative forest-fire protection, which permitted Federal participation in a much more effective way than had previously been possible with the relatively small appropriations made available under the Weeks Law.

Cooperation in Reforestation and Management. The act authorized an annual appropriation of not more than $100,000 to enable the Secretary of Agriculture to cooperate with the various states in providing nursery stock for the establishment of windbreaks, shelterbelts, and farm wood lots upon denuded or nonforested lands, on not to exceed a 50–50 basis. Although the authorized appropriation was small, it was highly significant in extending Federal cooperation with the states into a new and important field.

Another authorization was for an annual appropriation of not more than $100,000 to enable the Secretary of Agriculture to cooperate with the various states or, in his discretion, with other suitable agencies, in bringing educational assistance and technical advice to the owners of farms in establishing, improving, and renewing wood lots, shelterbelts, windbreaks, and other valuable forest growth, on not to exceed a 50–50 basis. This provision extended Federal cooperation to still another new field, perhaps of even greater importance than assistance in the growing of planting stock.

Liberalization of Acquisition Program. The act amended the Weeks Law of March 1, 1911, by authorizing and directing the Secretary of

Agriculture to recommend for purchase such forested, cutover, or denuded lands within the watersheds of navigable streams as in his judgment may be necessary for the regulation of the flow of navigable streams or for the production of timber. The National Forest Reservation Commission could approve of purchases only after the receipt of reports by the Secretary of Agriculture and the Director of the Geological Survey showing that the control of such lands by the Federal government would promote or protect the navigation of streams, or by the Secretary of Agriculture showing that such control would promote the production of timber thereon.

These provisions are among the most important in the act. They removed the restriction in the Weeks Law that purchases must be confined to the "headwaters" of navigable streams, and they permitted the purchase of lands for timber production as well as stream-flow protection. Although the proviso that purchases must be "within the watersheds of navigable streams" was retained, presumably as a constitutional safeguard, Congress had evidently become convinced that such purchases could be justified not only by its power to regulate interstate commerce but also by its power "to lay and collect taxes . . . to provide for the common defense and general welfare of the United States." That either act will hereafter be challenged on the ground of unconstitutionality seems most unlikely.

Extension of National Forests. The act authorized the Secretary of Agriculture to accept gifts of forest lands to be added to the national forests, subject to such reservation by the donor of the present stand of merchantable timber or of mineral or other rights, for a period not exceeding twenty years, as the Secretary may find to be reasonable, provided that all reservations of property rights, easements, and benefits shall be subject to taxation under the laws of the states in which lands are located. The total acreage involved in gifts has been small, but some of them have been of considerable value, particularly for research purposes.

It authorized the Secretary of Agriculture to ascertain the location of public lands chiefly valuable for stream-flow protection or for timber production which could be economically administered as parts of national forests and to report his findings to the National Forest Reservation Commission, which in turn, if it agreed with the Secretary's conclusions, was to present its findings to Congress through the President. The Secretary located a considerable area of unreserved public land which he believed could advantageously be added to the national forests, but because of opposition by the Secretary of the Interior, no action was taken on his recommendations.

Finally, the act authorized the President to establish as national forests any lands within government reservations other than national parks, national monuments, reservations for phosphate and other mineral deposits or water-power purposes, and Indian reservations, which in the opinion of the Secretary of the Department administering the area and the Secretary of Agriculture are suitable for the production of timber, provided that military reservations thus established as national forests should remain subject to the unhampered use of the War or Navy Department for military purposes.

Within a year, thirteen military reservations in the Eastern United States totaling some 210,000 acres were established as national forests and five additional military reservations totaling some 144,000 acres were added as districts to adjoining national forests. The arrangement did not work out as satisfactorily as had been hoped, and by 1929 all but two of the first group of reservations had been removed from national-forest status. Military uses and timber production often proved to be incompatible.

Significance of Act. The Clarke-McNary Act justified Greeley's description of it as "the greatest step forward in forestry in America since the Weeks Act of 1911." It greatly strengthened that act in the fields of forest acquisition and cooperative forest-fire protection; it extended the principle of Federal financial cooperation with the states to the important fields of reforestation and forest management; and it authorized studies of the difficult problems of forest taxation and forest insurance. Supplementary legislation was, however, needed for the support of other research, and the appropriations authorized had to be actually made, and in due time materially increased, in order to carry out effectively the constructive policies embodied in the act.

McNary-Woodruff Act. The original Snell bill of 1920 had authorized an annual appropriation of 10 million dollars for five years for the purchase of lands for national forests. When the Clarke-McNary Act was approved four years later, it greatly liberalized the provisions of the Weeks Act with respect to such purposes but carried no authorization. In order to remedy this omission, a bill authorizing the appropriation of 40 million dollars over a ten-year period, in accordance with a proposal made by George D. Pratt, president of the American Forestry Association, was introduced in Congress on December 20, 1924. It was thought that the amount suggested would make possible the acquisition of that part of the 6 million acres in the southern Appalachian Mountains and White Mountains which had not yet been bought, and an additional 5 million acres in the Lake states and the pine region of the South.

This proposal had the vigorous support of both the American Forestry

Association and the National Forestry Program Committee. It was before Congress in one form or another until the passage, on April 30, 1928, of the McNary-Woodruff Act introduced by Senator McNary of Oregon and Representative Roy O. Woodruff of Michigan. That act was somewhat of a letdown in that it authorized the appropriation of 2 million dollars for the fiscal year 1929 and 3 million dollars each for the fiscal years 1930 and 1931—a total of only 8 million dollars. The act also provided that, except for the protection of the headwaters of navigable streams or the control and reduction of floods therein, not more than 1 million acres of land were to be purchased in any one state.

Actual appropriations amounted to only 1 million dollars for 1929 and 2 million dollars each for 1930 and 1931, or a total of 5 million dollars as against the hoped-for 40 million dollars. An additional special appropriation of 1 million dollars was made for the purchase of the so-called Waterville tract comprising some 23,000 acres of virgin spruce and other valuable timber in the White Mountains of New Hampshire. Even these amounts permitted a substantial stepping up of the acquisition program and aroused considerable interest in the extension of national forests in the East.

McSweeney-McNary Act. During 1925 and 1926 a thorough study of the current status of forest research was made by a special committee of the Washington Section of the Society of American Foresters headed by Earle H. Clapp, Chief of the Branch of Research in the Forest Service. The exhaustive report of the committee was published in 1926 by the American Tree Association for the Society of American Foresters. It served as the basis for a bill authorizing a comprehensive program of forest research introduced by Representative John R. McSweeney of Ohio on March 3, 1927, at the close of the Sixty-ninth Congress. The bill was printed and used for educational purposes during the following months by the National Forestry Program Committee, which had sponsored it at the request of the Society of American Foresters.

The bill was reintroduced in the next Congress by Representative McSweeney and Senator McNary in December, 1927, and was enacted into law within less than six months on May 22, 1928, a remarkable record for speed. The broad scope of the act is indicated by its instructions to the Secretary of Agriculture

to conduct such investigations, experiments, and tests as he may deem necessary . . . to determine, demonstrate, and promulgate the best methods of reforestation and of growing, managing, and utilizing timber, forage, and other forest products, of maintaining favorable conditions of water flow and the prevention of erosion, of protecting timber and other forest growth from fire, insects, disease, or other harmful agencies, of obtaining the fullest and most effective use of forest lands, and to determine and promulgate the economic considerations for the management of forest land and the utilization of forest products.

Research Authorizations. The act authorized an annual appropriation of not more than $3,375,000 for research in these fields, regardless of the bureau by which the work was conducted, during the next ten years. Thereafter it authorized such annual appropriations as might be necessary to carry out the provisions of the act. Of this amount, 1 million dollars was for the support of research at the regional forest experiment stations enumerated in the bill, and $1,050,000 for research at the Forest Products Laboratory.

In addition, the act authorized an annual appropriation of not more than $250,000 to enable the Secretary of Agriculture to cooperate with states, private owners, and other agencies "in making a comprehensive survey of the present and prospective requirements for timber and other forest products in the United States, and of timber supplies, including a determination of the present and potential productivity of forest land therein, and of such other facts as may be necessary in the determination of ways and means to balance the timber budget of the United States." No time limit was set for the completion of the survey, but the total appropriation was limited to 3 million dollars—a restriction which was later removed.

The McSweeney-McNary Act focused attention on the fundamental importance of forest research in providing the basis for effective forest management and utilization and laid a solid foundation for the enviable record of accomplishment subsequently made by the Forest Service and other bureaus in the Department of Agriculture. It deserves to be ranked as one of the few major pieces of legislation in the development of Federal forest policy.

End of National Forestry Program Committee. During the remainder of 1928 the National Forestry Program Committee participated actively, and with reasonable success, in efforts to obtain adequate appropriations to implement the authorizations in the three acts which it had sponsored —Clarke-McNary, McNary-Woodruff, and McSweeney-McNary. This was the last of its activities. As its chairman later said, "The committee never formally disbanded. It just ceased to function in 1929 when the particular undertaking for which it was set up was accomplished." Few organizations have accomplished as much in as short a time.

Knutson-Vandenberg Act. The Knutson-Vandenberg Act of June 9, 1930, was the outcome of an effort to speed up reforestation and improve silvicultural practice on the national forests. It set up a fiscal program for national-forest planting by authorizing an appropriation of not more than $250,000 for the fiscal year 1932, $300,000 for 1933, and $400,000 for 1934 and each year thereafter for the operation of nurseries and the establishment and maintenance of plantations.

It also authorized the Secretary of Agriculture to require a special deposit by purchasers of national-forest timber for such tree planting, seed sowing, or forest-improvement work on cutover areas as may be needed to improve the future stand of timber. This provision has permitted much desirable silvicultural work on cutover areas that would otherwise have been impossible. Any unused funds revert to the Treasury of the United States and not to the depositor.

Shipstead-Nolan Act. The Shipstead-Nolan Act of July 10, 1930, grew out of a strong movement to preserve the natural beauty of the innumerable lakes in northern Minnesota and so far as practicable to maintain the wilderness character of the region. It withdrew from entry all public land in parts of northern Cook, Lake, and St. Louis Counties, Minnesota; required the Forest Service to conserve for recreational use the natural beauty of the lakes in the Superior National Forest, including the reservation from cutting of all trees along their shore lines; and provided that there should be no further change in the natural or existing water level of any lake or stream without specific act of Congress.

Minnesota enacted a similar law applicable to state lands in the same region. These acts constituted the first steps toward the later development of roadless areas and the establishment in cooperation with Canada of the Quetico-Superior International Peace Memorial.

Recreation and the National Forests. They provide merely one illustration of the increasing emphasis being placed by the Forest Service on recreational use of the national forests. By the middle of the 1920's it had become clear that the Service must formulate a definite policy for the handling of recreational activities in order to take care of the ever-mounting number of visitors (more than 11 million in 1924), to meet public criticism of the current lack of policy, and to avoid the danger that large areas of national-forest land might be transformed into national parks where recreation was the major activity.

Recreation gradually came to be regarded not as a minor use of the national forests, a by-product, but as a major use demanding recognition and administrative attention similar to that accorded other major uses. The attitude was well expressed in a letter of instructions dated September 27, 1932, to regional foresters from Robert Y. Stuart, who had succeeded Greeley as Forester on May 1, 1928, following the latter's resignation to become secretary-manager of the West Coast Lumbermen's Association:

The importance of recreational use as a social force and influence must be recognized and its requirements must be met. Its potentialities as a service to the American people, as the basis for industry and commerce, as the foundation of the future economic life of many communities, are definite and beyond question. Its rank in National Forest activities will in large degree be a major one and in a limited degree a superior one. It will in many situations constitute a use of natural resources coordinate and occasionally paramount to their industrial conversion into commercial commodities, and as a recognized form of use of natural resources it deserves and should receive the same relative degree of technical attention and administrative planning that is now given to the other forms of utilization.

Protection of the water supply and production of timber still remain the only two purposes for which national forests can be legally created. Congress has, however, never objected to their use "for all proper and lawful purposes" that do not interfere with these primary objectives. Consequently a wide variety of other uses has always been permitted, among which grazing and recreation rank particularly high.

Under the policy promulgated by Stuart, an attempt has been made to meet the legitimate wants of recreationists of every variety—transient motorists, overnight campers, summer residents, hunters, fishermen, skiers, and lovers of the unspoiled wilderness—with due regard to their health, comfort, and safety, and in coordination with other uses of the national forests. The latter has not always been easy, since watershed management, timber management, wildlife management, and recreation management are sometimes not wholly compatible.

Central States Forestry Congress. The growing effort to get more forestry practiced in the woods found regional expression in the organization of the Central States Forestry Congress, which held its first meeting in Indianapolis in 1930. Eleven states were represented at the meeting. The constitution adopted at that time stated that the object of the congress was "to call public attention to the part forests play in the life of our communities and the need for protecting, improving, and extending the forests of the Ohio and Central Mississippi valleys so that they may best serve our people."

The congress met annually in a different state each year through 1936, when it quietly dissolved. Its resolutions stressed particularly the desirability of enlarging the area of forest land in public ownership and of centralizing all Federal activities dealing with organic natural resources in the Department of Agriculture. Although the congress served a useful purpose in providing a forum for the discussion of problems of common interest in a large and important region, it is doubtful whether it exercised any major influence on the development of forest policy.

Trouble on the Range. In 1916 the Secretary of Agriculture announced a plan to increase the very low fees then charged for grazing stock on national forests in order to bring them more nearly in line with the true value of the privilege. Substantial increases were made in 1917 and 1919 after consultation with representatives of the livestock industry. In the latter year, for the first time, a large number of permits were issued for a period of five years, with the understanding that grazing fees would not be increased during the five-year period.

These increases did not satisfy Congress. Consequently in 1920 an attempt was made to insert in the agricultural appropriations bill a rider requiring the Forest Service to impose charges equal to commercial rates, which were still substantially higher than those on national forests. Although the attempt failed, it forced the Forest Service to undertake an intensive study of the subject, with the expectation that any further increases which might be found justifiable would be put into effect in 1924. The study was made by C. E. Rachford, whose report showed that in general national-forest grazing fees would have to be raised some 60 or 70 per cent to equal what the stockmen were willing to pay for the use of similar land in private ownership.

It had first been planned to put the new schedule of fees based on the Rachford appraisal into effect with the grazing season of 1925. However, because of the severe depression which the livestock industry had been suffering since 1921, the Secretary of Agriculture announced that the new schedule would become effective in 1925 only in cases where it carried a reduction in fees and that no increases would be made prior to 1927. Then, if there were a substantial improvement in business conditions, the Secretary reserved the right to determine the fees to be charged. Certain changes in grazing regulations were put into effect beginning with 1925, the most important of which was an increase in the period for which permits might be issued from five to ten years.

Stockmen's Demands. All of these developments combined to stir up a violent controversy over the whole problem of grazing on the national forests. Although the stockmen admitted the fairness of Rachford's study and of the Secretary's actions, they wanted no increase in fees if it could possibly be avoided. That attitude is understandable in view of the fact that economic conditions had been so bad as to lead to a remission of grazing fees by Congress in the years 1921, 1925, and 1926.

The stockmen were obsessed by the idea that they must have even more security than that afforded by ten-year permits. Many were insistent on permanent easements for existing permittees. They wanted the right to appeal to the courts from administrative decisions, and they wanted allotments made on the basis of area rather than number of stock. Emotion as well as economics played a prominent part in the thinking of the more radical elements in the industry, who took the lead in pressing demands which the government had no choice but to oppose.

The uproar caused Secretary of Agriculture W. M. Jardine in 1925, at the suggestion of the Forest Service, to have an impartial review of the Rachford report made by Dan D. Casement, a prominent stockman of Manhattan, Kansas. It also led the Senate in the winter of 1925 to adopt a resolution authorizing its Committee on Public Lands and Surveys "to investigate all matters relating to the national forests and the public domain and their administration, including grazing lands in the forest reserves and other reservations of lands withdrawn from entry."

A subcommittee, under the chairmanship of Senator R. M. Stanfield of Oregon, concentrated its attention chiefly on grazing in the national forests. It held hearings throughout the West at which it not only permitted but encouraged the malcontents to air their grievances, real and imaginary, against the Forest Service. Incidentally, the Senator himself had a grievance since his stock-grazing permit had recently been canceled because of misrepresentation.

Grazing Policies under Fire. Even before the subcommittee completed its investigation, Senator Stanfield introduced a bill applying both to the national forests and to the unreserved public domain which included virtually all of the provisions recommended by the most radical of the stockmen. It was opposed by the Secretary of Agriculture and the Secretary of the Interior, who made some alternative suggestions that Senator Stanfield incorporated in the original bill as amendments. The chief items relating to national forests provided for legal recognition of grazing as a subordinate use; continued regulation of grazing along existing lines; charging of fees that are reasonable in view of all the circumstances affecting use of the range; issuance of ten-year permits having a contract status; preference to certain classes of present permittees, with opportunity for the admission of new qualified applicants; and establishment of local grazing boards to cooperate in administering the act and in deciding appeals from decisions of forest officers.

The attack on Forest Service grazing policies by certain members of the livestock industry and of Congress brought a prompt counterattack by forestry and conservation organizations, notably the American Forestry Association and the Society of American Foresters. The secretary of the former conducted a vigorous publicity campaign to defeat the demands of the stockmen, while the latter appointed a special committee to study and report on the matter.

Two members of the committee (H. H. Chapman and Henry Schmitz) opposed not only the legislation sought by the stockmen as presented in the original Stanfield bill, but also that proposed by the Department of Agriculture. They took the position that the Secretary of Agriculture already has complete authority to regulate grazing on the national forests under the act of 1897 and that legal recognition of grazing even as a secondary use is both unnecessary and highly dangerous. In their judgment, it would eventually give the stockmen the easements or servitudes which they were seeking and which once established would be as difficult to extinguish here as they have been in other countries.

The third member (A. W. Sampson) favored legislation along the lines recommended by the Department of Agriculture on the ground that it would give stockmen the additional security in the use of the range to which he felt they were entitled. The executive council of the Society of American Foresters unanimously supported the position taken by the majority of the committee and stood ready if necessary to oppose publicly the passage of any legislation dealing with grazing on the national forests.

Stalemate and Fee Adjustments. The upshot of the matter was that Congress took no action. When it became clear that the extreme demands of the stockmen would not be approved, they lost interest and made no real effort to push through the more moderate proposals made by the Secretary of Agriculture. On the other hand, the Department was not sufficiently eager for legislation on the subject to insist on favorable consideration of its recommendations in the face of strong popular and professional opposition to them. Subsequent attempts to obtain legislation on the subject have similarly failed of success (through 1955).

The controversy over grazing fees was settled in 1927 by administrative action by the Secretary of Agriculture, after personal conferences with representatives of the two national livestock associations. The settlement provided for the gradual establishment of a schedule of maximum and minimum fees based upon the range appraisals by the Forest Service and Casement's recommendations, with no further increases during the ten-year period beginning in 1935 unless there should be a material change in the factors on which fees are based. The average increase was about 38 per cent for cattle and about 55 per cent for sheep.

Grazing on the Unreserved Public Domain. The situation with respect to grazing on the unreserved public domain was far different from that on the national forests. On the latter it was already being effectively administered by the Secretary of Agriculture, with real question as to whether additional legislation might not only be unnecessary but undesirable. On the former there was no administration whatever, and the urgent need for legislation of some sort was unquestioned. The Secretary of Agriculture and the Secretary of the Interior were in full agreement as to the basic principles that should underlie a leasing system, adoption of which they repeatedly urged on Congress. Each felt, however, that jurisdiction over the lands should be placed in his Department, and this difference of opinion may have been in part responsible for the failure of Congress to take any action.

When Herbert Hoover became President and Ray Lyman Wilbur Secretary of the Interior in the spring of 1929, the condition of the public range lands was steadily deteriorating, with no prospect of improvement so far as Congress was concerned. In the belief that something had to be done, they decided to support the oft-made proposal that these lands be turned over to the states. Such a move, they felt, could not make the situation worse than it already was, and might improve it.

Secretary Wilbur opened up the subject in a speech at Boise, Idaho, on July 9, 1929:

The forests must be protected or harvested constructively, overgrazing must be stopped, and experts in plant life and water conservation must be our guides. . . . It seems to me that it is time for a new public-land policy which will include transferring to those States willing to accept the responsibility the control of the surface rights of all public lands not included in national parks or monuments or in the national forests. With sound State policies based on factual thinking it may eventually develop that it is wiser for the States to control even the present national forests.

A month later President Hoover followed up this suggestion in a letter read at a conference of the governors of the public-land states held at Salt Lake City, August 26 to 27, 1929:

It may be stated at once that our Western States have long since passed from their swaddling clothes and are today more competent to manage much of these affairs than is the Federal Government. Moreover, we must seek every opportunity to retard the expansion of Federal bureaucracy and to place our communities in control of their own destinies. . . . It is not proposed to transfer forest, park, Indian, and other existing reservations which have a distinctly national as well as local importance. Inasmuch as the royalties from mineral rights revert to the Western States, either direct or through the reclamation fund, their reservation to the federal control is not of the nature of a deprival.

Committee on Conservation and Administration of the Public Domain. The President then went on to propose the appointment of a commission to study the whole problem of the disposal of the remaining unreserved public lands. This proposal was approved by Congress on April 10, 1930, in an act authorizing the President to appoint a Committee on the Conservation and Administration of the Public Domain. The Committee consisted of twenty members, with James R. Garfield, Secretary of the Interior under President Roosevelt, as chairman, and with the Secretaries of the Interior and of Agriculture as ex officio members. W. B. Greeley was the only forester on the Committee.

Probably to the surprise of President Hoover and Secretary Wilbur, public reaction to the proposed grant of range lands to the states was far from enthusiastic even in the West. Opposition from the Eastern states was to be expected because of their equity in the lands as a national heritage, and from conservationists generally because of their fear that state ownership would lead to private ownership and would in the long run result in even more serious deterioration than was taking place under Federal ownership. This fear was intensified by Secretary Wilbur's hint that the same treatment might later be applied to the national forests.

That the West should be inclined to look a gift horse in the mouth was hardly to be anticipated, particularly in view of past efforts to obtain state control of public lands. Opposition was based primarily on two grounds—that the lands were so run down as to be more of a liability than an asset and that mineral rights were not to be included in the grant. As Governor Dern of Utah put it, "The States already own, in their school-land grants, millions of acres of this same kind of land, which they can neither sell nor lease, and which is yielding no income. Why should they want more of this precious heritage of desert?" Still another objection was that the grant would materially lessen the substantial contributions received by the states from the Federal government for the building and maintenance of highways on the basis of the proportion of land in each state in Federal ownership.

Guiding Principles. In spite of the obvious reluctance of much of the West to accept a gift of doubtful value and with a heavy string attached to it (reservation of mineral rights), the Committee on Conservation and Administration of the Public Domain included the proposed grant in its recommendations, which were submitted to the President on January 16, 1931. These recommendations were based on the following general policies:

1. That all portions of the unreserved and unappropriated public domain should be placed under responsible administration or regulation for the conservation and beneficial use of its resources.

2. That additional areas important for national defense, reclamation purposes, reservoir sites, national forests, national parks, national monuments, and

migratory-bird refuges should be preserved by the Federal Government for these purposes.

3. That the remaining areas, which are chiefly valuable for the production of forage and can be effectively conserved and administered by the States containing them, should be granted to the States which will accept them.

4. That in States not accepting such a grant of the public domain responsible administration or regulation should be provided.

5. We recognize that the Nation is committed to a policy of conservation of certain mineral resources. We believe the States are conscious of the importance of such conservation, but there is a diversity of opinion regarding any program which has for its purpose the wise use of those resources. Such a program must of necessity be based upon such uniformity of Federal and State legislation and administration as will safeguard the accepted principles of conservation and the reclamation fund. When such a program is developed and accepted by any State or States concerned, those resources should be transferred to the State. This is not intended to modify or be in conflict with the accepted policy of the Federal Government relating to the reservation stated in conclusion No. 2 above.

Recommendations of the Committee. The Committee made twenty special recommendations based on these five general principles. First of all, it recommended that Congress pass an act granting to the public-land states all the unreserved, unappropriated public domain within their respective boundaries, subject to the reservation to the United States of all mineral rights. All grants should be impressed with a trust for the administration and rehabilitation of the lands concerned. Unreserved lands in any state which does not by legislation within ten years accept the proposed grant should be organized by the government into "national ranges."

State boards should be created consisting of five members, one designated by the President of the United States, one by the Secretary of the Interior, one by the Secretary of Agriculture, and two by the state, to decide what lands should be added to or withdrawn from existing national forests and to recommend additions to or withdrawals from other reservations. The present ratio of Federal participation in the construction of Federal-aid highways should be continued for a period of ten years. Interstate compacts should be used as far as practicable in matters not requiring Federal intervention.

The two final recommendations deserve quotation in full:

It is the conclusion of the committee that as to agricultural and grazing lands, private ownership, except as to such areas as may be advisable or necessary for public use, should be the objective in the final use and disposition of the public domain.

In order to provide for a more effective administration of the public domain and the various reservations and areas now under the control of the Federal Government and to promote the conservation of natural resources, it is recommended that the Congress be asked to authorize the President to consolidate and coordinate the executive and administrative bureaus, agencies, and offices created for or concerned with the administration of the laws relating to the use and disposition of the public domain, the administration of the national reservations, and the conservation of natural resources.

Greeley Withholds Signature. Greeley neither signed the report of the committee nor submitted a minority report. He stated later that he believed

it would be sound policy to transfer to the states prepared to accept them most of the unreserved lands whose primary value is for grazing, for the following reasons:

1. This is apparently the only practical solution of the paramount necessity of placing such lands under some responsible custody and administration.

2. Most of the Western States are now administering large areas of similar land, scattered through the unreserved Public Domain; and as a matter of state development and progress, they might wisely be entrusted with the care and administration of the intermingled and surrounding public lands.

In spite of this general agreement with the major recommendation of the Committee, he withheld his signature because he believed that all public lands of evident importance for the conservation of stream flow or the control of destructive erosion, including range lands, should so far as practicable be placed within national forests, regardless of the nature of the cover. He also believed that the allocation of the public lands to various forms of disposal, through the medium of the proposed state boards, should take the form of a real classification, without time limit, and should be subject to confirmation by the President. Finally, he questioned the soundness of the vague recommendation concerning the possible future transfer of "certain mineral resources" to the states.

Congress Fails to Act. Bills embodying the recommendations of the Committee were promptly introduced in Congress and extensive hearings held on them. Widespread opposition was expressed for the reasons already indicated. Another bill which would have granted mineral rights as well as surface rights to the states was equally strongly opposed by those who shared Greeley's views that in the mineral field the national interest was paramount.

Although the Committee's activities resulted in no legislative or administrative action, they served a useful purpose in arousing national interest and in clarifying the issues involved. The discussions in and out of Congress made it evident that the states would not accept the unreserved range lands in their current deplorable condition without transfer also of the mineral rights, and that such transfer stood no chance of approval. The only alternative was for the Federal government to tackle the job of stopping further deterioration and promoting rehabilitation, which everyone agreed must be done. The way was prepared for the Taylor Grazing Act approved three years later.

Society of American Foresters Committee on Forest Policy. Prior to the appointment of the Committee for the Application of Forestry in 1919, the Society of American Foresters had confined its activities mainly to technical problems in the management and utilization of forests and forest products. The activities of that committee and the society's participation in the grazing controversy involved it inextricably in the consideration of social, economic, and political problems.

It was only natural therefore that in 1928 the society should appoint a committee to formulate a statement of the principles on which the profession believed that a comprehensive forest policy for the nation should rest. After three years of careful study under the chairmanship of Barrington Moore, the committee prepared such a statement, which in May, 1931, was approved by an overwhelming majority of the 839 members who voted on it.

The statement emphasized the fact that private owners have "an obligation to handle their forests in such a manner as to prevent the public injuries which follow destructive exploitation and failure in fire protection," but added that "the industrial and general welfare of the country requires a direct participation of the public in the protection, development, and continuance of the forests." It approved public assistance to private owners in the control of fire, insects, and disease; in reforestation and forest management; in research and educational activities; in tax reform; and in stabilization of the forest industries.

"The public has the responsibility to exercise such control over the exploitation of private forests as may be necessary to prevent injury to the community at large," but such control should be applied through local and regional boards on which forest owners are represented. Federal cooperation in the field of public control was to be limited to advice and financial cooperation and to the encouragement of interstate compacts. Private forests located in zones of major importance for watershed protection should be subject to such regulatory measures as the public interest requires, with the public bearing a larger share of protection costs and road building than on areas where less onerous regulatory measures are needed. A greatly enlarged system of public forests at the Federal, state, and local level was advocated.

Government Reorganization. In the fall of 1930, while the statement of society policy was still in preparation, the committee prepared a plan for reorganizing the conservation work of the Federal government which was approved by the Council of the Society and transmitted to President Hoover. This plan proposed to assign to the Department of Agriculture responsibility for all activities dealing with organic resources and to the Department of the Interior responsibility for all activities dealing with inorganic resources. Under this setup the Forest Service, the National Park Service, the Biological Survey, the Bureau of Fisheries, and administration of forest and range lands in the unreserved public domain would be in the Department of Agriculture.

On June 30, 1932, Congress authorized the President to reorganize the executive departments of the government in the interest of economy and efficiency by "grouping, coordinating, and consolidating executive and administrative agencies of the Government, as nearly as may be, according to major purposes." Five months later (December 9, 1932) President Hoover issued proclamations reorganizing the Department of Agriculture and the Department of the Interior along lines considerably different from those suggested by the society.

Both plans left the Forest Service in the Department of Agriculture, which was the main objective of the society's recommendation. President Hoover, however, transferred the General Land Office from Interior to Agriculture, left the National Park Service in Interior, and left the Bureau of Fisheries in Commerce. On January 19, 1933, the House of Representatives disapproved all of President Hoover's reorganization proposals, none of which therefore became effective. The foresters, however, had made clear their determined opposition to the removal of the Forest Service from the Department of Agriculture.

State Forestry Progresses Slowly. Passage of the Weeks Act greatly stimulated the organization of state departments of forestry and expansion of their activities in the field of fire control. In 1911 only eleven states cooperated with the Federal government under the provisions of the act. By 1924, when the Clarke-McNary Act was passed, the number had risen to twenty-eight, and by 1933 to thirty-eight. The area involved, however, did not increase as rapidly as the number of states; in 1933 it was estimated that only a little more than half of the forest area in private and state ownership needing protection from fire was actually under organized protection. Progress was slowest in the Southern states. Marked improvement in the efficiency of fire-control activities accompanied the increase of area covered.

Expansion in other state forestry activities was much less striking. For the most part, the period from 1911 to 1933 was still one of beginnings.

Fairly large percentage increases in appropriations, particularly for fire protection, were not uncommon, but these started from so low a figure that the amounts involved were far from impressive. Several states enlarged considerably the area in state forests, and the Clarke-McNary Act stimulated somewhat cooperation with private owners in reforestation and forest management. A few states. notably Michigan, Wisconsin, Oregon, and Washington, passed optional yield-tax laws, of which relatively few owners took advantage.

One highly significant development was the organization in 1920 of the Association of State Foresters. This was an outgrowth on a national scale of the Association of Eastern Foresters, which had been in operation since 1908. The new association served a most useful purpose not only in providing a clearinghouse for the exchange of views and information but also in promoting cooperation among the states and between the states and the Federal government. Its strength has grown with the years.

Private Forestry Handicapped by Economic Conditions. For private forestry, even more than for state forestry, the period was still one of beginnings. More and more owners were studying the practicability of improved forest practices on their own lands, but economic conditions were not yet ripe for putting thinking and talking into action on any extensive scale. Prospects for doing so were considerably lessened by the depression from which the lumber industry suffered during the 1920's, nearly a decade before other major industries except agriculture were seriously affected.

Hopes nevertheless ran high that private forestry was about to materialize in a big way. In 1924, Greeley expressed the belief that "commercial forces are now placing a powerful pressure behind the practice of forestry by private landowners." That autumn a well-attended National Wood Utilization Conference was held at Washington to stimulate efficiency in the use of forest products and was followed by establishment of a National Committee on Wood Utilization.

In 1925, Greeley pointed to "convincing evidence" that "the possibility of growing successive crops of timber on private land as a business has aroused the interest and is receiving the attention of the forest industries throughout the country." A year later he again called attention to the changing attitude of the forest industries and timberland owners and their willingness "to weigh carefully, as business men, the methods of forestry."

In 1927, however, Greeley reported:

There have been no marked changes during the year with respect to the growing of timber on privately owned lands, and there is little prospect of any sudden or striking change. Forest-land practices are undergoing a slow and gradual evolution, and some time must elapse before the results can be appraised in specific terms. . . .

Among the difficulties which stand in the way are the current depression in lumber markets, the financial obligations imposed by existing investments in plants or timberland, the uncertainty as to the future course of taxation in respect both to merchantable timber and young forest growth, and the burden of raw material wastes still carried by most forest industries of the United States.

Two years later Stuart confirmed this realistic appraisal. Even before the advent of the great depression, he stated:

Economic conditions have not reached the point at which the remaining stands are being sufficiently husbanded, with due regard to their unique quality and to future requirements. Beyond this, the pressure to liquidate holdings of virgin timber is operating as a powerful deterrent to the reorganization of our forest economy on the basis of timber growing in place of forest exploitation.

Economic conditions had indeed become so bad that President Hoover was asked to initiate a study of the facts in the hope that a national attack on the problem might relieve the industry's distress. A decidedly pessimistic view of the situation was taken by Lieut.-Col. George P. Ahern in two publications vividly entitled "Deforested America" (1928) and "Forest Bankruptcy in America" (1933), in which he pointed out state by state the small amount of forestry actually being practiced by private owners in spite of the considerable amount of attention being given the subject and optimistic reports in certain quarters as to the progress being made.

Conference on Commercial Forestry. On November 16 to 17, 1927, an important Conference on Commercial Forestry was arranged by the Chamber of Commerce of the United States to help promote interest in and the actual practice of forestry by private owners. It was held in Chicago under the auspices of the chamber's Natural Resources Production Department, which had been organized in 1921, and had recently conducted a study showing that 174 companies owning 21 million acres of forest land were "practicing some sort of forestry."

The conference itself expressed its gratitude to the Chamber of Commerce for having afforded "the first auspicious opportunity which the timber owners and forest industries of the country have ever had to voice their attitude toward a national problem that more directly affects and concerns them than any other group in the national community. As a rule, forestry conferences have been gatherings of non-owners intent upon telling the owners how to administer their properties."

The discussions brought out a mixture of optimism and pessimism.

Many specific instances of progress were cited, but there was general agreement that conversion from timber mining to timber cropping was not an easy process and that really satisfactory accomplishment in this direction lay in the future rather than in the past. Everett G. Griggs, president of the St. Paul & Tacoma Lumber Company, made the point that "the time is about past when commercial forestry has to be sold to the forest owner as a desirable thing. He believes it, or wants to. The time has come to sell it to the public that governs the conditions to make it possible. I believe we can do that, too. But it will take some effort and we must make an effort ourselves."

Conclusions of the Conference. The formal conclusions adopted by the conference indicate clearly its confidence, its concern, and above all its recognition of the need for public cooperation.

We are glad to report that despite adverse conditions, unparalleled in any other progressive country, an amazing advance in forestry has been made in the United States within the last 25 years and especially within the last 10 years. . . . But from all sides, come reports of a discouraging struggle against conditions, which can be rectified only by public cooperation, through full recognition of individual and collective public responsibility.

To continue reforestation, to expand it to meet national and community requirements, the following measures are necessary:

1. Adequate protection against forest fires.
2. Equitable and stable taxation.
3. Full technical and economic information through research.
4. Complete recognition by the people of state and federal responsibility.

An important feature of the conference was the submission of a report on the subject of forest-fire insurance by a special committee appointed by the president of the Chamber of Commerce of the United States. The committee was unique in that it consisted of five members representing timberland owners and five members representing stock and mutual forest-fire-insurance companies. The report aroused considerable interest but led to no concrete results.

Appointment of Timber Conservation Board. Just about a year after the Conference on Commercial Forestry, Wilson Compton recommended to the directors of the National Lumber Manufacturers Association that President-elect Hoover be asked to appoint a Timber Conservation Board for the purpose of making a comprehensive investigation of the acute problems confronting the lumber industry. Such a request was made in April, 1930, by a distinguished group representing various public and private agencies.

On November 12 of the same year President Hoover announced the appointment of a Timber Conservation Board of thirteen members, with Robert P. Lamont, Secretary of Commerce, as chairman. In essence, its purpose was to study "the economic problems and consequences of overproduction in the forest industries" and to propose "sound and workable programs of public and private effort, with a view to securing and maintaining an economic balance between production and consumption of forest products and to formulating and advancing a deliberate plan of forest conservation."

The Board organized on January 7, 1931, and appointed an advisory committee of twenty-two members, with R. Y. Stuart as chairman. The advisory committee promptly divided itself into seven subcommittees, which proceeded to make careful studies of the subjects assigned to them. One of the valuable outcomes of these studies was a report by the Forest Service on *The Forest Situation in the United States,* which brought up to date the information contained in the *Capper Report* of 1920. Another report of lasting value was one by David T. Mason on the place of sustained yield in forest management, with special reference to its economic advantages.

Activities of the Board. Early results of the Board's activities were an order by President Hoover on May 14, 1931, to curtail national-forest timber sales as an aid both to the commercial situation and to forest conservation, and publication on July 1 of the first of the reports, still (1955) being issued, by the Special Lumber Survey Committee appointed to analyze current lumber consumption. On June 10 to 11, 1931, the Board held an open hearing at which representatives of the wood-using industries made comprehensive and convincing statements of the critical situation in their respective industries.

Great stress was laid, particularly by the lumber industry, on the dangers of overproduction which had been chronic for the past fifteen years and which had been brought on by a large excess of plant capacity, heavy carrying charges on stumpage, declining lumber markets, and other factors beyond the industry's control. If this overproduction were allowed to continue, it threatened economic chaos. Among other ill effects, it tended to promote destructive lumbering and premature and wasteful exploitation of the forest; it had a deadening effect on land values and the practice of forestry; it added to insecurity of employment; and it was driving forest land from the tax rolls.

Remedies suggested included withdrawal from sale of government timber, establishment of regional production quotas, sanctioning, under proper public control, of reasonable trade agreements to control production, protection against unfair competition of imported lumber, and frank recognition of the fact that no timber famine is in sight. Much emphasis was placed on the burden of carrying the immense amount of standing timber which the government had allowed private owners to acquire. One method suggested for reducing the burden was to permit private owners to donate their timberlands to the government, with retention of cutting rights, and heavy purchases of timberlands by the government were almost universally approved.

Many of the claims made by the timber industry were borne out by the findings of the subcommittee on privately owned timber, logging, manufacturing plants, and distributing facilities. The report of that subcommittee showed that the industry had been in a depressed condition even before the depression. This was indicated by the large percentage of concerns reporting no net income, by the low ratio of net income to gross income and to capital stock and surplus, and by the low ratio of current assets to current liabilities. Although bonded and other indebtedness was moderate, the industry was carrying a load of mature timber and of nonrestocking land; it had a mill capacity far in excess of the amount that it had ever been possible to utilize even in the most active years; and its method of handling its forest properties necessitated a heavy depletion charge.

These items, the subcommittee stated, involved capital and carrying charges that might readily become disastrous in periods of economic stress. They pointed to sustained-yield forest management as the basic remedy for the difficulties of the wood-using industries. Meanwhile, immediate relief must be brought about by other measures such as those urged by the industries.

Recommendations and Results. On June 18, 1932, the Board made its recommendations to the President. They included practically all of the items commonly proposed as major parts of any comprehensive national forest policy. Prompt application of sustained-yield management to all publicly owned or controlled forest land was urged, and study of its possibilities on privately owned lands, with consideration of legislation necessary to permit cooperative sustained-yield management.

"Consideration of control of production under competent Federal supervision" was recommended, as well as continued study of interstate compacts as "practicable and desirable means of advancing the cause of timber conservation, controlling timber cutting, and establishing and enforcing State production quotas." The expansion of Federal and state acquisition of timberlands was declared to be sound public policy.

Although it is difficult to trace either legislative or administrative action directly to recommendations made by the Timber Conservation Board, it performed a valuable service in analyzing the difficult economic and technical problems faced by industry and in promoting close cooperation between public and private agencies in the solution of those problems. The task was continued a year later by the group set up to carry out Article X of the Lumber Code promulgated under authority of the National Industrial Recovery Act.

National Conference on Land Utilization. Matters of interest to private timberland owners as well as to the general public were considered at a National Conference on Land Utilization called by the Secretary of Agriculture and the Association of Land Grant Colleges, which met at Chicago November 19 to 21, 1931. The preamble to the report of the Committee on Summaries and Conclusions began as follows: "Our Federal and State land policies have, in the main, encouraged the rapid transfer of public lands to private ownership with little regard given to the uses to which the land was best adapted or to the demand for its products. The economic and social difficulties in agriculture which are widely recognized at present, are in considerable degree traceable to the effects of these policies."

As a staff member of the National Lumber Manufacturers Association remarked, these sentences "could equally well be used for the introductory remarks by the Timber Conservation Board by substituting for the word 'Agriculture' the words 'Forest Industries,' particularly in its proposed consideration of the advisabilities and possibilities of extensive reacquisition by the federal government of surplus privately owned timber reserves in the West."

The recommendations adopted by the conference urged the organization of the unreserved forage lands in the public domain "into public ranges to be administered by a Federal agency in a manner similar to and in coordination with the national forests." The conference also emphatically recommended that Federal and state agencies develop a coordinated program of land utilization for the extensive areas of idle or misused agricultural and forest lands.

"We believe it to be a sound policy that before we undertake to retain or acquire land for public ownership every reasonable effort should be made to remove the conditions that discourage forms of private utilization not inconsistent with public welfare." Reform of taxation and consolidation of scattered holdings were among the adjustments suggested. At the same time the conference favored public acquisition of lands that could not be utilized profitably by private individuals.

All of the recommendations constituted sound doctrine so far as the forestry thinking of the day was concerned. The conference arranged for the appointment of two important continuing committees—a National Land Use Planning Committee and a National Advisory and Legislative Committee on Land Use.

Copeland Report. On March 27, 1933, the Secretary of Agriculture, Henry A. Wallace, transmitted to the Senate a report of more than 1,650 pages entitled *A National Plan for American Forestry*. It had been prepared in response to a resolution introduced the previous year by Senator Royal S. Copeland of New York and thus acquired the name *Copeland Report*, by which it is universally known.

The Department interpreted the resolution as a mandate to prepare a coordinated plan for the utilization of our forest land so as to make "all of its timber and other products and its watershed, recreational, and other services available in quantities adequate to meet national requirements." This it proceeded to do in the most complete and comprehensive report of its kind yet prepared. The character and extent of our forest resources, their past utilization, and future programs to assure the attainment of maximum economic and social benefits from them were all considered at length.

Findings and Recommendations. The main findings were summarized as follows:

1. That practically all of the major problems of American forestry center in, or have grown out of, private ownership.
2. That one of the major problems of public ownership is that of unmanaged public lands.
3. That there has been a serious lack of balance in constructive efforts to solve the forest problem as between private and public ownership and between the relatively poor and the relatively good land.
4. That the forest problem ranks as one of our major national problems.

The main recommendations, "as the only assured means of anything approaching a satisfactory solution of the forest problem," were for:

1. A large extension of public ownership of forest lands, and
2. More intensive management on all publicly owned lands.

The proposed extension of public ownership was to involve the acquisition by the states of an additional 90 million acres of land and by the Federal government of 134 million acres—a total of 224 million acres. Part of this area would be abandoned agricultural land, but most of it would be cutover and other forest land. The size of the proposed acquisition program can be better appreciated when it is remembered that at the time only 11 million acres of commercial forest land were in state ownership and only 88 million acres in Federal ownership. The prospects that the country would embark on so ambitious an enterprise were always dim and became increasingly so as recovery from the depression diverted attention from the "burden" of private ownership.

Although the body of the report dealt with the desirability of state and Federal regulation of cutting on private lands, that approach was not suggested in the Secretary of Agriculture's letter of transmittal as one of the major methods of maintaining the productivity of forest lands. The Secretary did recommend a substantial increase in public aid to private owners, in which the Federal government and the states should join forces. When F. A. Silcox succeeded Stuart as Chief Forester, following the untimely death of the latter in the fall of 1933, he revived with vigor the campaign for public control.

The *Copeland Report* was, and still is, an encyclopedia of valuable information. It contained a wealth of facts relating to the forest resources of the country, in the realms both of technique and policy, and presented programs of action in many fields that continue to be provocative of constructive discussion. It constituted an outstanding contribution to American forestry literature.

Summary. Shortly after the First World War the Forest Service and the Society of American Foresters proposed greatly enlarged programs of forestry for the nation, with special reference to lands in private ownership. The proposal for public regulation of cutting on private lands, which was included in both programs, led to much controversy as to whether such regulation should be exercised by the Federal government or the state governments.

The result was a stalemate on that issue. However, under the leadership of the National Forestry Program Committee, four major pieces of legislation were enacted—the General Exchange Act of 1922, the Clarke-McNary Act of 1924, the McNary-Woodruff Act of 1928, and the Mc-Sweeney-McNary Act of 1928. These acts strengthened the programs for the acquisition of national forests and for cooperation with the states in forest-fire control initiated by the Weeks Act of 1911; extended Federal cooperation with the states to the fields of reforestation and forest management; and established a sound and comprehensive program for forest research.

Vigorous controversy over management of range lands in national forests was precipitated by pressure on the part of Congress in 1920 for increased grazing fees and by subsequent demands on the part of the livestock industry for privileges that the government opposed. None of the numerous bills dealing with the subject were approved by Congress; but in 1925 the Secretary of Agriculture increased the period for which grazing permits might be issued from five to ten years, and in 1927 he approved a gradual increase in grazing fees.

Discussion of grazing on the national forests was accompanied by attempts to bring about management of range lands in the unreserved public domain, which were in deplorable condition as a result of long-continued overgrazing. After several bills to improve the situation failed of passage in Congress, Secretary of the Interior Wilbur proposed that the lands be turned over to the states. This action led to the appointment of a Committee on the Conservation and Administration of the Public Domain, which repeated Wilbur's proposal. The states, however, showed little enthusiasm for accepting a gift that might prove more of a liability than an asset, and nothing came of this or other proposals made by the committee.

State departments of forestry increased in number and in strength, but except in fire control their development was disappointing as compared with that in the Federal field. Private forestry made some progress in all parts of the country, but interest and planning exceeded accomplishment in the woods. Economic conditions were on the whole not favorable for the widespread adoption of improved forest practices by private owners. Toward the end of the period, the depression from which the lumber industry had been suffering during most of the 1920's became so acute that the President appointed a Timber Conservation Board to devise means of alleviating the economic and social ills resulting from overproduction, consequent industrial instability, and wasteful logging. In both state and private forestry the period was essentially one of preparation for greater and more effective activity in the years to come.

REFERENCES

Cameron, Jenks: "The Development of Governmental Forest Control in the United States," chap. 11, Johns Hopkins Press, Baltimore, 1928.

Committee on Conservation and Administration of the Public Domain (James R. Garfield, Chairman): "Report," Washington, 1931.

Graves, Henry S.: A Policy of Forestry for the Nation, *Jour. Forestry*, 17:901–910, 1919.

Greeley, William B.: "Forest Policy," chap. 15, McGraw-Hill Book Company, Inc., New York, 1953.

Hosmer, Ralph S.: The National Forestry Program Committee, 1919–1928, *Jour. Forestry*, 45:627–645, 1947.

Peffer, E. Louise: "The Closing of the Public Domain," chaps. 10–11, Stanford University Press, Stanford, Calif., 1951.

Pinchot, Gifford: The Lines Are Drawn, *Jour. Forestry*, 17:899–900, 1919.

Society of American Foresters, Committee for the Application of Forestry (Gifford Pinchot, Chairman): Forest Devastation—A National Danger and a Plan to Meet It, *Jour. Forestry*, 17:911–945, 1919.

U.S. Department of Agriculture, Forest Service: "Timber Depletion, Lumber Prices, Lumber Exports, and Concentration of Timber Ownership" ("Capper Report"), Report on S. Res. 311, 66th Cong., 2d Sess., 1920.

————: "A National Plan for American Forestry" ("Copeland Report"), 2 vols., S. Doc. 12, 73d Cong., 1st Sess., 1933.

CHAPTER 10

Conservation under the New Deal

The New Deal administration that took office on March 4, 1933, found the country in a state of widespread and acute economic distress. The banks were closed, unemployment was at an all-time high, agriculture and industry were faced by bankruptcy. In the lumber industry, for example, estimated production had shrunk from 39 billion board feet in 1929 to 14 billion feet in 1932, or a decrease of 64 per cent. Improved forest practices seemed beyond the reach even of timberland owners who would have liked to adopt them.

Another Roosevelt. Under these circumstances the new President, Franklin Delano Roosevelt, followed the lead of his cousin Theodore in making the conservation of natural resources one of the outstanding features of his administration. Many Presidents have given conservation their support, but only the two Roosevelts elevated it to a position of major importance. T.R. started the crusade for wise use of natural resources; F.D.R. revived it. In between the two was a period of steady but unspectacular progress.

Franklin Roosevelt's previous experience prepared him even better for leadership in the field than did that of his predecessor. Himself a forest owner, he had long practiced forestry on his estate at Hyde Park, New York, so that he had firsthand knowledge of the technical and economic problems involved in the art and business of forest management. How keenly this activity interested him is indicated by his own description of his occupation in "Who's Who in America" as a "tree grower."

As a youthful member of the New York Legislature he served as chairman of the Senate Committee on Forestry and consistently supported constructive measures in forestry and related fields. As governor of the state, he successfully championed a program for the purchase, reforestation, and management of nearly half a million acres of abandoned farm land. It is not surprising that as President his interest and enthusiasm

247

led him to include conservation of natural resources as an important part of every program fathered by his administration. During the worst depression and the worst war in the history of the country he made conservation one of the most vital issues of the day because he recognized clearly the indispensability of an abundant supply of natural resources both in peace and in war.

Emergency Conservation Work. The dramatic revival of conservation as a major national issue was foreshadowed in July 2, 1932, in Roosevelt's acceptance of the nomination for the Presidency.

Every European nation [he stated] has a definite land policy and has had one for generations. We have not; having none, we face the future of soil erosion and timber famine. It is clear that economic foresight and immediate employment march hand in hand in the call for the reforestation of these vast areas.

In spite of ridicule from political opponents and others, he continued during the campaign to advocate land improvement as a means of relieving unemployment, and in less than three weeks after his inauguration he asked Congress for legislative authority to proceed with the program. Ten days later the Emergency Conservation Act of March 31, 1933, had become a law. A record for speed!

The act authorized the President to employ unemployed citizens on works of a public nature "for the purpose of relieving the acute condition of widespread distress and unemployment now existing in the United States, and in order to provide for the restoration of the country's depleted natural resources and the advancement of an orderly program of useful public works." The program was to be conducted on Federal or state lands but could be extended to county, municipal, and private lands for the control of fires, insects, disease, and floods. Research in forest management and wood utilization was also authorized.

Civilian Conservation Corps. Duration of the act at first was limited to two years, after which it was continued by annual appropriations until 1937. The Civilian Conservation Corps (C.C.C.) was then formally established for a period of three years by the act of June 28, 1937. The purpose of the act was stated to be not only to provide employment in conservation of the natural resources of the country, but also to provide vocational training, to which ten hours per week might be devoted. Some educational work had been a part of the program from the beginning, but after 1937 it was systematized and expanded. Further continuations kept the Corps alive until June 30, 1943, when it was finally liquidated by Congress.

Enrollment was at first open to young men between the ages of eighteen and twenty-five who were unmarried, unemployed, and had dependents. Later the age limits were set at seventeen to twenty-three; enrollees had to be unemployed and in need of employment, but were not required to have dependents. Compensation was $30 per month, of which at first $25 and later $22 had to be assigned to dependents, if any, with slightly higher pay for a few in supervisory positions. Provision was also made for employment of a limited number of Indians, of veterans of the First World War, and of "local experienced men."

General supervision over the program was exercised by a director, Robert Fechner, with the assistance of several Federal Departments. The Department of Labor handled recruiting; the War Department operated the camps and ran the educational program; and the Department of Agriculture and the Department of the Interior directed the field activities of the men. In 1939 the Corps was placed in the Federal Security Agency.

Camps of 200 men each were located on Federal and state lands throughout the entire country, and to a less extent on private lands. The first camp, on the George Washington National Forest in Virginia, was occupied on April 5, 1933, and actual work in the woods started on April 17. When the program was at its peak in 1935 there were 520,000 enrollees in 2,652 camps, of which about half were forestry camps. Altogether the Corps gave employment to approximately 3 million men at a cost of some 2½ billion dollars.

Activities and Accomplishments. Never before had there been a comparable enterprise for the simultaneous building up of young men and of natural resources. No conceivable activity that would improve the latter was overlooked. The Director of the Corps reported more than 150 major lines of work that might be classed under the general headings of reforestation, forest protection and improvement, soil conservation, recreational developments, range rehabilitation, aid to wildlife, flood control, drainage, reclamation, and emergency rescue activities. In forestry alone, the Forest Service estimated that 730,000 man-years were devoted to such activities as reduction of fire hazard, construction of firebreaks, actual fire fighting, timber-stand improvement, tree planting, and the building of roads, trails, bridges, telephone lines, lookout towers, and other permanent improvements. Similar results were accomplished in other fields.

It is true that the quality of the work was not always topnotch, that some of it could be classed as "busy work," and that it was expensive. What else could be expected of an emergency program, directed largely by inexperienced supervisors and carried out by unskilled youths many of whom had never been in the woods before, the primary aim of which was relief of unemployment? The marvel is not that the program had its technical and financial shortcomings, but that these were not more serious. Taking everything into consideration, both the quantity and the quality of the work accomplished were impressive.

Even more impressive was the success of the program on the human side. Young men who had become penniless, hopeless, sometimes desperate, through no fault of their own but simply because there were no jobs to be had, were given an opportunity not only to support themselves through orderly and useful work but to help out their equally needy families. The results in physical, mental, and spiritual growth were amazing.

A highly important by-product of the C.C.C. was a great enlargement of the purchase program for the acquisition of national forests. In order to make available more Federal land in the Eastern United States on which the Corps could usefully pursue its activities, President Roosevelt in May, 1933, allocated 20 million dollars of emergency funds for the purchase of forest lands under the Weeks Act of 1911 and the Clarke-McNary Act of 1924. Subsequent allocations in 1934 and 1935 brought the total made available for land purchases to $44,534,500. That sum was 76 per cent greater than all of the appropriations for acquisition made by Congress from 1911 to 1932. It resulted in the establishment of nearly sixty new purchase areas and in the acquisition of 7,725,000 acres, or two and a half times as much as during the preceding twenty-two-year period.

Taken as a whole, the C.C.C. program proved to be one of the most constructive and most popular of all of the New Deal projects. In addition to achieving its primary objective of relieving unemployment, it gave some three million young men a new start and a new outlook on life, educated the general public as to the importance of natural resources in the national economy, expanded Federal ownership of forest lands, and accomplished much in the restoration and improvement of our land and water resources.

Other Relief Projects. Large sums were made available for forestry and other conservation activities through the Federal Emergency Relief Act of May 12, 1933, the National Industrial Recovery Act of June 16, 1933, and the Works Relief Act of April 8, 1935. The Public Works, Civil Works, Transient Relief, Drought Relief, Works Progress, and National Youth agencies seemed to have almost unlimited funds which conservation agencies were not slow to tap.

Coupled with the still larger amounts provided for the C.C.C. in this particular field, these funds made possible progress in resource development that would not have taken place for decades in the ordinary course of events. For this, the imagination, interest, and vigor of the President were largely responsible. Credit must also be given to the conservation agencies, and particularly to the Forest Service and the National Park Service, for having long-range plans that assured wise use of the windfalls resulting from the emergency.

Shelterbelt Project. One distinctive, highly publicized, and decidedly controversial use of emergency relief funds was for the Shelterbelt, or Prairie States Forestry, Project. The unprecedented drought of 1934 in the Plains area of the Middle West brought about a condition of acute distress which urgently demanded relief measures. To meet the situation the President allocated to the Forest Service an initial 1 million dollars of emergency funds for large-scale shelterbelt planting, which he hoped would provide not only temporary employment but lasting benefits through the amelioration of adverse natural conditions.

In its original form the project envisaged the planting of shelter strips 10 rods in width at intervals of 1 mile in a belt 100 miles wide extending through the Prairie states from Canada to Texas, so located as to intercept the prevailing winds. In practice this formula, which was severely criticized for its mathematical rigidity, was modified to meet the biological facts of tree life. Planting was actually done not in accordance with a geometrical pattern but where climate and soil were such as to give the trees a reasonable chance of survival and where the protection of a shelterbelt was particularly needed. The obvious necessity for research to solve the many problems involved in afforestation under the difficult conditions existing in the Plains region led Congress on June 15, 1936, to authorize the establishment of a Great Plains Forest Experiment Station in addition to the stations listed in the McSweeney-McNary Act of 1928.

The Shelterbelt Project confounded its critics by being on the whole surprisingly successful, although relatively expensive. High survival and high cost went together. Planting had to be done with more than usual care and followed by cultivation and protection for several years to assure the establishment of the plantation. On June 30, 1942, the Prairie States Forestry Project was terminated as an emergency activity and transferred from the Forest Service to the Soil Conservation Service. Up to that time 18,600 miles of shelterbelts had been planted in cooperation with some 30,000 farmers and with the states of North Dakota, South Dakota, Nebraska, Kansas, Oklahoma, and Texas. Since then, the encouragement of shelterbelt planting has been an integral part of the farm-planning activities of the Soil Conservation Service.

Timber-salvage Operations. Another form of emergency expenditure by the Federal government in the field of natural resources followed the New England hurricane of September, 1938. That storm blew down some three billion board feet of timber and created a fire hazard and a salvage problem that private owners and the states alone were unable to handle. Congress promptly appropriated $500,000 for the reduction of forest-fire hazard on the White Mountain National Forest and 5 million dollars for hazard reduction outside of the national forests in New England. The latter amount had to be matched by the states or other political subdivisions.

The work outside of the national forests was directed by the New England Forest Emergency Organization, which reported that by the spring of 1939 more than 14,000 men were engaged in the task. Of these, more than 11,000 were employed by the Civilian Conservation Corps and the Works Progress Administration. So successful was the work that fire losses following the blowdown were kept within normal bounds.

Assistance in salvaging the enormous amounts of down timber, which could not be handled rapidly enough in the course of ordinary logging operations to prevent serious deterioration, was rendered by the Northeastern Timber Salvage Administration (N.E.T.S.A.). This agency was organized and administered by the Forest Service as agent for the Federal Surplus Commodities Corporation. It bought logs, put them into wet storage in ponds or sawed them into lumber, and sold the logs or lumber as rapidly as practicable without disrupting the market. Funds were supplied by a loan of some 6 million dollars to the Surplus Commodities Corporation (of which N.E.T.S.A. was a part) from the Disaster Loan Corporation, organized in 1937 as a subsidiary of the Reconstruction Finance Corporation. More than 92 per cent of the loan was repaid from sales of logs and lumber.

At the peak of the project 241 sawmills were operating under government control. During its four years of activity, the Timber Salvage Administration handled more than 660 million board feet of logs and 35,000 cords of pulpwood. It enabled some 13,500 landowners, mostly farmers, to obtain fair prices for wind-thrown timber which they would otherwise have had to sell for a fraction of its value or leave in the woods to rot. Most of the salvaged material went directly or indirectly into war use.

A similar but smaller salvage operation was conducted during 1944 and 1945 on an area of about 310,000 acres in eastern Texas where the pine had been seriously damaged by a severe ice storm in January, 1944. The job of salvaging the damaged timber, which had to be harvested promptly in order to avoid almost complete loss, was much greater than local industries could handle alone. Another loan from the Disaster Loan Corporation to the Federal Surplus Commodities Corporation, this time for 3 million dollars, enabled the Forest Service to harvest some 80,000 cords of pulpwood and 11 million board feet of sawlogs before deterioration made further salvage operations impossible. Most of the logging was done by prisoners of war, and practically all of the salvaged material was made available for war needs.

River-basin Development. One of the earliest and most radical of the many unorthodox proposals made by President Roosevelt during his first few months in office was that for the unified development of all of the resources of the Tennessee River Valley. In a special message to Congress he said:

It is clear that the Muscle Shoals development is but a small part of the potential usefulness of the entire Tennessee River. Such use, if envisioned in its entirety, transcends mere power development: it enters the wide fields of flood control, soil erosion, afforestation, elimination from agricultural use of submarginal lands, and distribution and diversification of industry. In short, this power development of war days leads logically to national planning for a complete river watershed involving many states and the future lives and welfare of millions. It touches and gives life to all forms of human concerns.

I, therefore, suggest to the Congress legislation to create a Tennessee Valley Authority—a corporation clothed with the power of government but possessed of the flexibility and initiative of a private enterprise. It should be charged with the broadest duty of planning for the proper use, conservation, and development of the natural resources of the Tennessee River drainage basin and its adjoining territory for the general social and economic welfare of the Nation. . . .

Many hard lessons have taught us the human waste that results from lack of planning. Here and there a few wise cities and counties have looked ahead and planned. But our Nation has "just grown." It is time to extend planning to a wider field, in this instance comprehending in one great project many States directly concerned with the basin of one of our greatest rivers.

This in a true sense is a return to the spirit and vision of the pioneer. If we are successful here we can march on, step by step, in a like development of other great natural territorial units within our borders.

Tennessee Valley Authority. Here at last was a concrete proposal for actually doing something about the admonitions of the Inland Waterways Commission, the National Waterways Commission, Theodore Roosevelt, Gifford Pinchot, and many others that every river system should be developed as a unit. Congress cooperated by passing the act of May 18, 1933, creating the proposed Tennessee Valley Authority (T.V.A.). The board of directors of the corporation was to consist of three members appointed by the President, by and with the advice and consent of the Senate, for staggered terms of nine years each. An interesting requirement was that "all members of the board shall be persons who profess a belief in the feasibility and wisdom of this Act."

The President was also instructed to recommend to Congress from time to time such legislation as he deemed proper to carry out the general purposes of the act and for the special purpose of bringing about in the valley:

(1) the maximum amount of flood control; (2) the maximum use of said Tennessee River for navigation purposes; (3) the maximum generation of electric power consistent with flood control and navigation; (4) the proper use of marginal lands; (5) the proper method of reforestation of all lands in said drainage basin suitable for reforestation; and (6) the economic and social well-being of the people living in said river basin.

It would be difficult to imagine a more comprehensive program. The Authority has paid special attention to navigation, flood control, power development, and fertilizer production, but has not overlooked other aspects of land and water utilization and industrial development. Since it has no control over the extensive area administered by the Forest Service and the National Park Service or over lands in state and private ownership, it is in no position to enforce or even to make any "master plan" for the region.

Outside of the strategic areas which the Authority itself controls, it attempts to promote better management practices through cooperation with other agencies. In the field of forestry, in addition to planting and other silvicultural practices on its own lands, chiefly surrounding reservoir sites, the Authority through its Division of Forestry Relations conducts research and educational activities aimed at bringing about better forest, wildlife, and watershed management, and more effective utilization of forest products throughout the region.

Proposals for Other Authorities. Roosevelt's vision that the T.V.A. might set a precedent that would make it possible to "march on, step by step, in a like development of other great natural territorial units" has been shared by others. Bills have been introduced in Congress for creating a Missouri Valley Authority and a Columbia Valley Authority, and even for dividing the entire country into nine regions each of which would have its own regional authority.

These proposals command vigorous support and equally vigorous opposition, in each case on both theoretical and practical grounds. Whatever the outcome of the debate, the T.V.A. has made it impossible to ignore the urgent problem of substituting unified for piecemeal development of regional and national resources. Whatever method or methods may eventually be adopted for achieving this result, its experience will aid in determining the most effective forms of organization and the most helpful relationships between Federal, state, local, and private agencies.

N.I.R.A. and the Lumber Code. One of the declared purposes of the National Industrial Recovery Act of June 16, 1933, was "to conserve natural resources." In accordance with the spirit of this declaration the Code of Fair Competition for the Lumber and Timber Products Industries approved by the President on August 19, 1933, contained the following statement headed "Conservation and Sustained Production of Forest Resources":

Art. X. The applicant industries undertake, in cooperation with public and other agencies, to carry out such practicable measures as may be necessary for the declared purposes of this Code in respect of conservation and sustained production of forest resources. The applicant industries shall forthwith request a conference with the Secretary of Agriculture and such State and other public and other agencies as he may designate. Said conference shall be requested to make to the Secretary of Agriculture recommendations of public measures, with the request that he transmit them, with his recommendations, to the President; and to make recommendations for industrial action to the [Lumber Code] Authority, which shall promptly take such action, and shall submit to the President such supplements to this Code, as it determines to be necessary and feasible to give effect to said declared purposes. Such supplements shall provide for the initiation and administration of said measures necessary for the conservation and sustained production of forest resources, by the industries within each Division, in cooperation with the appropriate State and Federal authorities. To the extent that said conference may determine that said measures require the cooperation of Federal, State or other public agencies, said measures may to that extent be made contingent upon such cooperation of public agencies.

Conference and Agreement. Two conferences were held to carry out the provisions of this article, one in the fall of 1933 the other in the winter of 1934. A pall was cast over the first conference by the accidental death of the Chief Forester, Robert Y. Stuart, just as it was about to convene. He was succeeded on November 15 by Ferdinand A. Silcox, a forester with a long record of leadership in the Forest Service and in industrial and labor organizations, who participated in the second conference. Dean Henry S. Graves of the Yale School of Forestry presided over both conferences, which included approximately equal representation from the forest-products industries and the general public.

After some intensive committee work and some lively discussion, unanimous agreement was reached on the programs to be recommended for adoption by the Lumber Code Authority and by public agencies. Proposed public action included expansion of Federal and state forests; increased participation in the protection of forests from fire, insects, and disease; modification of the property tax as applied to forests; provision of forest-fire insurance and forest credit; more extensive cooperation with private owners in reforestation, forest management, and the marketing of forest products; and strengthening of research in forest protection, forest production, and utilization of forests and forest products. Since administration of the general property tax lies wholly with the states, the President addressed a letter to the governor of each state urging careful consideration of ways and means of modifying the tax to remove its deterrent effect on the practice of forestry by private owners.

Rules of Forest Practice. The program for industry took the form of an amendment to the Code of Fair Competition for the Lumber and Timber Products Industries which was approved by the President on May 23, 1934, as Schedule C—Forest Conservation Code. It provided that each division and subdivision of the industry having jurisdiction over forest-utilization operations should establish an agency to formulate rules of forest practice and to exercise general supervision over their application and enforcement.

The rules of forest practice were to include

practicable measures to be taken by the operators to safeguard timber and young growing stock from injury by fire and other destructive forces, to prevent damage to young trees during logging operations, to provide for restocking the land after logging if sufficient advance growth is not already present, and where feasible, to leave some portion of merchantable timber (usually the less mature trees) as a basis for growth and the next timber crop.

After approval by the Lumber Code Authority, the rules were to become effective on June 1, 1934. Thereafter violators were subject to punishment by fine and imprisonment. In order to permit flexibility each operator was encouraged to submit a specific plan of management for his own property, which, if approved by the agency, could be followed instead of the standard rules. Each agency was also instructed to encourage the application of sustained-yield forest management wherever feasible.

There was great variation in the speed with which rules of forest practice were adopted and put into effect in the different regions. In some regions, action was prompt; in others, there was evidence of stalling. Nowhere were the rules in force for as much as a year when the Supreme Court on May 27, 1935, unanimously invalidated the act on the grounds that it involved an unconstitutional delegation of legislative power, that it exceeded the power of Congress to regulate interstate commerce, and that it invaded the powers reserved exclusively to the states. As Justice Cardozo expressed it, "This is delegation running riot." The decision of course abrogated not only the rules of forest practice but also the provisions of the Lumber Code relating to hours of labor, minimum wages, control of production, cost protection, and fair-trade practices.

Aftermath. This experiment in self-regulation of cutting practices did not run long enough to give any fair indication of how effective it would have proved. The idea itself was so new and the practices recommended so foreign to the experience of most operators that considerable time would have been required to attain the goal of sustained production.

Voluntary action in this direction was strongly encouraged by several regional organizations, notably the West Coast Lumbermen's Association, the Western Pine Association, and the Southern Pine Association. Foresters were employed to place their activities on a sound technical basis; frequent conferences were held to consider ways and means of promoting better forest management; and several manuals of forest practice were issued for the guidance of timberland owners. The Pacific Northwest Forest Industries established a standing Forest Conservation Committee, which led in 1953 to organization of the Industrial Forestry Association.

The Forest Service, which had cooperated actively in the formulation of the rules of forest practice, also continued its efforts to be helpful under the new setup. Although Article X of the Lumber Code proved to be an abortive attempt to effect a revolution in forest practice on private lands, it was undoubtedly an educational influence of great value in hastening the evolution of forestry as a private enterprise.

Federal Resources Planning. One of the early actions of President Roosevelt's administration was the appointment on July 20, 1933, of a National Planning Board to assist the Administrator of Public Works in the preparation of the "comprehensive program of public works" required by the National Industrial Recovery Act. Among the functions of the Board were the development of "comprehensive and coordinated plans for regional areas" and the conduct of surveys and research concerning "the distribution and trends of population, land uses, industry, housing and natural resources."

By subsequent Executive orders the National Planning Board was replaced on June 30, 1934, by the National Resources Board, and the latter in turn was replaced on June 7, 1935, by the National Resources Committee. President Roosevelt's Reorganization Plan No. I, effective July 1, 1939, transferred the National Resources Committee and the Federal Employment Stabilization Office in the Department of Commerce to the Executive Office of the President and consolidated them under the name National Resources Planning Board. This agency, under its various names, issued a large number of reports, many of them voluminous, dealing with the country's natural resources of all kinds and with the broad subjects of regional and national planning. Its philosophy with respect to the solution of problems in these fields is illustrated by the following statement in its report of December 31, 1934: "The natural resources of America are the heritage of the whole Nation and should be conserved and utilized for the benefit of all of our people."

The publications of the Board constitute an impressive mass of factual information and of stimulating plans which have doubtless been used by many agencies and individuals. It is, however, difficult to identify any specific legislation that can be traced directly to the activities of the Board, which was discontinued by Congress in 1943. One of its principal services was to promote integrated thinking with respect to the interrelationships between all natural resources and between the different parts of entire regions.

Agricultural Conservation Program. The Soil Conservation and Domestic Allotment Act of 1936 authorized the Federal government to share with farmers and ranchers the cost of carrying out approved conservation practices. The primary objective of the program was the protection of the public interest in the nation's soil and water resources by helping landowners to carry out desirable practices which they otherwise would not adopt to the needed extent.

The first major forestry activity to receive assistance under this legislation was a special naval-stores conservation program administered by the Forest Service. Its purpose was to encourage better fire protection, turpentining, and cutting activities on turpentine farms. About 3,200 turpentine farmers have participated in the program, which has included about 75 per cent of the area cupped. Among other things it has resulted in reducing the number of trees below 9 inches in diameter tapped for turpentine from 28 per cent to less than 2 per cent of the total number.

Other forestry practices for which Federal payments have been made include tree planting; stand improvement by thinning, cleaning, and pruning; establishment and maintenance of windbreaks and shelterbelts; forest-site preparation; construction of firebreaks; and fencing for protection against grazing and for other purposes. Administration of the program covering these activities was transferred in November, 1953, from the Production and Marketing Administration to the Agricultural Conservation Program Service. From the beginning the Forest Service has been responsible for providing the necessary specialized technical assistance, developing specifications for approved forestry practices, and through state and local committees determining performance in meeting these specifications.

In 1953 the Federal share of the cost of approved forestry practices was approximately 1 million dollars under the "regular" program and $417,000 under the naval-stores program. Although payments under the regular program constituted only about ½ per cent of the total payments for improved agricultural practices, the program did tend to give forests a more prominent place in the thinking of the average farmer than they had previously occupied. That the amount is small relatively as well as absolutely is indicated by the fact that the annual needs for forest conservation practices on farms are estimated at no more than 2 per cent of the total needs for agricultural conservation practices.

Under the 1955 Agricultural Conservation Program promulgated on July 1, 1954, Federal cost sharing in the field of forestry is limited to tree planting; thinning; pruning of crop trees; release of desirable tree seedlings by cutting or otherwise; site preparation for natural reseeding; construction of permanent fences, excluding boundary and road fences; and approved naval-stores practices. In 1954 the definition of what constitutes a farm for purposes of Federal cost sharing was broadened by providing that "a farm may include or may consist entirely of woodland which is being operated for the production and sale of forest products."

"Rangeland means non-irrigated land, located in arid and semiarid areas, growing native grasses and forage plants primarily, and used for grazing by domestic livestock." For purposes of the program it is limited to areas west of the ninety-eighth meridian except as other specific areas may be approved by the Director of the Agricultural Conservation Program Service. Range-management practices to which Federal cost sharing has been extended, and which are included in the 1955 program, include artificial seeding; deferred grazing to permit natural reseeding; control of competitive plants; construction of reservoirs and wells; construction of stock trails to permit better distribution of grazing; construction of fireguards; and construction of drift fences. The published statistics commonly group expenditures for the improvement of pasture and range land, but it appears that in 1953 Federal contributions to the latter constituted about 1 per cent of all Federal payments.

Under the 1955 program, Federal cost sharing is not authorized for repairs or for upkeep or maintenance of any practice. In general, practices that have become a part of regular farming operations in a particular county are not eligible for cost sharing. The total Federal payment to any person with respect to farms, ranching units, and turpentine places is limited to $1,500 for 1955.

Taylor Grazing Act. The long controversy over the disposition of range lands in the unreserved public domain was settled, for a while at least, by passage of the Taylor Grazing Act of June 28, 1934. The act was fathered by Representative Edward T. Taylor, previously an advocate of session to the states, who became an equally strong supporter of the leasing plan after Congress failed to act on the recommendations of the Committee on the Conservation and Administration of the Public Domain.

As originally introduced, the bill had the full support of both the Secretary of Agriculture and the Secretary of the Interior. During its passage through Congress it was so greatly modified, largely through amendments introduced by Senator McCarran of Nevada, that the Secretary of Agriculture not only withdrew his support but urged its veto by the President as no longer a conservation measure. Chief among his many objections, which were based on advice from the solicitor of the Department, were that the bill did not give the Secretary of the Interior full authority to regulate grazing on the lands involved; that it partially abdicated Federal control over these lands in favor of the states; that it granted permanent rights to present users of the range; that it was limited to only a small part of the total area over which control should be exercised; that its enforcement provisions were weak; and that at best the phrase "pending final disposal" meant that it was a stopgap measure.

The Secretary of the Interior, on the advice of the solicitor of his Department, believed that these objections were either invalid or unimportant and urged signature of the bill by the President. In view of the diametrically opposite view of the Secretaries and solicitors of the two Departments, the President sought the advice of the Attorney General, who sided with the Department of the Interior. The President thereupon signed the bill, with an accompanying statement lauding it as an exceptionally constructive piece of conservation legislation.

Establishment and Administration of Grazing Districts. As finally approved, "in order to permit the highest use of the public lands pending its [*sic*] final disposal," the act authorized the Secretary of the Interior to establish as grazing districts not more than 80 million acres of vacant, unappropriated, and unreserved public lands which in his judgment were chiefly valuable for grazing and raising forage crops. It further authorized him to make rules and regulations for the occupancy and use of the districts and the development of the range; and it provided that "any willful violation of the provisions of this Act or of such rules and regulations thereunder after actual notice thereof shall be punishable by a fine of not more than $500."

The Secretary was authorized to issue permits for use of the grazing districts, for periods of not more than ten years, on payment of reasonable fees, with preference to certain classes of owners.

No permittee complying with the rules and regulations laid down by the Secretary of the Interior shall be denied the renewal of such permit, if such denial will impair the value of the grazing unit of the permittee, when such unit is pledged as security for any bona fide loan. . . . Nothing in this Act shall be construed as restricting the respective States from enforcing any and all statutes enacted for police regulation, nor shall the police power of the respective States be, by this Act, impaired or restricted, and all laws heretofore enacted by the respective States or any thereof, or that may hereafter be enacted as regards public health or public welfare, shall at all times be in full force and effect: *Provided, however,* that nothing in this section shall be construed as limiting or restricting the power and authority of the United States.

All lands in grazing districts were open to hunting and fishing under the laws of the states and the United States and to prospecting and patenting under the mineral laws.

Provision was made for the acceptance of gifts of land within grazing districts and for the exchange of lands with states or private owners on an equal-value basis. Lands more valuable for the production of agricultural crops than for forage could be opened to homestead entry by the Secretary of the Interior in tracts not exceeding 320 acres in size.

The act of August 30, 1890, authorizing the Secretary of the Interior to sell isolated tracts of unreserved public land not including more than 160 acres was amended to authorize him (1) to sell any isolated tract of not more than 760 acres at public auction for not less than its appraised value and not more than three times its appraised value, and (2) to sell not more than 160 acres of land which is mountainous or too rough for cultivation, whether isolated or not, to owners of adjoining land.

Authorization was granted to the Secretary to lease for grazing purposes, under such conditions as he might prescribe, any unreserved public lands so situated as not to justify their inclusion in a grazing district. These are known as "Section 15" lands.

The President was authorized to place under national-forest administration, in any states where he had the power to create national forests, any unappropriated public lands lying within watersheds forming a part of the national forests which in his judgment could best be administered in connection with existing national forests, and to place under the administration of the Department of the Interior any lands within national forests principally valuable for grazing which, in his opinion, could best be administered under this act.

Disposition of Receipts. All moneys received from grazing districts, except those on ceded Indian lands, were to be deposited in the Treasury of the United States as miscellaneous receipts, but 25 per cent of such receipts was made available, when appropriated by Congress, for the construction, purchase, or maintenance of range improvements, and 50 per cent of the receipts was to be paid to the states to be expended as the Legislature might prescribe for the benefit of the counties in which the grazing districts were located.

In the case of ceded Indian lands, 25 per cent was made available for range improvement and 25 per cent for the benefit of public schools and public roads in the counties concerned. The remaining 50 per cent was to be deposited to the credit of the Indians pending final disposition under applicable laws, treaties, or agreements.

Cooperative funds received for the protection, administration, and improvement of grazing districts were not subject to disposal by any of these methods but were to be used in their entirety for the purpose for which they were made available.

Early Amendments of the Act. The act of June 26, 1936, increased the maximum allowable area of grazing districts to 142 million acres. It also authorized the exchange of lands with states on either an equal-value or equal-area basis. When the exchange was on the basis of equal area, mineral rights must be reserved by the government and could be reserved by the state. The President was authorized, with the advice and consent of the Senate, to select a Director of Grazing; and the Secretary of the Interior was authorized to appoint such assistant directors and other employees as might be necessary to administer the act, provided that every appointee must have been for one year a bona fide citizen or resident of the state in which he was to serve.

An amendment of June 23, 1938, authorized the Secretary of the Interior to lease at rates determined by him any state, county, or private land chiefly valuable for grazing purposes within the exterior boundaries of a grazing district. Such leases were not to run for more than ten years, and the fees charged for grazing privileges on the leased land were not to be less than the rental paid by the United States for them. All moneys received in the administration of leased lands were made available, when appropriated by Congress, for the leasing of lands under the act and were not to be distributed in the manner provided for other receipts.

Still another amendment, adopted on July 14, 1939, provided for the establishment within each grazing district of an advisory board of five to twelve local stockmen elected by the users of the range but appointed by the Secretary of the Interior, who might also on his own initiative appoint one wildlife member on each board. Advisory boards were to offer advice on applications for grazing permits and on all other matters affecting the administration of the Taylor Grazing Act within their respective districts. Except in an emergency, the Secretary of the Interior must request the advice of the advisory board prior to the promulgation of any rules and regulations affecting the district.

Later amendments of the act will be discussed in Chapter 11.

Grazing on National Forests. In contrast to the drastic reversal of policy with respect to the handling of unreserved range lands brought about by the Taylor Grazing Act, little change in the administration of grazing on the national forests took place during the 1930's. Although there was a steady undercurrent of criticism on the part of certain members of the livestock industry, this criticism did not become acute until the middle forties.

In 1936, in response to a request from the Senate initiated by Senator Norris of Nebraska, the Forest Service submitted a comprehensive report entitled *The Western Range,* which discussed in detail the technical, administrative, and legislative problems involved in the utilization of all range lands west of an irregular line through the prairie states. Its publication, as a Senate document, led to a rather sharp exchange of compliments between the Secretary of the Interior and the Secretary of Agriculture as to which Department was the authority in that particular field.

Indian Forest and Range. During the 1930's and 1940's the timber and forage resources on Indian reservations continued to be handled by the Forestry Branch of the Indian Service with no major changes in policy; timber sales increased materially; and substantial improvements were effected by the C.C.C. Timber and range surveys, for example, were made on about 5,500,000 acres of Indian land; insect- and disease-control operations were conducted on 700,000 acres; some 80,000 acres of forest land were planted or otherwise improved; range revegetation was carried out on 300,000 acres; noxious weeds were eradicated on 250,000 acres; and prairie dogs and other destructive rodents were largely eliminated from 13,300,000 acres of range and agricultural lands.

The act of June 18, 1934, was a highly constructive piece of legislation. It directed the Secretary of the Interior to make rules and regulations for managing Indian forestry units on the principle of sustained yield; for restricting the number of livestock grazed on Indian range units; and for protecting the range from deterioration, preventing soil erosion, and assuring full utilization of the range. Decentralization of administration was promoted in 1946 by action of the Secretary of the Interior in delegating to the Commissioner of Indian Affairs authority to approve sales of timber from Indian lands up to 40,000 M board feet and to adjust stumpage prices on these sales; and the following year the Commissioner authorized the regional offices to approve sales up to 15,000 M board feet.

Federal Aid for State Forests. The Fulmer Act of August 29, 1935, was a response to the request of many states, particularly in the South, for financial assistance from the Federal government in the acquisition of state forests. It authorized an appropriation of 5 million dollars for the purchase of lands approved by the National Forest Reservation Commission, which would be administered by the respective states under plans of management approved by the Secretary of Agriculture. States taking advantage of the act must provide for the employment of a state forester, who shall be trained and of recognized standing. They must also, by June 30, 1942, provide for the reversion to the state, or to a political unit thereof, of title to tax-delinquent lands which are more suitable for public than for private ownership and which in the public interest should be devoted primarily to the production of timber crops and/or to the maintenance of forests for watershed protection. The original purchase price of the lands was to be returned to the government, without interest, by paying to it currently 50 per cent of the gross proceeds from the management of the forests.

Although this legislation was directly in line with the recommendation of the *Copeland Report* of 1933 for a large increase in state ownership of forest land, Congress has never made any appropriation to put it into effect. Failure to implement the act has caused considerable resentment on the part of the states, several of which passed the necessary legislation to enable them to take advantage of its provisions. The Forest Service claims that it has urged the Bureau of the Budget to include appropriations under the act in its recommendations to Congress, but without success. Although the law is still on the statute books, it can probably be regarded as a dead letter. Austin F. Hawes, State Forester of Connecticut, referred to it in 1941 as "the greatest disappointment of the last decade."

Forests and Flood Control. The Flood Control Survey Act of June 22, 1936, recognized the influence of vegetative cover on water runoff and floods by providing that thereafter Federal investigations and measures for retarding waterflow and preventing soil erosion on watersheds should be under the jurisdiction of the Department of Agriculture. In September of the same year a well-attended Upstream Engineering Conference, in which forestry and range management figured prominently, was held in Washington, D.C., to emphasize the importance of this phase of flood and erosion control.

Amendments of the 1936 act in 1937 and 1938 provided that the Secretary of Agriculture should make runoff and erosion surveys on watersheds on which the Secretary of War has been authorized to make flood-control surveys; and the comprehensive flood-control act of 1944 continued his duties in this field. Substantial appropriations have been made to the Department of Agriculture under these and other acts both for research and for actual upstream control of runoff and erosion.

Federal-State Cooperation in Forest Management. Pressure from both Federal and state agencies for enlargement of the Federal-aid program in the field of forest management initiated by the Clarke-McNary Act of 1924 finally resulted in passage of the Cooperative Farm Forestry Act, or Norris-Doxey Act, of May 18, 1937. The purpose of the act was "to aid agriculture, increase farm-forest income, conserve water resources, increase employment, and in other ways advance the general welfare and improve living conditions on farms through reforestation and afforestation in the various States and Territories."

In furtherance of these objectives Congress authorized an annual appropriation of $2,500,000 to enable the Secretary of Agriculture, in cooperation with the land-grant colleges and universities and state forestry agencies,

to produce or procure and distribute forest trees and shrub planting stock; to make necessary investigations; to advise farmers regarding the establishment, protection, and management of farm forests and forest and shrub plantations and the harvesting, utilization, and marketing of the products thereof; and to enter into cooperative agreements for the establishment, protection, and care of farm- or other forest-land tree and shrub plantings.

There was no legislative requirement for state matching of Federal funds.

The first appropriation under the act was for the fiscal year 1940, in the amount of $300,000. Over-all administration of the act was at first lodged in the Soil Conservation Service, with active participation by the Forest Service and the Extension Service. Owners of farms in which forests were distinctly a secondary crop were normally served by the Soil Conservation Service, while owners of farms in which forests were the major crop were served by the Forest Service. The distinction between "farm forests" and "forest farms" was, however, often difficult to draw, and in 1945 administration of the program was transferred to the Forest Service in order to consolidate in the Department's principal forestry agency the cooperative and other work in that field.

Strictly educational, as contrasted with service, activities were generally handled by the Extension Service. From the beginning, the program was administered so far as practicable through state agencies such as the soil-conservation districts, the state departments of forestry, and the state extension services, and after 1948 entirely so.

The act was replaced in 1950 by the Cooperative Forest Management Act, which will be discussed in the next chapter.

Oregon and California Railroad Lands. When title to the lands included in the O. and C. and the Coos Bay Wagon Road grants was revested in the United States, Congress provided for the disposal of the lands with no regard to the conservative management of the forests, which constituted their chief value. Steadily mounting public dissatisfaction with this situation finally led to passage of the act of August 28, 1937, which completely reversed the previous policy.

The new act provided that such portions of these lands

as are or may hereafter come under the jurisdiction of the Department of the Interior, which have heretofore or may hereafter be classified as timberlands, and power-site lands valuable for timber, shall be managed . . . for permanent forest production, and the timber thereon shall be sold, cut, and removed in conformity with the principal [*sic*] of sustained yield for . . . providing a permanent source of timber supply, protecting watersheds, regulating stream flow, and contributing to the economic stability of local communities and industries, and providing recreational facilties [*sic*].

Pending studies to determine the annual productive capacity of the lands, Congress limited the annual cut to 500 million board feet. The Secretary of the Interior was authorized to establish management units, within each of which the sustained-yield principle would apply. Cooperative agreements with other Federal agencies, with state agencies, and with private owners were authorized to secure coordinated administration of O. and C. and intermingled lands both in fire control and timber management.

Grazing was authorized where it would not interfere with the production of timber or the other purposes of the act. The Secretary of the Interior could classify and open to homestead entry or purchase under the provisions of the Taylor Act any lands that in his judgment are more suitable for agricultural use than for afforestation, reforestation, streamflow protection, recreation, or other public purposes. He was also authorized "to perform any and all acts and to make such rules and regulations as may be necessary and proper" for administering the act. In formulating forest-practice rules and regulations he was further authorized, which presumably means encouraged, to consult with the Oregon State Board of Forestry, representatives of timber owners and operators, and other persons or agencies interested in the use of the lands.

Disposition of Receipts and Other Provisions. Receipts were to be deposited in a special "Oregon and California land-grant fund," of which 50 per cent was payable immediately to the counties concerned, plus an additional 25 per cent after payment of back taxes due under the act of July 13, 1926, and of reimbursable Federal charges against the fund. All of these charges have since been paid off, so that the counties now receive 75 per cent of the gross receipts, less the costs of building access roads. The remaining 25 per cent is available for administration of the lands in such annual amount as is determined by Congress. Any part of this 25 per cent not used for administrative purposes is covered into the general funds of the Treasury. Receipts from Coos Bay Wagon Road lands were at first treated in the same way as those from the O. and C.

lands, but since 1939 the counties have received an ad valorem payment in lieu of taxes, which is determined by applying the local tax rates to the appraised value of the lands.

An act of April 8, 1948, reopened the revested lands in both grants, except power sites, to mineral exploration and entry, to which they had been closed since 1937. The general mining laws applied, with the important exceptions that claimants must obtain cutting permits for all timber used in the development of their mines; that when claims are patented the government retains title to all the timber on the area, including future growth; and that all mining claims must be recorded with the Federal government.

More than 2 million acres of the revested lands have been classified by the Department of the Interior as timberlands. Substantial sales of timber are made regularly from these lands on the sustained-yield basis required by law. Much less than the 25 per cent of the receipts available for administration has actually been appropriated by Congress for that purpose. An advisory board was soon set up to assist the General Land Office in the management of the timberlands, and in 1948 local or district advisory boards were also established.

Controverted Lands. When title to the O. and C. lands was revested in the United States in 1916, there were some 465,000 acres in odd-numbered sections in the indemnity limits to which the company had not received patent, for which no application for patent was pending, and which had been included in national forests by presidential proclamation between 1892 and 1907. The Department of Agriculture promptly raised the question as to whether these were lands to which the company was entitled to receive patent and which should therefore be transferred to the jurisdiction of the Department of the Interior. The question was answered in the negative by the Commissioner of the General Land Office in 1919, and again by the Assistant Secretary of the Interior in 1923. Doubt as to the correctness of these decisions, however, left the lands in a highly confused and unsatisfactory status.

Following passage of the act of 1937 the Department of the Interior proceeded to make three sales of timber on the lands in question, over which both Departments at that time claimed jurisdiction. Shortly thereafter the Forest Service advertised a sale of timber on one of the controverted sections, and its right to do so was promptly challenged by the Department of the Interior.

These actions obviously created an impossible situation, which presently led to an agreement between the two Departments that the Forest Service would continue to administer the lands but that all receipts from them would be placed in a special fund until the status of the lands should be definitely determined. A year later (1940) the Attorney General ruled that in his opinion "disturbance of the continued administration of these lands by the Department of Agriculture as part of the national forest reserves would not be warranted under existing law."

The matter was later reopened with a new Attorney General, who declined to withdraw the opinion rendered by his predecessor. The Secretary of the Interior and the eighteen Oregon counties in which the lands are located thereupon attempted to obtain legislation effecting the desired transfer of jurisdiction. Their efforts finally resulted in the act of 1954, which will be discussed in the next chapter.

Northern Pacific Land Grant. In 1929 Congress passed an act authorizing adjudication of the claim of the Northern Pacific Railway Company for damages because it was denied the right to select 2,800,000 acres within the indemnity limits of its 1864 land grant in Wyoming, Montana, Idaho, and Washington. Most of the disputed area lay within national forests and included much valuable timberland.

The case was a complicated one, and the suit dragged on until 1940, when the United States Supreme Court supported the most important contentions of the government. The final outcome was to confirm Federal title to the controverted lands without any payment by the United States. In addition, the railroad company relinquished certain other lands and made cash settlement for still other lands erroneously patented to it.

National Parks, Monuments, and Forests. Under authority of the act of March 3, 1933, authorizing the President to reorganize the executive branch of the government, Roosevelt, on June 10, 1933, placed all national monuments under the jurisdiction of the National Park Service in the Department of the Interior. The change affected sixteen national monuments in national forests and areas under withdrawal for national-forest purposes which had been previously under the jurisdiction of the Department of Agriculture. It also led the Chief of the Forest Service, in his annual report for 1934, to emphasize the fundamental distinctions between national parks and national monuments, on the one hand, and national forests, on the other hand. He concluded: "National forest administration is integral, and separate jurisdiction over either limited areas or specific activities, if carried out in any large way, would lead to chaos."

Fear that large additions to the national parks might seriously impair the entire national-forest system induced the Chief to discuss the principles involved at greater length in his report for 1936:

The physical characteristics of national parks and national forests are in many ways similar. Both embrace interesting and sometimes unique geological and organic examples of the operation and effect of natural laws, possessing high inspirational, educational, and recreational values. The basic differences relate to the form of administration through which the American people can derive from a given area the maximum social and economic benefits.

Sometimes the intrinsic values involved justify maintaining the area inviolate as a permanent "museum" piece deserving of national concern, and demand its administration exclusively as a source of scientific knowledge, education, inspiration, and recreation. In other cases the best public interest may require that the area be so managed as to derive from it a coordinated series of benefits and uses, proper balance being maintained between the intangible services of scientific, spiritual, and recreational character and the tangible services to industry, commerce, and the general economy. A national-park status may be given to the "museum" areas, but not to areas where the principles of management most in the public interest are incompatible with those necessary for national parks truly meriting the name.

Unsettled questions of boundary adjustments between national parks and national forests therefore find their origin in economic rather than administrative circumstances, and should be determined accordingly. . . . If careful study and analysis of all factors show the national-park status to promise the highest and most permanent social benefit, the status should be promptly established. Otherwise, it should not be. Principles are involved, not a mere jurisdictional rivalry. There should be no compromise that would break down the standards and the principles of management which distinguish national parks from national forests.

New Parks and Monuments. Creation of the Olympic National Park by Congress on June 29, 1938, removed 648,000 acres from the Olympic National Forest. Two years later President Roosevelt, acting under authority granted him by Congress, transferred an additional 187,000 acres from the national forest to the national park. The Forest Service felt that the area of highly productive timberlands included in the withdrawals was unnecessarily large, while timber interests regarded them as so serious a blow to the economy of the region that they have consistently but unsuccessfully attempted to have the size of the national park reduced.

Following the creation of the Olympic National Park, an interdepartmental committee consisting of two members each from the Department of Agriculture and the Department of the Interior was established to facilitate consideration of matters of policy and administration of mutual interest. During the first year of its existence the committee agreed upon reports to Congress and the Bureau of the Budget concerning more than thirty legislative proposals, including the transfer of certain national-forest areas to the Rocky Mountain National Park in Colorado and the Glacier Bay National Monument in Alaska.

King's Canyon National Park in California was created by Congress on March 4, 1940. It consists almost wholly of land formerly in the Sequoia National Forest and the Sierra National Forest. Although there was some opposition to the establishment of the park, the superb scenery of the area was generally regarded as sufficiently unique to justify national-park status.

A major controversy was precipitated by President Roosevelt's proclamation establishing the Jackson Hole National Monument in Wyoming in 1943. The area involved included some 170,000 acres of national-forest land and some 33,000 acres which had been purchased by John D. Rockefeller, Jr., for donation to the government. It did not include the neighboring wildlife refuge, which continued under the administration of the Fish and Wildlife Service.

Immediate and vigorous protest against the President's action was made on the ground that if grazing were prohibited, as is usually the case in national monuments, this would seriously cripple an important local industry and would materially reduce payments to the counties in lieu of taxes; that the community would lose the taxes on the lands purchased by Rockefeller; and that establishment of so large a monument against the wishes of the local residents constituted a dangerous precedent that might be followed elsewhere. Secretary Ickes attempted to meet the criticism by promising to continue the grazing rights of present permittees during their lifetime and the lifetime of their heirs and by recommending that part of the receipts from national parks and national monuments be paid to the counties in lieu of taxes. He also reminded the objectors that national monuments attract tourists in large numbers, to the economic advantage of local communities.

Sufficient pressure was exerted on Congress to get it to pass a bill the next year (1944) abolishing the monument. This was pocket-vetoed by the President. Bills were also introduced to rescind the power of the President to establish national monuments and to require the consent of the state concerned before a national monument could be established within its borders. None of these were approved. Congress did, however, prevent development of the area by including in the annual appropriation acts for the Department of the Interior a prohibition against any expenditure on the monument except for the minimum necessary for protection and maintenance. The final chapter was written in 1950, when Congress incorporated most of the Jackson Hole National Monument in the Grand Teton National Park.

Recreation in the National Forests. In accordance with its basic "multiple-use" program for the balanced use of all national-forest resources, the Forest Service has long recognized recreation as an important use. This has been particularly true since Stuart's instructions of 1932 to regional foresters to regard it as a major use.

Congress similarly recognized the recreational values inherent in forest wildlife by the act of June 23, 1933, authorizing the President to create a game refuge within the Ouachita National Forest in Arkansas. The next year it made the legislation general by authorizing the President, upon recommendation of the Secretary of Commerce and the Secretary of Agriculture and with the approval of the Legislature of the state concerned, to establish fish and game sanctuaries or refuges within any of the national forests. Such lands continue to be a part of the national forests and may be used for other purposes not inconsistent with their use as sanctuaries. The Secretary of Agriculture and the Secretary of Commerce were jointly authorized to make all needful rules and regulations for their administration, including the control of predatory animals, provided the jurisdiction of the states is not altered or changed without the approval of their Legislatures.

In addition to the more conventional provisions for forest recreation, the Forest Service has established some 78 wilderness areas containing about 14 million acres, in which only the most primitive forms of transportation are available. Outstanding in this respect is the roadless area in the Superior National Forest in Minnesota, which is in line with the aim of the Shipstead-Nolan Act of 1930 to preserve the unique recreational assets of the region. An active program for the acquisition of lands which should be in Federal ownership has been under way since 1947.

On December 12, 1949, President Truman issued a proclamation establishing an air-space reservation within which airplanes could not land except in case of emergency, and over which through planes must fly at an altitude of not less than 4,000 feet. The validity of the proclamation was upheld by a Federal district court in a test case in 1952.

State Forestry. Progress in state forestry during the decade preceding the Second World War continued to be slow but sure. Technical direction of the work of C.C.C. camps located on state lands constituted a major activity. Accomplishments in forest-fire control, forest management, and the construction of permanent improvements went far beyond what would have been possible with regular appropriations.

Another important development was in the field of assistance to woodlot owners under the Cooperative Farm Forestry Act of 1937, to which reference has already been made. Control of forest fires improved materially, although large areas remained without organized protection, particularly in the South. In some states the forfeiture of tax-delinquent timberlands added substantially to the area of state forests.

New York in 1931 embarked on an ambitious program for the acquisition and reforestation of submarginal agricultural lands outside of the Forest Preserve in the Adirondack and Catskill Mountains. A constitutional amendment adopted that fall (Article VII, Section 16) mandated the Legislature to appropriate 20 million dollars over a period of fifteen years for the purchase and planting of approximately a million acres.

Because of the financial situation the Legislature failed to appropriate the specified amounts; and in 1938 a constitutional convention replaced the 1931 provision relating to reforestation by a new section (Article XIV, Section 1), approved by the electorate on November 8. This section stated that "wildlife conservation and reforestation are hereby declared to be policies of the State," but did not include any reference to annual appropriations. As of 1953, 553,000 acres of submarginal farm lands in thirty-four counties had been acquired as state forests at a cost of well over 2 million dollars, and the great bulk of the plantable area had actually been planted.

Private Forestry. Economic conditions during this period were not such as to encourage any marked or general improvement in forest practice by private owners. The lumber industry was particularly hard hit by the Great Depression. Lumber production fell off by 64 per cent during the three years from 1929 to 1932; stumpage prices were low; the future was uncertain. The pulp and paper industry, on the other hand, continued to expand, particularly in the South and West, and steadily increasing production led many companies to give serious thought to ways and means of assuring a permanent and adequate supply of pulpwood.

As has already been indicated, Article X of the Lumber Code focused attention not only on the need for private forestry but on specific measures to bring it into being throughout the country. After any legal compulsion had been removed by the Supreme Court's invalidation of the National Industrial Recovery Act, several regional associations continued their efforts to formulate workable rules of forest practice and to obtain their adoption by private owners.

William L. Hall, timberland owner and consulting forester in Arkansas, summed up the situation in 1939 as follows: "Good forest practice by private landowners is not yet sufficiently evident in the United States to be impressive to all, [but] the idea of good forest practice is advancing and is steadily strengthening its hold on landowners."

Industry Conferences. On April 7 to 9, 1937, the organized lumber industry held a national conservation conference of private and public representatives at Washington, to review the situation and recommend action for its improvement. Special attention was given to industrial forest practices and to public timber acquisition and disposal. The conference accepted "the continuous production, or sustained yield, of forest resources as the ultimate objective of our industries."

It favored the formulation of rules of forest practice, but did not directly advocate their enforcement by state legislation. It recommended continuation of the general policy of "withholding from sale all Federal timber including Indian timber not needed to maintain existing operations, unless contrary to the best economic and social interests of the Indians." Other subjects covered by the six committees into which the conference was divided included forest protection, forest taxation, forest credit, forest research, and farm woodlands and other small ownerships.

A month later (May 3, 1937) representatives of a majority of the pulpwood-using companies in the South, meeting at New Orleans, Louisiana, adopted a "Statement of Conservation Policy for the Southern Pine Pulpwood Industry." In this statement, "It was agreed that all land, including non-company land, must be cut over in a manner which will maintain and build up the forest growing stock" and that each pulpwood operator would employ the necessary qualified personnel to ensure proper compliance with this agreement. Rules of forest practice were subsequently prepared for each of the three subregions, and the Southern Pulpwood Conservation Association was organized to direct their enforcement.

On July 30, 1937, the National Lumber Manufacturers Association through its board of directors urged each of its regional affiliates to continue the development of suitable forest-practice rules and methods. It further advocated that, "when confirmed by the tested experience of the region as effective in furthering its general purposes of continuous forest production and sustained yield operation of forest lands, such rules and methods of forestry practice be considered as matters for State legislation."

Summary. President Roosevelt's inauguration in 1933 was followed by a burst of conservation activities unequaled since early in the century. Many of the programs undertaken were emergency measures designed to relieve unemployment and to help the nation get back on its economic feet. Notable in this group of activities were the C.C.C., the Shelterbelt Project, the retirement of submarginal agricultural lands, and the formulation of rules of forest practice under the National Industrial Recovery Act.

Other important measures that were more the result of the normal evolution of forest and range policy than of the current emergency included the inauguration of range management on the unreserved public domain under the Taylor Grazing Act of 1934; the reclassification of the revested O. and C. timberlands and the application to them of sustained-yield forest management under the act of 1937; and the extension of Federal-state cooperation in the improvement of wood-lot management under the Cooperative Farm Forestry Act of 1937.

Progress in the field of State forestry resulted largely from the activities of the C.C.C. on state lands and the increased assistance to farmers made possible by the Cooperative Farm Forestry Act. At the same time substantial progress was made in fire protection, reforestation, and the acquisition of state forests.

In spite of unfavorable economic conditions, private owners both collectively and individually paid more attention than ever before to the possibilities of improved forest management. Rules of forest practice were prepared under the legal pressure exerted by Article X of the Lumber Code prior to invalidation of the National Industrial Recovery Act by the Supreme Court, and subsequent efforts were made both at the regional and the national level to get improved practices voluntarily adopted by private owners. Actual accomplishment in the woods was not conspicuous, but the foundation was laid for the rapid progress that took place during the forties.

REFERENCES

Butler, Ovid: "Youth Rebuilds—Stories from the C.C.C.," American Forestry Association, Washington, 1934.

————: The Oregon Checkmate, *Amer. Forests,* 42:157–162, 196–197, 1936.

Conference of Lumber and Timber Products Industries with Public Agencies on Forest Conservation, *Jour. Forestry,* 32:275–307, 1934.

Greeley, William B.: "Forests and Men," chaps. 7–8, Doubleday & Company, Inc., New York, 1951.

Peffer, E. Louise: "The Closing of the Public Domain," chaps. 12–14, Stanford University Press, Stanford, Calif., 1951.

Robbins, Roy M.: "Our Landed Heritage," chap. 24, Peter Smith, New York, 1950.

CHAPTER 11

Forestry Comes of Age

Forestry as a profession in the United States can be regarded as dating back to 1900, when the Society of American Foresters was organized. The actual practice of forestry on an extensive scale started in 1905 following the transfer of the forest reserves to the Department of Agriculture. Forestry on Indian lands generally began in 1910 and on O. and C. revested lands in 1937, in both cases under the administration of the Department of the Interior.

By 1940 Federal forestry was well established. State forestry was on firm ground in some regions and in some fields, notably fire control. Private forestry was gradually assuming increasing importance, although the total area under good management was still relatively small. From then on there was a decided change in the situation, with marked progress on all fronts, and particularly on the part of state governments and private owners.

President Urges Congressional Study of Forest Problems. On March 14, 1938, President Roosevelt sent a special message to Congress recommending the appointment of a joint committee to conduct a study of the forest problem which he hoped would form the basis for essential legislation at its next session.

Forests [said the President] are intimately tied into our whole social and economic life. . . . The forest problem is therefore a matter of vital national concern, and some way must be found to make forest lands and forest resources contribute their full share to the social and economic structures of this country, and to the security and stability of all our people. . . . The public has certain responsibilities and obligations with respect to private forest lands, but so also have private owners with respect to the broad public interests in those same lands. . . .

The fact remains that, with some outstanding exceptions, most of the States, communities, and private companies have, on the whole, accomplished little to retard or check the continuing process of using up our forest resources without replacement. This being so, it seems obviously necessary to fall back on the last defensive line—Federal leadership and Federal action. Millions of Americans are today conscious of the threat. Public opinion asks that steps be taken to remove it.

Among the specific items that the President asked Congress to consider was "the need for such public regulatory controls as will adequately protect private as well as the broad public interests in all forest lands." This proposal was in line with the policy that Silcox had advocated with increasing vigor since becoming Chief of the Forest Service on November 15, 1933. In his annual report for 1937 he presented the orthodox "three-point program" of the Forest Service, consisting of public ownership and management, public cooperation with private owners, and public regulation. The latter he referred to as "a margin of sovereignty over private forest lands" to "protect broad vital public interests. Private owners who recognize social obligations inherent in forest-land management will also be protected, by such regulation, from owners who might otherwise continue ruthless exploitation."

Joint Congressional Committee on Forestry. Congress acceded to the President's request by appointing a Joint Committee on Forestry consisting of five members from each House. The Committee was directed to investigate the present and prospective situation with respect to the condition, ownership, and management of forest land in the United States "as it affects a balanced timber budget, watershed protection, and flood control and other commodities and social benefits which may be derived from such lands." Among other things, the Committee was specifically instructed to ascertain the adequacy and effectiveness of present activities in forest protection and of cooperative efforts between the government and the states; other measures necessary to ensure timber cropping on private lands; the need for extending public ownership; and the need for public regulatory control.

The Committee held public hearings at eight places in different parts of the country, at which testimony was presented by representatives of practically every group having an interest in forests, forest products, and forest services. Much of the discussion dealt with cooperative and regulatory relationships between the public and private owners. The Forest Service took the position that "the crux of the Nation's forest problem lies in commercial forest lands in private ownership, and the uncontrolled exploitation of these lands."

Position of Forest Service. Earle H. Clapp, who served as Acting Chief of the Service from the date of Silcox's sudden death on December 20, 1939, until the appointment of Lyle F. Watts as Chief on January 8,

1943, presented the views of the Forest Service in his annual report for 1940. After stating that the Forest Service preferred Federal to state regulation, he summarized the compromise plan which it had submitted to the Committee. Since the principles on which that plan was based were consistently adhered to by the Service during Watts' administration, they justify quoting:

1. That States shall have the opportunity to administer regulation, with a reasonable but definite period of 5 to 7 years within which to pass State legislation and apply it.

2. That the Federal Government shall contribute on a 50–50 basis to the cost of such State administration.

3. That State legislation and standards of enforcement shall be satisfactory to the Federal Government, with mandatory provision that Federal financial assistance in regulation be withdrawn if enforcement proves unsatisfactory.

4. That if requested by a State to do so, or if after the formative period the State does not undertake it or does not attain satisfactory standards, the Federal Government shall be authorized to administer enforcement within such State or States.

Committee's Report. The Committee submitted its report to Congress on March 24, 1941. The report was divided into three parts—findings of fact, recommendations, and statistics, based chiefly on information supplied by the Forest Service as a result of the forest survey that it had been conducting since 1930. The Committee concluded that

the President was justified in emphasizing the importance and seriousness of the forest situation, and the need for prompt and definite remedial action. . . . For what has happened to our forest land, timber landowners are responsible because of their improvident treatment of the forest, the public, for its indifference, and the government for allowing it to continue.

Commercial forest lands in private ownership were found to constitute the nation's major forest problem. The solution of that problem, in the judgment of the Committee, must be based on two fundamental principles: "Private ownership of forest land carries with it certain obligations that in the public welfare private owners must help redeem. . . . Public interests inherent in all forest lands carry with them certain obligations which the public must redeem."

Recommendations and Results. The specific recommendations offered for the application of these principles were as follows:

1. Increase of the authorization for Federal cooperation in fire control to 10 million dollars "provided the respective States pass legislation pro-

viding for proper State, county, and district fire protection and regulations governing minimum forestry practices to be administered as approved by the Secretary of Agriculture." Withdrawal of Federal assistance in fire control and regulation would be mandatory if state standards and enforcement proved unsatisfactory, and discretionary with respect to other forest cooperative funds. Private owners should be given full opportunity to participate in formulating requirements, with the right of appeal through nongovernmental boards or other channels for review and reconsideration of requirements. States should be given a reasonable period of perhaps three to five years within which to pass and apply satisfactory legislation.

2. Extension of the Clarke-McNary Act to provide for cooperative protection against forest insects and diseases on private and state-owned forest lands.

3. Extension of Federal cooperation in reforestation to all landowners and authorization of the amounts needed for that purpose.

4. Expansion of Federal financial cooperation in forest management under the Cooperative Farm Forestry Act of 1937.

5. Passage of "legislation authorizing the establishment of cooperative sustained units to enable sustained-yield management of intermingled public and private holdings conditional upon management and woods practices approved by the Secretary of Agriculture." Within such units, timber could be sold without competitive bids at not less than its appraised value.

6. Provision for a forest-credit system to make long-term loans at low interest rates to private forest and naval-stores operators through facilities already available. "These loans should be conditional upon both sound forest practice and sound investment."

7. Encouragement of farm-forest cooperatives, including Federal financial aid in building and operating forest industries and woodworking plants.

8. Passage of legislation providing for leases and cooperative agreements with private forest landowners, communities, institutions, and states.

9. Extension and intensification of research in all phases of forestry and wood utilization.

10. Amendment of the State Forest Acquisition Act of 1935 (Fulmer Act) by making it applicable to institutions, communities, and subdivisions of the state, and by increasing the authorized appropriation to 10 million dollars.

11. Acceleration of the program for the acquisition of land for national forests, with special emphasis on the blocking up of present scattered areas.

12. Provision of an equitable system of financial contribution to local governments in lieu of taxes on forest land removed from the tax rolls through Federal acquisition.

13. Provision for more adequate protection against fire, insects, and diseases on national forests.

14. More intensified management of timber, forage, wildlife, recreation, and watershed resources on national forests.

15. Authorization of an annual appropriation of $750,000 for early completion of the forest survey of the United States.

16. Investigation of the apparent monopolistic purchasing of pulpwood and price fixing of paper and other pulp products.

The activities of the Committee had great educational value. Its report also helped materially to secure passage in 1944 of the Sustained-yield Forest Management Act, the increased authorization for fire control under the Clarke-McNary Act, and the increased authorization for the forest survey. It also paved the way for the constructive amendments of the Clarke-McNary Act in 1944 and 1949 and the passage of the Cooperative Forest Management Act in 1950. It marked a decided step forward in the development of forestry in this country, and particularly in the strengthening of cooperative relationships between the Federal government, the states, and private owners.

"Cooperative Forest Restoration Bill." The Committee's recommendations concerning public control of cutting on privately owned lands did not result in any Federal legislation on the subject, although several bills dealing with it were presented to Congress. The one which attracted most attention and around which discussion largely centered was introduced on November 13, 1941, by Senator Bankhead of Alabama (S. 2043). It became known both as the "cooperative forest restoration bill" and as the "omnibus forestry bill" because of the broad scope of Federal activities covered by its eleven titles.

Title I was headed "State Regulation of Private Forest Lands." It authorized any state to submit to the Secretary of Agriculture a "plan for the conservation and proper use of privately owned forest lands within its borders"; and it authorized the Secretary to cooperate financially in the administration of any state plan meeting his approval. To be acceptable a plan must comply with certain requirements, including provision "for the adoption and enforcement of rules of forest practice . . . which will prevent forest destruction and deterioration and maintain forest lands in a reasonably productive condition." With the approval of the appropriate state agency, working plans for individual properties might be substituted for compliance with the rules of forest practice.

The bill prohibited transportation "in commerce" of "any forest product produced contrary to any provisions of any State plan." It also instructed the Secretary of Agriculture, whenever he found it necessary or appropriate, to require a certificate of clearance as a prerequisite to the movement of forest products "in commerce" from the state or region concerned. Violation of either of these provisions was made punishable by a fine of not more than $10,000.

Should any state within three years fail to adopt a satisfactory plan or should it fail to enforce adequately the provisions of an approved plan, the Secretary was directed to withdraw financial assistance to that state under the Clarke-McNary Act. He was further authorized, with the approval of a National Forestry Board of twelve members to be appointed by the President, to withdraw or curtail any assistance being extended "in furtherance of any flood-control, erosion-control, or other program, the purposes of which, in his judgment, are related to the purposes of this title, and which, in his judgment, would be adversely affected by the absence or inadequate administration of any State plan."

The bill made no provision for direct Federal control in case a state should fail to act, as had been done in an earlier bill introduced by Representative Pierce of Oregon. Although none of the proposals dealing with public control were approved by Congress, their influence was felt in regulatory legislation enacted by several of the states, beginning with Oregon in 1941.

War, Wood, and Livestock. With the outbreak of the Second World War, wood became a critical war material. Wood and wood derivatives were needed for housing, ships, wharves, airplanes, trucks, boxes and crates, paper and paper products, explosives, and a host of other uses. Colonel F. G. Sherrill, who was chief of the Materials and Equipment Section of the Army Corps of Engineers, said that lumber is "the most vital material for the successful prosecution of the war" and that the lumber industry "is the most important war industry in the country."

Civilian consumption of wood had to be substantially curtailed, and the combined efforts of Federal and state agencies and private industries were required to meet military needs. Research by the Forest Products Laboratory, the Timber Engineering Company, and other agencies was stepped up to a new high. At the Laboratory, courses were given for the training of more than 3,600 inspectors of packaging, wood aircraft, wood-ship construction, and the manufacture of veneer and plywood.

In spite of depleted personnel, the cut from national forests was greatly increased, including the shipment of Sitka spruce for aircraft construction from the Tongass National Forest in Alaska. The production of lumber, pulpwood, and other forest products by small owners in the East was stimulated by the Timber Production War Project conducted cooperatively by the Forest Service and state forestry organizations. Industry, in spite of handicaps imposed by shortages of labor and equipment, unfavorable weather, and transportation difficulties, responded loyally and effectively to the demand for suddenly increased production.

That demand emphasized the fact that the situation with respect to the timber supply was far from satisfactory. As the Acting Chief of the Forest Service put it:

We now have to search out isolated remnants of timber for specialty use. For timber of high quality we have to depend altogether too largely on the remaining virgin stands on the Pacific coast. This involves long and costly hauls to industrial centers at a time when transportation facilities are already severely taxed. And many of these consuming centers have large areas of forest land near by, potentially capable of good timber production, but now producing only a fraction of what they could.

The immediate result of the pressure for more and more wood under conditions that brooked no delay, and that at best were unfavorable to adherence to high standards of forest practice, was to worsen an already unsatisfactory situation. Forest depletion was accelerated rather than checked. On the other hand, the long-term result of the pressure was probably beneficial, since it forced more general recognition than had previously existed of the need for vigorous action both by government and industry to bring about improved forest practices if the forests are to make the contribution of which they are capable to the country's safety in war and prosperity in peace. Legislation by Congress and a campaign by industry to speed up the practice of forestry by private owners soon proved that the lesson had been at least partially learned.

Wartime requirements for meat, wool, and hides similarly emphasized the depleted condition of much of the Western range and the need for improved range management if the supply of livestock was to be increased or even maintained. Special effort was made to avoid the overgrazing which had caused serious depletion during the First World War and to speed up the research on which sound management must rest.

Cooperative Forest-fire Control. Federal cooperation with the states in the protection of forests from fire was greatly strengthened by the act of June 28, 1944. This act increased the maximum appropriation of $2,500,000 authorized by the Clarke-McNary Act to $6,300,000 for the fiscal year 1945, with additional annual increases up to 9 million dollars for the fiscal year 1948 and thereafter. A new feature incorporated in an act passed later in the same year was the authorization of the Secretary of Agriculture to spend not more than 1 million dollars a year during the existing emergency for preventing and suppressing forest fires on critical areas of national importance without requiring an equal expenditure by states and private owners.

Even more substantial increases were made by the act of October 26, 1949, which authorized an appropriation of 11 million dollars for the fiscal year 1950 with additional annual increases up to 20 million dollars for the fiscal year 1955 and thereafter. This legislation was based on the theory that the national interest in the protection of forests from fire is such as to justify the assumption by the Federal government of approximately half of the estimated cost of adequate protection on state and private lands. Actual appropriations have, however, never exceeded about a fourth of the total cost.

Allocations are made to the individual states primarily on the basis of the estimated cost of adequate protection and the expenditures made by the state itself, subject always, of course, to the matching requirement. In 1953 the Forest Service was cooperating with forty-three states and the Territory of Hawaii in the protection of 368,692,000 acres of forest land, leaving more than 58 million acres still without organized protection.

Protection from Insects and Diseases. Forest insects and forest diseases, like forest fires, are no respecters of property lines. That public agencies therefore have a responsibility to assist in preventing and reducing damage by them has been recognized by governmental participation in the control of the chestnut blight, the white pine blister rust, the spruce budworm, the hemlock looper, and many other enemies of the forest. In 1940 the Lea Act specifically provided for Federal cooperation in the protection of forest lands from white pine blister rust, irrespective of ownership, except that on state or private lands Federal expenditures must be at least matched by state or local agencies and/or individuals and organizations.

There was no general legislation on the subject until 1947, when the Forest Pest Control Act declared it to be the policy "of the Government of the United States independently and through cooperation with the governments of States, Territories, and possessions, and private timber owners to prevent, retard, control, suppress, or eradicate incipient, potential, or emergency outbreaks of destructive insects and diseases on, or threatening, all forest lands irrespective of ownership." It authorized the appropriation of "such sums as the Congress may from time to time determine to be necessary" to enable the Secretary of Agriculture to achieve these purposes.

The Secretary is specifically authorized to conduct both surveys and actual control measures, either independently or in cooperation with state and local agencies and private owners. Operations on Federal lands not administered by the Department of Agriculture must be with the consent of the agency having jurisdiction or they may be conducted by that agency with funds made available by the Secretary of Agriculture. No funds can be expended on non-Federal lands "until such contributions toward the work as the Secretary may require have been made or agreed upon in the form of funds, services, materials, or otherwise."

With losses from forest pests now exceeding those from forest fires, the act constitutes an essential step toward their effective control.

Cooperation in Forest Management. Federal cooperation with the states in various phases of forest management under the Clarke-McNary Act of 1924 and the Norris-Doxey Act of 1937 was handled through both the state extension services and the state departments of forestry. In order to clarify the respective responsibilities of these agencies, the Association of State Foresters and the Association of Land Grant Colleges in 1948 agreed upon a statement of policy allocating primary responsibility for educational work to the state extension services and for technical-service assistance to the state departments of forestry. The closer working relationships between the two groups resulting from this agreement materially strengthened state activities in farm forestry and also helped to bring about the enactment of two important pieces of legislation expanding Federal participation in that field.

An act of October 26, 1949, amended the Clarke-McNary Act by increasing the authorization for Federal cooperation in the production and distribution of forest-tree planting stock ("CM-4") from $100,000 to $1,000,000 for the fiscal year 1950, with additional increases up to $2,500,000 for the fiscal year 1953 and thereafter. It also removed the limitation on the distribution of such stock to farmers only.

The same act also increased the authorization for Federal cooperation in educational service to landowners in forest management ("CM-5") from $100,000 to $500,000. Cooperation in that field continued to be limited to farmers, but was broadened to include assistance in the harvesting, utilizing, and marketing of forest products. The work is handled by the state extension services.

The Cooperative Forest Management Act of August 25, 1950, authorized the Secretary of Agriculture "to cooperate with State Foresters or equivalent officials . . . for the purpose of encouraging the states, territories, and possessions to provide technical services to private forest land owners and operators, and processors of primary forest products with respect to the management of forest lands and the harvesting, marketing, and processing of forest products. . . ." Such cooperation is to be provided in accordance with a plan agreed upon by the Secretary and the state forester.

The act authorized an annual appropriation of $2,500,000. Apportionment of the available funds is made by the Secretary of Agriculture after consultation with a national advisory committee of not less than five state foresters selected by a majority of the state foresters. Federal expenditures are not to exceed those of the state for the same purpose.

It was further specified that the provisions of the act and of the plan are to be carried out in such manner as to encourage the utilization of private agencies and individuals furnishing services of the type covered by the act. The increasing availability of such services is indicated by the organization in 1948 of the Association of Consulting Foresters. Its objectives are to assure forest owners of competent professional service through maintenance of high standards of performance by consulting foresters and to promote the most economic and scientific management of forest resources.

This act, which repealed the Cooperative Farm Forestry Act of 1937 as of July 1, 1951, marked a major advance in cooperative relations between the Federal government and the states.

Sustained-yield Forest Management Act. Recognition of the fact that the stability of certain communities dependent primarily on the manufacture of forest products can be assured only by the steady and continuous flow of raw materials from the forest led Congress to pass the sustained-yield forest management act of March 29, 1944. This act, which followed the general pattern set by the O. and C. act of 1937, authorized the Secretary of Agriculture and the Secretary of the Interior to establish either (1) cooperative sustained-yield units consisting of Federal forest land and private forest land or (2) Federal sustained-yield units consisting only of Federal forest land, whenever the stability of communities primarily dependent upon Federal stumpage cannot be maintained through the usual timber-sale procedures.

In such cases the Secretaries were authorized to enter into agreements with private forest landowners, with each other, and with other Federal agencies providing for the coordinated management of forest lands for the purposes of (a) promoting the stability of forest industries, of employment, of communities, and of taxable forest wealth; (b) providing for a continuous and ample supply of forest products; and (c) securing the benefits of forests in maintenance of water supply, regulation of stream flow, prevention of soil erosion, amelioration of climate, and preservation of wildlife.

Provision was made for the sale of Federal stumpage in sustained-yield units to cooperating landowners or to responsible purchasers within communities dependent on Federal stumpage, without competitive bidding, at prices not less than the appraised value of the timber. In the case of cooperative sustained-yield units involving private land, the agreement may run for as long as 100 years, and timber on the private land must be utilized at a rate and under methods of cutting and harvesting approved by the appropriate Secretary. Provision is made for the periodic adjustment of stumpage prices as warranted by changing market conditions.

So far only one cooperative sustained-yield unit has been established. The cooperative agreement is with the Simpson Logging Company, runs for 100 years, and aims to stabilize the towns of Shelton and McCleary, Washington. Altogether 111,466 acres of Federal land and 158,760 acres of company land are included. Several other similar agreements have been proposed, but their approval has been blocked by the vigorous opposition of smaller operators and certain labor groups.

Federal sustained-yield units have been established by the Department of Agriculture as follows:

Vallecitos Unit of 73,000 acres on the Carson National Forest in New Mexico, 1948

Flagstaff Unit of 905,000 acres on the Coconino National Forest in Arizona, 1949

Grays Harbor Unit of 116,583 acres on the Olympic National Forest in Washington, 1949

Big Valley Unit of 82,185 on the Modoc National Forest in California, 1950

Lakeview Unit of 488,010 acres on the Fremont National Forest in Oregon, 1950 and 1951

Prior to 1953, action looking to the establishment of Federal sustained units was initiated by the Forest Service. A change in policy effected in that year now places the initiative on the communities that would be benefited by such units. Final decision with respect to their establishment still rests with the Secretary of Agriculture.

No cooperative or Federal sustained-yield units have been established on the revested O. and C. lands under the 1944 act. In 1946 and 1947, however, the Secretary of the Interior, acting under authority of the 1937 act, established twelve "master units" consisting of smaller sustained-yield units, the product of which should logically go to definite marketing areas. It is normally required that all timber cut from the master units be given primary manufacture within these marketing areas.

Modification of Federal Income Tax. In 1944 an important amendment to the Federal income tax law permitted any profit from the sale or cutting of timber held for more than six months to be reported as a long-term capital gain. Prior to that time, any profit from the sale of timber was subject to the capital-gains tax, which is levied at half the applicable income tax rate and never exceeds 25 per cent; while any profit from the harvesting of timber by the owner himself was subject to the straight income tax rate, which may run as high as 92 per cent. By removing the discrimination against the owner who does his own cutting, the amendment tended to encourage the stability of ownership which is so important for sustained-yield forest management.

The same amendment also permitted the forest owner to choose between reporting such items as silvicultural operations and temporary improvements as current expenses deductible from current income or as capital investments. The owner must, however, continue to use whichever method he chooses; shifts from one method to the other are not permissible. Expenditures for planting, for nonexpendable equipment, and for permanent improvements continue to be classified as capital investments and cannot be deducted from current income.

Federal Credit for Forest Management. Until 1953, national banks were not allowed to make loans secured by forest lands as collateral, since such lands were classified as unimproved property not acceptable as security. This situation, which made it difficult and often impossible for forest owners to obtain credit for the improvement of their timberlands, was partially remedied by passage of an act authorizing national banks to lend up to 40 per cent of the appraised value of forest lands properly managed in all respects. Loans may run for not more than ten years, with amortization at the rate of 10 per cent a year.

Attack on the Grazing Service. The cordial relations that had existed between the Grazing Service and the livestock industry were marred in 1939 by the refusal of certain Nevada cattlemen to pay fees for temporary revocable licenses pending the acquisition of sufficient information to justify the issuing of ten-year permits. A ruling by the Nevada Supreme Court restraining the Grazing Service from interfering with the free use of the range by the plaintiffs was reversed in 1941 by the United States Supreme Court.

Meanwhile Senator McCarran of Nevada in 1940 introduced a resolution severely criticizing the administration of the public lands and proposing a searching investigation of the situation. The investigating committee appointed the following year was headed by McCarran and continued its activities until 1947, when its final report was submitted.

Controversy at first centered largely on the question of grazing fees. In 1941 the Grazing Service proposed a graduated increase in fees for the use of grazing districts that on the average would have about tripled the nominal fee of 5 cents per animal unit per month established in 1936. The outbreak of war delayed action on this proposal, but in 1944 a new

Director of Grazing, Clarence L. Forsling, formerly chief of the Branch of Research in the Forest Service, attempted to put it into effect. At the same time the House Committee on Appropriations suggested higher fees, and the next year (1945) indicated its belief that fees should be increased or appropriations decreased. The Grazing Service thus found itself in the same position that the Forest Service had occupied twenty years earlier, with Congress and users of the range pushing it in opposite directions.

Appropriations and Grazing Fees. When it appeared unlikely that the Interior Department would increase fees, Congress adopted the other alternative. It reduced the already meager appropriation for salaries and expenses in the Grazing Service by 44 per cent for the fiscal year 1947, and by an additional 11 per cent the following year. The result was so seriously to cripple administration of the grazing districts that, as Director Forsling expressed it, the reduced appropriations were inadequate to permit anything but "licensed abuse" of the range.

The livestock industry was not averse to Federal expenditures that would build up the range. On the other hand, it favored financial starvation of the Grazing Service until it could be brought to terms on the matter of grazing fees, including a reduction in the percentage of such fees allocated to local communities.

The deadlock was broken by inclusion in the Interior Department Appropriations Act of August 6, 1947, of an amendment drastically revising the distribution of receipts from grazing specified in the Taylor Grazing Act. The amendment provided that permittees should thereafter be charged a grazing fee and a range-improvement fee; that 12½ per cent of the grazing fee, instead of the previous 50 per cent, should be paid to the counties; and that all of the range-improvement fee should be retained by the government. The purpose of the change was to make a larger proportion of the receipts from grazing districts available for their administration and improvement and thereby to obviate the need for any material increase in grazing fees. The counties were to continue to receive 50 per cent of the grazing fees from unreserved lands not in grazing districts, and their share of the receipts from ceded Indian lands was changed from 25 per cent to 33⅓ per cent.

On May 1, 1947, Secretary Krug had raised the grazing fee from 5 to 8 cents per animal unit per month. Immediately following passage of the act of August 6, he designated 6 cents of the new fee as an outright grazing fee and 2 cents as a range-improvement fee. Substantial increases in appropriations for the Grazing Division have subsequently been made, but the amount available is still far from adequate for effective management. Fees were raised to 11 and 4 cents, respectively, in 1955.

Other Policies. The act of August 6, 1947, also required the Secretary of the Interior to "take into account the extent to which such districts yield public benefits over and above those accruing to the users of the forage resources for livestock purposes." In effect, this ambiguous language apparently legalized the stockmen's claim that grazing fees should be based on the cost of administration and not on the value of the forage. It obviously raised the question as to why the same principle should not be applied to other Federal lands and to other resources.

The appropriation for the Bureau of Land Management for the fiscal year 1948 carried the wording: "For the administration of the public lands and their resources . . . including their protection, use, maintenance, improvement, development, and disposal." The inclusion for the first time of the words "and disposal" was undoubtedly a result of a campaign started by the stockmen in 1946 to provide for the sale to private owners of public lands which had been allotted to them for use in connection with their private lands. The proposal did not have the complete support of the livestock industry, was vigorously opposed by conservation groups, including notably those interested in recreational use of the public lands, and did not receive serious consideration in Congress.

An act of May 28, 1954, removed the limit of 142 million acres on the total area that might be included in grazing districts.

Forest Service under Fire. The Forest Service did not escape involvement in the campaign started by the stockmen to bring about the sale to existing permittees of public lands chiefly valuable for grazing and to give them greater control over the administration of such lands. In May, 1947, the House of Representatives authorized an investigation by its Committee on Public Lands of any phase of the public-land situation. The task was assigned to a subcommittee which became known as the Barrett committee because of the fact that it was headed by Representative Barrett of Wyoming. Like its predecessor, the McCarran committee, it held hearings throughout the West at which every opportunity was offered for the airing of grievances against both the Forest Service and the Bureau of Land Management, in which administration of the grazing districts was then located.

The stockmen's proposals and criticisms met with strenuous opposition as well as support. This opposition came not only from the bureaus under attack but even more effectively from the general public. Disapproval was especially strong from the recreation, forestry, and reclamation groups, and many of the stockmen themselves did not follow the lead of their association officers. The Idaho Legislature went so far as to pass a formal resolution objecting to the proposed transfer of title.

Recommendations of Barrett Committee. After completion of the hearings, the Barrett committee presented to the Secretary of Agriculture six specific proposals concerning the administration of grazing on the national forests. Four of these proposals he accepted without question as being virtually in accord with existing policy. The Secretary did not feel that he could surrender to "impartial appeal boards" his responsibility for final administrative decisions or that he could agree to a three-year moratorium on livestock reductions on national-forest ranges.

The committee's formal report was submitted to the House of Representatives in August, 1948. It expressed approval of the multiple-use policy of the Forest Service and recommended against the sale or transfer of national-forest lands. Although stating that it was unalterably opposed to overgrazing, the committee repeated its recommendation of a three-year moratorium on reduction in numbers of livestock. It recommended that Congress recognize grazing, recreation, and wildlife as basic uses of the national forests and that grazing advisory boards be given legal status. It strongly urged larger appropriations for reseeding and other range improvements, with substantial results in subsequent appropriation measures.

Granger-Thye Act. Three provisions of the Granger-Thye Act of April 24, 1950, were undoubtedly due primarily to the activities of the Barrett committee:

1. The act authorized the appropriation of grazing fees from national forests in amounts equivalent to 2 cents per animal month for sheep and goats and 10 cents per animal month for other kinds of livestock, for artificial revegetation, construction and maintenance of range improvements, control of range-destroying rodents, and eradication of poisonous plants and noxious weeds, "in order to protect or improve future productivity of the range."

2. It authorized the election of grazing advisory boards for each national forest or administrative subdivision upon request of a majority of the grazing permittees. These boards consist of three to twelve members who are grazing permittees in the area concerned, plus one wildlife member appointed by the state game commission or corresponding body. Appeals and proposed changes in regulations must be referred to the boards for advice, and they may also initiate recommendations concerning the administration of grazing on the national forests. If the Secretary disapproves of any recommendation made by a board, his reasons for doing so must be stated in writing.

3. It authorized the Secretary of Agriculture, upon such terms and conditions as he may deem proper, to issue permits for the grazing of

livestock on national forests for periods not exceeding ten years and renewals thereof.

The two latter provisions did little more than give legal recognition to policies already in effect. Although the act made no change in the basic purposes for which national forests may be established, as specified in the act of 1897, it undoubtedly strengthened the position of grazing permittees. To some extent, therefore, it did exactly what was feared and opposed by those who fought the enactment of any legislation on the subject in the 1920's.

As a result of continued agitation for further legislation, detailed provisions for the administration of grazing on the national forests were added by the Senate to the 1954 agricultural support and marketing bill but were eliminated in conference.

National Forest Advisory Council. Another important development, which resulted in part from the criticism of grazing administration, was the creation in May, 1948, of a National Forest Board of Review of three members to advise the Secretary of Agriculture concerning problems arising in connection with use of the national forests. One of its first assignments was to make a study of the grazing situation on the Roosevelt National Forest in Colorado. Its conclusions were that watershed protection, recreation, grazing, and timber uses should be favored in that order of priority; that closure of limited areas to grazing was justified; and that "there should be a clear and general understanding that it is not a major policy of the Forest Service to exclude grazing on any except such particularly unsuitable, limited areas." It also recommended the creation of a Colorado Forest Resource Board to assist the Forest Service in formulating local policies and in promoting the fullest use of all resources consistent with good management.

At the request of the American National Live Stock Association and the National Wool Growers Association, the Board in January, 1950, reviewed Forest Service policies relating to transfer adjustments and grazing trespass. It recommended retention of the existing policy for transfer readjustments, with clarification of procedure, and the incorporation in the grazing manual of the provisions for the handling of trespass.

In 1950 the name of the Board was changed to National Forest Advisory Council. During 1951 and 1952 it made a thorough study of the application of the mining laws to the national forests, as a result of which it suggested several important changes in basic legislation.

Mining and the National Forests. Mineral claims have long been the cause of much difficulty in the administration of the national forests created from the public domain. These forests are open to unrestricted mineral exploration and the patenting of valid claims under the act of May 10, 1872, except for the specific minerals covered by the Mineral Leasing Act of 1920 and its amendments. Patenting is, however, not required, and until 1955 a claimant could continue indefinitely to control the use of all of the resources on his claim by spending $100 a year for development work. If the claim is patented, the timber, forage, and other resources go with the title.

This situation made it possible for a person to hold, or to get patent to, a claim that is more valuable for other purposes than for any mineral it may contain. That such practice was common is shown by a Forest Service estimate that as of January 1, 1950, less than 15 per cent of the patented claims in the Western national forests had been or were commercially successful mines and that less than 3 per cent of the unpatented claims were producing ore in commercial quantities. The unpatented claims covered an area of 7,100,000 acres, with a timber value of $57,251,000.

Use of the mining laws to obtain control of, or title to, nonmineral values has resulted not only in loss to the public of those values but also in serious interference with national-forest administration. There are thousands of claims containing little or no mineral of commercial value, both patented and unpatented, which are holding up timber sales, road construction, recreational developments, and other activities that are essential features of multipurpose management. The apparent remedy of canceling invalid claims is ineffective because the process is a long, costly, and difficult one and because there is nothing to prevent the filing of a new claim on virtually the same area.

Many measures have been proposed to prevent abuse of the mining laws without hampering the legitimate miner. These include the granting of title to the minerals only; permitting use of the surface by the claimant only to the extent necessary to extract the mineral, with logging and grazing under applicable national-forest practices; restricting to five or ten years the period during which a claim may be held without applying for patent; canceling new claims at the end of three years if acceptable evidence of the occurrence of valuable mineral deposits has not been obtained; requiring new claims to be recorded at the Federal district land office; applying the provisions of the Mineral Leasing Act of 1920 to nonmetalliferous minerals such as sand, gravel, and pumice.

Congress Acts to Curtail Abuses. Bills embodying many of these proposals in various forms were before Congress for several years, but not until July 23, 1955, was any action taken on them. Congress then passed a law containing the following provisions:

1. It broadened the scope of the Materials Disposal Act of July 31,

1947, by removing common varieties of pumice, pumicite, and cinders as well as sand, stone, gravel, and clay from the operation of the mining laws on both reserved and unreserved public lands (except national parks, national monuments, and Indian lands), and on land-utilization projects acquired under the Bankhead-Jones Farm Tenant Act of 1937. Hereafter these minerals (and also vegetative materials) may be disposed of only under rules and regulations prescribed by the secretary of the department having jurisdiction over the lands in question.

2. It limited the use of unpatented mining claims hereafter located under the mining laws to prospecting, mining, or processing operations and uses reasonably incident thereto. Any cutting of timber for use in the development of the claim, other than to provide clearance, must be in accordance with sound principles of forest management.

3. It authorized the Federal government to manage and dispose of the timber, forage, and other nonmineral resources on unpatented claims hereafter located under the mining laws, provided that such use shall not endanger or materially interfere with mining operations. If such management deprives the claimant of timber needed in the development of his claim, adequate supplies for this purpose must be made available to him without charge from other Federal lands.

4. It established a procedure whereby the right to timber and other nonmineral, surface resources on existing, inactive, mining claims may be canceled or waived.

These provisions will do much to improve a situation that has previously been a serious handicap in the management of the national forests and other Federal lands. At the same time it will protect the interests of the legitimate miner and will not interfere with the location and development of valid mining claims. Other safeguards against abuses can be added later if experience proves that they are needed.

Miscellaneous National-forest Legislation. Partly because of the increased hazard resulting from the war, the Secretary of Agriculture was authorized in 1944 to pay rewards for information leading to arrest and conviction for violating laws and regulations relating to fires in or near national forests.

An act of August 7, 1947, made the mineral leasing laws, with certain exceptions, applicable to lands acquired by the United States, including those acquired under the Weeks Act and the Clarke-McNary Act.

Further strengthening of the protection of the national forests was afforded by an act of February 10, 1948, which provided that "whoever, without lawful authority or permission, shall go upon any national-forest land while it is closed to the public by or under authority of a regulation of the Secretary of Agriculture made pursuant to law" shall be subject to a fine of not more than $500 and to imprisonment for not more than six months.

An act of March 30, 1948, permitted "the use and occupancy of national-forest lands in Alaska for purposes of residence, recreation, public convenience, education, industry, agriculture, and commerce, not incompatible with the best use and management of the national forests, for such periods as may be warranted but not exceeding thirty years and of such areas as may be necessary but not exceeding eighty acres."

The Anderson-Mansfield Act of October 11, 1948, authorized increased appropriations for the reforestation and revegetation of the forest and range lands of the national forests at a rate that would permit the planting of 4 million acres in fifteen years. Appropriations have so far fallen considerably short of authorizations.

The Granger-Thye Act of 1950 amended the act of August 11, 1916, which had authorized the Secretary of Agriculture to require deposits from timber purchasers for disposing of brush and other debris resulting from cutting operations on the national forests, so as to cover any unused balance into miscellaneous receipts, forest-reserve fund, instead of refunding it to the depositor. The same act also authorized the Secretary of Agriculture to sell forest-tree seed and nursery stock to states and political subdivisions thereof and to public agencies of other countries at not less than the actual or estimated cost.

Congress Acts on Controverted O. and C. Lands. After many years of delay, during which legislation on the subject was almost continuously under consideration, Congress in 1954 finally settled the jurisdictional controversy over the revested O. and C. grant lands in the indemnity strip. An act of June 24 declared these lands "to be revested Oregon and California railroad grant lands," which "shall continue to be administered as national-forest lands by the Secretary of Agriculture."

It further stipulated that revenues hereafter derived from these lands and revenues heretofore derived and placed in a special deposit shall be disposed of as provided in the act of August 28, 1937. This means that the eighteen counties involved will get 75 per cent of the gross receipts, less cost of access roads. In order to facilitate administration and accounting, the Secretary of Agriculture was authorized to designate in each county an area of national-forest land of a value substantially equal to that of the controverted land in the county, from which revenues will be disposed of under the 1937 act in lieu of revenues from the controverted land.

Another provision of the act directed the Secretary of the Interior and the Secretary of Agriculture, within two years, to exchange administrative jurisdiction over national-forest and intermingled and adjacent (within 2 miles) O. and C. lands of approximately equal aggregate value, for the purpose of facilitating administration. So far as practicable, exchanges within each county must be on the basis of equal area as well as equal value, unless otherwise agreed to by the county.

Unreserved Public Lands. Congress in 1913 had authorized the sale of fire-killed timber on the unreserved public lands, but no authority to sell live timber existed until 1944, when the Secretary of the Interior was authorized to dispose of timber and other material on such lands during the period of hostilities. An act of July 31, 1947, gave the Secretary permanent authority to dispose of materials such as sand, stone, timber, and other forest products on the public lands of the United States "if the disposal of such materials (1) is not otherwise expressly authorized by law, including the United States mining laws, (2) is not expressly prohibited by the laws of the United States, and (3) would not be detrimental to the public interests."

Adequate compensation was required, although free use was permitted for other than commercial or industrial purposes or for resale. Sales of more than $1,000 must be made to the highest bidder after advertising and of less than that amount in such manner as the Secretary might prescribe. Receipts were to be disposed of in the same manner as receipts from the sale of public lands. The act stated specifically that it did not apply to national forests, national parks, national monuments, or Indian lands.

Developments in Alaska. Utilization of the forest resources of Alaska, nearly all of which are owned by the Federal government, has been slow in developing. There has been a small sawmill industry in the coastal forests of the territory for the last fifty years, but economic factors prevented their use on any large scale until 1953 and 1954, when the Alaska Plywood Corporation and the Ketchikan Pulp Company opened a new era by commencing production in a big way. Other prospective sales of national-forest timber give promise of still further development of the extensive forests of the region.

In the interior of Alaska is a vast empire of 125 million acres of forest land, some 40 million acres of which are considered to be of commercial density. The total stand in the combined commercial and woodland areas is estimated at 170 billion board feet, with an average annual cut of only about 20 million feet. The toll exacted by fire has, however, been exceedingly heavy. It is believed that since 1890 some 80 per cent of the area has been burned at least once. Not until 1940, when Congress appropriated $37,500 for the control of fires on the unreserved public lands, was any real attempt made to improve the situation. By 1954, the appropriation for this purpose had increased to a meager $242,000, which, with additional emergency funds, amounted to only about 2 mills per acre.

Timber management of the interior forests on anything even approaching an adequate scale still remains to be inaugurated. With an average of one forester to ten million acres of forest land, their effective protection and administration is obviously impossible. The present and potential value of the forests of the interior, for other purposes as well as for timber production, justifies much more attention than they have so far received.

An act of August 30, 1949, provided for the sale at public auction of public lands in Alaska not within national parks, national monuments, national forests, Indian reservations, or military reservations. When classified by the Secretary of the Interior as suitable for industrial or commercial purposes, including the construction of housing, such lands may be sold in tracts of not more than 160 acres to any bidder who furnishes satisfactory proof that he has bona fide intentions and the means to develop the tract for his intended use. The act somewhat resembles the act of March 30, 1948, relating to national forests, but provides for outright sales instead of leases.

Forest Appraisals. The nationwide Forest Survey authorized by the McSweeney-McNary Act of 1928 was initiated in 1930. This action constituted the first attempt to bring up to date, with adequate field work, the figures on the forest resources of the country which had been obtained by the Bureau of Corporations some twenty years earlier. Initial emphasis was placed on the Douglas fir region in the Pacific Northwest because of the location there of the heaviest remaining stands of virgin timber.

Progress reports for different regions presented the results of the survey as rapidly as they became available. Estimates of area, volume of standing timber, growth, and drain for the entire country were supplied to the Joint Congressional Committee on Forestry and were published in its 1941 report, although field work in many regions still remained to be completed. The marked excess of drain over growth shown by the figures undoubtedly contributed to the emphasis placed by the committee on the need for public control of cutting on private lands.

In 1944 Congress amended the original provisions of the McSweeney-McNary Act by authorizing an annual appropriation of $750,000 to complete the initial survey of forest resources, with the stipulation that total appropriations for this purpose should not exceed $6,500,000. It also authorized an additional appropriation of $250,000 a year to keep the survey current. In 1949 these figures were increased to $1,000,000, $11,000,000, and $1,500,000, respectively.

Simultaneously with the continuation of the survey a reappraisal of the nation's forest situation was made by the Forest Service during 1945 and 1946. The reappraisal showed, among other things:

That the country's forest growing stock was inadequate and poorly distributed.

That drain and growth were about equal in terms of cubic-foot volume, but that drain exceeded growth by about 50 per cent in terms of board-foot volume.

That in several regions low-grade hardwoods were increasing at the expense of the more valuable conifers.

That the management of forests in private ownership was on the whole, but with some conspicuous exceptions, much less satisfactory than of those in public ownership.

That the crux of the forest problem lies in the improved management of the small, nonindustrial holdings which comprise three-fourths of the total area of commercial forest land in private ownership.

Substantial agreement with these findings resulted from a concurrent forest-resource appraisal conducted by the American Forestry Association, independently but in close cooperation with the Forest Service, the states, and private agencies.

As of the fall of 1953, 72 per cent of the total forest area of the country had been covered by the Forest Survey, and 55 per cent of the area first surveyed during the 1930's had been resurveyed. In connection with these continuing activities, the Forest Service in 1952 undertook a special "timber resource review" for the purpose of presenting a comprehensive, up-to-date picture of the situation for the country as a whole. Active cooperation of public and private agencies was sought both in the planning and the conduct of the review.

American Forest Congresses. Following the completion of its forest-resource appraisal, the American Forestry Association sponsored an American Forest Congress at Washington, October 9 to 11, 1946. The gathering was the first of the sort to be held since the American Forest Congress of 1905. Its discussions centered around a preliminary statement of forest policy prepared by a Forest Program Committee which had met at Higgins Lake, Michigan, the preceding July.

The directors of the association then prepared a thirty-point "Program for American Forestry," which on February 1, 1947, was submitted to the entire membership for a referendum vote. The program was "based primarily on the need for education of forest owners in tree growing and its economic possibilities and on the need for more effective teamwork between private owners, the states and the federal government in realizing these possibilities." It was approved by more than 98 per cent of those voting.

The Fourth American Forest Congress was held in Washington, October 29 to 31, 1953, to revise the 1947 program in the light of changing conditions. Again the discussions centered around a proposed program prepared by a second "Higgins Lake Committee," the majority of whose membership was identical with that of the previous committee.

The program set up three immediate goals to be sought by constructive action in the fields of forest ownership, forest-land management, multiple use on all forest lands, education and assistance to forest owners, and research. These goals are: "(1) To meet the essentials of forest protection; (2) to improve the national timber crop sufficiently to balance the forest budget and establish a growth reserve; and (3) to attain the maximum of economic and social services from our forests by realistic application of the multiple-use principle to their management." The program, as revised by the directors of the association in the light of the discussions at the congress, was submitted to the entire membership for action in 1954 and was approved in its entirety by 93 per cent of those voting.

Mid-Century Conference on Resources for the Future. Another national conference of outstanding importance was held at Washington, December 2 to 4, 1953, under the sponsorship of Resources for the Future, a nonprofit corporation supported by the Ford Foundation. The Mid-Century Conference on Resources for the Future dealt with the major problems connected with the utilization and conservation of natural resources of all sorts, among which forests and range lands play an important part. It was divided into eight sections and attracted an attendance of some 1,500 people from all walks of life.

Since the purpose of the meeting was to identify problems and issues rather than to formulate policies, no votes were taken and no resolutions adopted. Its deliberations will serve as a guide to public and private agencies and to Resources for the Future in developing its own program.

President's Policy Commissions. In 1949 President Truman created the President's Water Resources Policy Commission of seven members, to which he assigned the task of developing a comprehensive water policy for the American people. The following year the Commission submitted a massive report in three volumes containing seventy recommendations covering all phases of the subject.

Two of these recommendations dealt somewhat indirectly with the influence of forests on runoff and erosion:

Watershed management should be included as a principal objective in the planning and development of basin programs, with large enough allotments of funds to enable soil conservation, range management, and forest agencies to undertake activities which will bring economically controllable deterioration of the land under control within a reasonable period of time. . . . Conservation storage of flood waters in the soil, underground, and in surface reservoirs on tributaries and upper reaches of rivers should be a principal factor in planning and development of river basin programs.

Forest influences were, however, discussed at greater length in the body of the main report.

The President's Materials Policy Commission of five members was appointed by President Truman in 1951. Its monumental five-volume report published in 1952 considered the country's present and prospective position with respect to the extent and availability of natural resources of all kinds and their probable adequacy to meet anticipated needs.

In the field of forests and forest products, with which it dealt at length, the Commission recommended large increases in current expenditures by Federal and state governments to provide more effective protection of the forests from fire, insects, and disease; a greatly expanded program of research; additional technical assistance to small-scale forest owners and wood processors; and more intensive management of the national forests. It also proposed the establishment of a national system of forest credit and forest insurance and the replacement of the ad valorem tax on forest properties by the yield tax whenever practicable. On the controversial subject of public control of cutting on private lands it recommended that for the next five years the Federal government assist the states in establishing such control and that if at the end of that time serious gaps remain in the state system of control, the Federal government establish minimum cutting regulations.

Reorganization of Federal Executive Departments. Assignment of jurisdiction over activities dealing with forests and other natural resources to Federal executive departments and bureaus has given rise to prolonged and often heated argument. Although none of the sweeping reorganizations proposed from time to time have been put into effect, several changes of importance were made during the period under consideration.

The first of these changes came in 1933, when President Roosevelt, acting under general authority granted him by Congress, placed all national monuments under the jurisdiction of the Department of the Interior. This action, as has previously been indicated, removed the relatively few monuments in national forests and military reservations from the administration of the Department of Agriculture and the War Department, which had previously exercised jurisdiction over them under the provisions of the American Antiquities Act of 1906.

On May 9, 1939, President Roosevelt used the blanket authority granted him by the Reorganization Act of April 3, 1939, to transfer the Bureau of Biological Survey from the Department of Agriculture and the Bureau of Fisheries from the Department of Commerce to the Department of the Interior. At the same time the Secretary of the Interior was made chairman of the Migratory Bird Conservation Commission, a position formerly held by the Secretary of Agriculture. On April 12, 1940, the President combined the two bureaus into the present Fish and Wildlife Service. A few days later soil-conservation activities on lands under the jurisdiction of the Department of the Interior were transferred to that department from the Soil Conservation Service in the Department of Agriculture.

On July 6, 1946, President Truman combined the General Land Office and the Grazing Service to form the Bureau of Land Management in the Department of the Interior.

Hoover Commission. The most ambitious attempt to bring order out of chaos in the executive departments and independent agencies resulted from the act of July 7, 1947, establishing the Commission on Organization of the Executive Branch of the Government, commonly known as the Hoover Commission from the name of its chairman, former President Herbert Hoover. Three of the twenty-four advisory committees, or "task forces," appointed by the Commission dealt at length with natural resources. These were the task forces on agricultural activities, on natural resources, and on public works.

With respect to forest and range lands, the three groups agreed that responsibility for administration of the national forests, O. and C. lands, and grazing districts should be centralized in the same department, and preferably in a single bureau. The task force on agricultural activities would, however, place the responsibility in the Department of Agriculture, while the task force on natural resources would place it in a new Department of Natural Resources, which would replace the Department of the Interior. The task force on public works at first recommended placing the responsibility in the Department of Agriculture, and later in a new Department of Works and Conservation.

The Hoover Commission itself (in 1949) recommended combination of Federal activities in forestry and range management in a Forest and Range Service in the Department of Agriculture. Three members of the Commission in a vigorous minority report recommended the establishment of a Department of Natural Resources, in which a Forest and Range Service would be one of the major units. Two other members of the

Commission, in a separate minority report, dissented vehemently from the recommendation of the majority that the Bureau of Reclamation and the civilian activities of the Army Corps of Engineers be combined in a new Water Utilization and Development Service in the present Department of the Interior.

Changes in Organization. Although neither the President nor Congress has taken any action on these recommendations, some minor changes in organization have been effected. Under reorganization plans submitted by President Truman and President Eisenhower the Secretary of the Interior and the Secretary of Agriculture were authorized to make such modifications in organization and administrative responsibilities as they believed would increase the efficiency of their respective departments. Acting under this authority, the Secretary of the Interior in 1950 grouped the units in his department under four assistant secretaries. Under the new arrangement the Bureau of Land Management, the National Park Service, the Fish and Wildlife Service, the Bureau of Indian Affairs, and the Office of Territories report to the Assistant Secretary for Public Land Management.

In the Department of Agriculture, the Secretary in 1953 adopted a somewhat similar arrangement under which the Forest Service, the Soil Conservation Service, the Agricultural Research Service, the Federal Extension Service, the Agricultural Conservation Program Service, and the Farmer Cooperative Service report to an Assistant Secretary in Charge of Federal-State Relations. He also transferred to the Forest Service the land-utilization projects acquired under the program for the retirement of submarginal lands formerly administered by the Soil Conservation Service and the research in forest entomology and forest pathology formerly handled by the Agricultural Research Administration. On the other hand, certain range research was transferred from the Forest Service to the Agricultural Research Service.

More Commissions. Two Federal Commissions, both of which dealt in part with activities in the field of natural resources, were established by Congress in 1953. A new Commission on Organization of the Executive Branch of the Government (Second Hoover Commission) was authorized to make recommendations not only on organization but also on policy in order "to promote economy, efficiency, and improved service in the transaction of the public business." The Commission on Intergovernmental Relations (Kestnbaum Commission) was authorized "to study the proper role of the Federal Government in relation to the States and their political subdivisions," including fiscal relations, "to the end that these relations may be clearly defined and the functions concerned may be allocated to their proper jurisdiction."

The Commission on Organization of the Executive Branch of the Government in its report on Real Property Management (June, 1955) quoted its task force on this subject as noting that "The Study of Federal rural lands by the first Hoover Commission's Task Force on Natural Resources was exhaustive and definitive . . . conditions affecting the administration of these lands have not changed materially in the intervening years." The only recommendations made by the Commission relating to the administration and management of forest lands, range lands, and other natural resources were as follows:

(a) That the President appoint a committee from the Federal and State Governments, and from forestry, agricultural, conservation, and mining interests, to make a study of Federal rural lands and the laws affecting them, and to make recommendations for their improved management.

(b) That after a thorough study, a uniform policy for all agencies involved in control of Federal rural lands be developed.

The Commission on Intergovernmental Relations in its report to the President (June, 1955) devoted a large share of the chapter on Natural Resources and Conservation to water resources. With respect to forest resources, it recommended that Federal grants-in-aid for state and private forestry cooperation be continued, with the further recommendation that funds now appropriated for reforestation purposes under Section 4 of the Clarke-McNary Act of 1924 be applied instead to cooperative forest management.

In the field of forest protection, the Commission suggested that Federal financial responsibility for cooperative forest-fire control should not exceed the present level of 25 per cent of the total cost of basic protection. It expressed the view that Federal activities under the Forest Pest Control Act of 1947 are not adequately financed, and suggested that Congress establish a special fund to be used in combating insect and disease infestations wherever and whenever they occur. Similar action by the states was suggested.

With respect to the broader aspects of Federal-state relations, the Commission expressed itself as follows:

Satisfactory solutions to problems of water and land management and development require establishment of a continuing medium through which a unified policy can be evolved.

The Commission therefore recommends establishment by the Congress of a permanent Board of Coordination and Review to advise the President and the Congress on a coordinated natural resources policy within the National Government and between it and the States. The Commission further recommends that each State designate an existing agency or establish a natural resource advisory council to coordinate State policies and administration and to facilitate cooperation with Federal agencies in planning, building, and operating natural resource projects. . . .

If its responsibilities are broadly defined, the proposed Board could study and make recommendations on Federal land ownership and its administration, and also on the use, development, and control of water—including urban problems. Both land and water problems need thorough exploration. Although a special study committee might be established for each, it would seem preferable to assign these interrelated problems to a single coordinating agency.

State Legislation on Cutting Practices. A major development in state forest policy started in 1941, when Oregon made use of its police power to regulate cutting by private owners of forest land. Prior to that year, in spite of extensive discussion of the subject, only five states—Nevada (1903), New Hampshire (1921), Louisiana (1922), Idaho (1937), and New Mexico (1939)—had passed laws attempting to control destructive cutting of forests in private ownership, and none of these were being effectively enforced. Since then, several states have enacted legislation dealing with cutting practices which gives promise of being more effective. These laws may be divided into two main groups depending on whether they rely primarily on compulsion or encouragement to attain improved management by private owners.

Oregon. The Oregon Forest Conservation Act of 1941 provides that "any person, firm, or corporation cutting live timber for commercial use from lands within the state of Oregon shall . . . leave reserve trees of commercial species deemed adequate under normal conditions to maintain continuous forest growth and/or provide satisfactory restocking to insure future forest growth." It further provides that "proper precaution shall be taken and every reasonable effort be made by the operator to protect residual stands and/or trees left uncut as a source of seed supply from destruction by fire or unnecessary damage resulting from logging operations."

The act states that the requirement for securing reproduction "shall be deemed to have been complied with in the forested areas situated east of the summit of the Cascade Mountains within the state of Oregon, if there shall have been reserved and left uncut all immature ponderosa pine trees less than 16 inches in diameter breast high outside the bark."

West of the summit of the Cascade Mountains the act requires that there shall be left uncut not less than 5 per cent of each quarter section (160 acres), well stocked with commercial tree species of seed-bearing size. This result may be accomplished (1) by leaving marginal long corners of timber between logged areas; (2) by leaving strips of timber along creeks, across valleys, or along ridges or natural firebreaks; or (3) by using staggered settings and reserving uncut settings long enough to reseed cutover areas. If none of these methods are feasible, seed trees of commercial species at least 18 inches in diameter breast high must be left at the rate of 2 trees per acre well distributed over the area cut.

In either region, any operator who desires to adopt other practical methods of attaining the objectives specified in the act may do so by submitting a plan of operation in writing thirty days prior to the commencement of cutting. If the state forester does not disapprove of the plan and if inspection shows that it has been carried out effectively, the operator will be given a release from complying with the specific provisions of the law. The highly desirable flexibility afforded by this provision has been taken advantage of by many of the larger operators. The state forester is empowered and directed to correct any undesirable condition caused by violation of the act at a cost to the operator of not more than $100 for each 40 acres.

Washington. Washington passed a very similar act in 1945. Its constitutionality was upheld in the state Supreme Court on February 18, 1949, on appeal from an adverse decision by the Superior Court of Pend Oreille County, with one justice dissenting. Justice Matthew W. Hill, who wrote the decision of the state court, commented: "We do not think that the state is required to stand idly by while its natural resources are depleted." The decision of the state Supreme Court was affirmed by the United States Supreme Court without comment on November 7, 1949. In both Oregon and Washington, the legislation had the general support of timberland owners, who were instrumental in obtaining its passage.

Mississippi. Mississippi's Forest Harvesting Act of 1944 requires the leaving of thrifty, uninjured, well-distributed seed trees in all commercial cutting operations. It gives detailed specifications about the number to be left in pine stands, including those worked for naval stores, in hardwood stands, and in mixed pine and hardwood stands. A fine of $25 to $50 is provided for violation of the law, and in addition the state forest commissioner is authorized to enter suit to enjoin any cutting in violation of its provisions.

Virginia. Virginia in 1948 passed a somewhat similar act requiring owners and operators cutting timber for commercial purposes on land on which loblolly pine and shortleaf pine constitute at least 50 per cent of the stand, to leave four healthy, wind-firm, cone-bearing pine trees 14 inches or more in diameter per acre. These trees cannot be removed until at least ten years after the cutting. A penalty of $2.50 is provided for the cutting of each tree required to be reserved.

Except in three counties, the governing body of the county must approve the act for it to become effective. It does not apply when, at the time of final cutting, there are as many as 500 loblolly pine and shortleaf pine seedlings 6 feet or more in height, or 300 seedlings 10 feet or more in height. Nor does it apply to any land for which a cutting or management plan has been prepared and approved by the state forester.

Maryland. Maryland's Forest Conservancy Districts Act of 1943 declares it to be "the policy of the state to encourage economic management and scientific development of its forests and woodlands to maintain, conserve and improve the soil resources of the state to the end that an adequate source of forest products be preserved for the people." As a step toward this goal it authorizes the Commission of State Forests and Parks to divide the state into forest-conservancy districts and in each district to establish a district forestry board of not less than five members. So far as practical, each board is to include a person representing each of the major types of forestry and woodworking interests and at least one person representing farm-woodland owners.

Each district board is charged with responsibility for formulating rules of forest practice, which after approval by the Commission of State Forests and Parks have the force of law. These rules serve as guides both to woodland owners and to the boards in determining the actual method of cutting to be used on specific properties. All owners must obtain the approval of the board before undertaking cutting operations and may submit working plans for the cutting and management of the forest.

On receipt of the application the board is required to advise the owner in writing "as to the most practical and satisfactory method of cutting the woodland covered by the application and to give its assent to do the cutting found best adapted thereto." Each board employs a district forester, whose duty it is, under the supervision of the state forester, to assist owners in every aspect of forest management. In addition to the provisions relating to cutting practices, the law requires every person engaged in a forest-products business to obtain a license from the commission.

California. California's Forest Practice Act of 1945 divides the state into four districts, each of which has its own forest-practice rules. These rules are prepared by a district forest-practice committee consisting of forest landowners and operators. Upon approval by the California State Board of Forestry they have the force of law. Forest owners may submit alternative plans of forest management which, after approval by the local committee and the state board, also have the force of law on the properties to which they apply.

According to the state forester, this form of regulation was adopted by the Legislature in the belief that "the industry itself could best determine what practical actions should be taken to leave the land in a productive condition after logging and to protect present and future forest crops from destruction by fire, insects and disease." Much emphasis has so far been placed on "educating the operators on their responsibilities to the rules and developing an effective system of administration and enforcement." There is no specific penalty for violation of the law, but the state forester can obtain a court injunction to shut down any operation for noncompliance with the rules of forest practice.

Massachusetts. Massachusetts, in 1943, was the first state to enact legislation dealing with cutting practices in which main reliance for improvement was placed on encouragement rather than compulsion. After declaring that "the public welfare requires the rehabilitation and protection of forest lands," the act states it

to be the policy of the commonwealth that all lands devoted to forest growth shall be kept in such condition as shall not jeopardize the public interests, and that the policy of the commonwealth shall further be one of cooperation with the land owners and other agencies interested in forestry practices for the profitable management of all forest lands in the interest of the owner, the public and the users of forest products.

The governor is empowered to appoint a state forestry committee consisting of representatives of farm–wood-lot owners, industrial-woodland owners, other woodland owners, and the general public, plus the director of the Division of Forestry ex officio. This committee is required to adopt rules of forest practice for each of the four regions into which it is authorized to divide the state. Selective cutting is emphasized in the law by directing the committee to require "that trees of desirable species and suitable size shall be retained uncut . . . except that provision shall be made for clear cutting in such individual cases as shall be approved by the director."

Any owner or operator who proposes to cut more than 40 M board feet, or 100 cords, in any calendar year is required to notify the director of the Division of Forestry of his intention. The director or his representative then examines the forest to be cut and assists the owner or operator in preparing and carrying out a plan of operations best calculated to conform to the forest practices adopted for the region. The director is further instructed to inspect the property both during the operation and upon its completion to determine whether the plan and practices are being followed and to report his findings in writing to the state committee.

A fine of $25 is provided for failure to give notice of intention to cut, but there is no penalty for failure to follow the plan approved by the director. The act is regarded as essentially a cooperative educational measure, and it is hoped that self-interest and public pressure will result in voluntary adoption of improved practices. In the words of one of its principal sponsors, it represents "an attempt at self-regulation by the forest growers rather than by public edict."

Vermont. Vermont's Forest Land Use and Management Act of 1945 requires the Department of Forestry to issue rules designed to prevent wasteful and dangerous forest practices and to provide information and service for all forest owners. Such owners are encouraged but not required to cooperate with the department. They are authorized to prepare their own management plans, which may be accepted by the state Conservation Board as constituting full compliance with the law if they meet the objectives of the rules and regulations issued by the state forester. All mills and all forest operators are required to file annual reports covering their operations.

New York. New York's Forest Practice Act of 1946 "is a voluntary, cooperative program, whose purpose is to help the private woodland owner practice forestry by making available to him the services of technically trained foresters from the Conservation Department." To this end it authorizes the Conservation Department to establish not more than twenty forest districts, of which fifteen have actually been established. In each district, a district forest-practice board consisting of three members from each county, of whom at least two must be forest landowners, sets up forest-practice standards applicable to the various forest types in the district. A state Forest Practice Board, composed of elected representatives of the district boards plus three ex officio members, passes upon these standards and coordinates the activities of the district boards.

For forest owners who agree to abide by the standards of forest practice the Conservation Department is required to provide "technical services in connection with forest operations including but not limited to, the marking of timber, marketing assistance, reforestation and silvicultural operations in immature stands." Free nursery stock may be furnished to cooperators in cases where planting is a part of the management plan approved by the district forester, who acts for the department and the district board in the execution of the law. Specific working plans for individual forest properties may be accepted provided the standards embodied in them are equivalent to those adopted by the boards.

The only penalty for failure to comply with the provisions of a cooperative agreement is discontinuance of the services of the Conservation Department. The response of forest owners has convinced the department that progress under the cooperative approach has been greater than would have resulted from any attempt to police the woods.

New Hampshire. New Hampshire's Forest and Taxation Act of 1949 (effective in 1950) replaced the general property tax on forest lands by a land tax and a yield tax of 10 per cent. The law also dealt with cutting practices by providing that the yield tax can be reduced to 7 per cent by compliance with forest-practice standards established by eight district forestry advisory boards and approved by the state forester with the advice and consent of the State Forestry and Recreation Commission. This provision had not been in effect long enough to determine its effectiveness when it was repealed by the Legislature in 1955. At the same time, the rate of the yield tax was increased to 12 per cent.

Yield and Severance Taxes. The yield tax is also used by twenty-five other states as a means of encouraging better forest practices, without conspicuous success. With the exception of Mississippi, these laws are all optional in character and vary widely in the requirements placed on the owner in return for the privilege of having the yield tax applied to his land. New Hampshire is the only state which has attempted to encourage improved management not only by applying the yield tax to all forest land, irrespective of its character or the wishes of its owner, but also by giving a substantial reduction in the rate of the yield tax to owners who voluntarily adopt state-approved standards of forest practice; and the latter provision was soon repealed.

Six states impose "severance taxes" for the privilege of removing timber and other forest products from the land. Such taxes are in addition to all other taxes. Their purpose is to raise revenue rather than to encourage better management, although in four states revenues from this source are devoted to forestry purposes.

Other State Activities. In addition to attempts to curb destructive cutting practices, state forestry organizations have increased greatly in size and effectiveness during the last quarter century, and particularly during the last ten years. This progress is strikingly illustrated by the fact that in the five years from 1945 to 1949 state forestry appropriations increased threefold and the number of state forestry employees doubled.

In the field of fire control, expenditures by states and by private owners under state supervision amounted to $26,636,876 in the fiscal year 1953, and by the Federal government on state and private lands to $8,960,230

—a total of $35,597,106. In addition, local governments and private timberland owners and operators spent over 16 million dollars for special fire-control equipment, improvements, and services. Although the forest area in need of but not receiving organized protection still amounted to more than 58 million acres, this figure represented a reduction of some 60 million acres in about ten years. More efficient methods of prevention, detection, and suppression of forest fires greatly increased the effectiveness of control operations. Increasing attention was paid to destructive insects and diseases, but little more than a good start has as yet been made toward reducing their depredations to a reasonable minimum.

In addition to the educational influence of state legislation dealing with forest-cutting practices, the states have emphasized programs of education and technical assistance to private owners on a steadily increasing scale. Most notable of these programs are the ones conducted in cooperation with the Federal government under the Norris-Doxey Act of 1937, and since 1950 under the Cooperative Forest Management Act. A study made by the American Forestry Association and published in 1951 under the title "The Progress of Forestry, 1945 to 1950" shows that during the fiscal year 1949 more than 17,000 owners were given on-the-ground assistance in the handling of nearly 1,770,000 acres of forest land at a cost of $573,882 to the states and $349,117 to the Federal government.

Much educational work with the owners of farm woodlands has also been done by the state extension services. Another form of assistance to private owners is through the distribution of planting stock from the 74 state-operated forest tree nurseries.

The area of forest land in state ownership increased greatly during the depression years of the 1930's, chiefly by the tax-delinquency route. With the return of prosperity, additions from this source practically ceased, and there was even some return of state-owned land to private ownership. In recent years the trend has been toward a gradual increase in state ownership as a result of purchase and exchange.

The growth of county, municipal, and school forests has been relatively more rapid, although the area involved is much smaller (about 4,400,000 acres in 1953). In 1949, some 30 per cent of the area in state and community forests in 29 states studied at that time was estimated to be under intensive management and 58 per cent under extensive management.

Six states have enacted legislation providing for the revegetation of lands damaged by strip mining of coal. West Virginia (1937), Indiana (1941), Illinois (1943), Pennsylvania (1945 and 1947), Ohio (1947), and Kentucky (1954) all require damaged land to be re-covered with trees or grass within a reasonable period after the completion of mining.

Surveys of State Forestry Administration. As a means of helping to strengthen state activities, surveys of state forestry administration were initiated in 1945 by a joint committee of the Society of American Foresters and the Charles Lathrop Pack Forestry Foundation, with the financial support of the foundation. The surveys, undertaken in each case at the request of the governor, ran through 1948 and covered nine states— Colorado, Idaho, Kentucky, Massachusetts, North Carolina, Ohio, Rhode Island, Tennessee, and West Virginia.

The following quotation from the committee's report for Idaho summarizes clearly the normal activities of a state department of forestry:

However begun, State forestry usually concerns itself with forest protection from fire, insects, and disease; with informing the public regarding forests, their values and needs; with acquiring and administering woodlands for production of wood and other values, including recreation; and with efforts to induce forest owners to manage their land intelligently and in the public good as well as for their own satisfaction and profit. Often there are other activities, but these are the most significant.

The reports of the committee dealt with improvements in administration under the particular conditions existing in each state studied, with emphasis in most cases on fire control, general education, state forests, reforestation, and assistance to and control of private owners. An increase in the area of state forests was recommended in eight of the nine states studied.

Northeastern Interstate Forest Fire Protection Compact. A previously untried device for increasing the efficiency of state activities in forest-fire control came into being in 1949 when Congress approved an interstate compact to promote more effective prevention and control of forest fires in the Northeastern United States. Among the means to be used for this purpose were (1) the development of integrated forest-fire plans, (2) the maintenance of adequate forest-fire-fighting services by the member states, (3) provision of mutual aid in fighting forest fires, and (4) establishment of a central agency to coordinate the activities of member states and perform such common services as they deem desirable. Permission to enact interstate compacts for the conservation of forests and the water supply had been granted by the Weeks Act of 1911 but never used.

The compact was ratified during the first year of its existence by all of the New England states and New York. It may be extended to include adjacent states on action by their respective legislatures and to adjacent provinces in Canada with the approval of Congress. None of the eligible states or provinces have as yet (1955) taken advantage of this provision.

The Northeastern Forest Fire Protection Commission created by the compact has been active in promoting cooperation among the signatory states. At its first meeting in July, 1950, it employed an executive secretary who was stationed first at Laconia, New Hampshire, and later at Chatham, New York. It also authorized the establishment of a uniform training program for the fire-fighting personnel in the several states, which has gone forward effectively under the direction of a Technical Committee composed of the seven state foresters. The program resulted in the preparation by the Commission and the Forest Service of a comprehensive manual on forest-fire-control organization that was published in 1954.

Another major activity has been preparation of a regional plan. It is essentially an action plan which provides for the pooling of regional resources in equipment and personnel in times of emergency. Maine received help from New York in 1952 in the form of power pumps and hose, and New Hampshire had similar aid from Maine and Massachusetts in 1953.

Other Interstate Compacts. The compact for the Northeastern states proved so successful that in 1954 two other regions followed the precedent which it had established. Congress on July 27 approved an interstate forest-fire-protection compact for the Southeastern states, and on August 24 for the South Central states. These compacts are similar to the one for the Northeastern states in general character but differ in some important details.

In a different field, the Southern Regional Education Compact of 1949 (consent of Congress not given and possibly not required) authorized the signatory states to establish, finance, and operate regional educational institutions. The Board of Control established by the compact has developed plans for the coordination of professional instruction and research in forestry at the schools of forestry in the Southern states, with particular reference to graduate work.

The Northwestern Region Educational Compact (not acted upon by Congress) also provides for the establishment, financing, and operation of programs of higher education at existing institutions or in new ones. It has not yet dealt with education in forestry.

Forestry on Private Land. During the 1940's the trend of the previous decade toward timber cropping in place of timber mining became more pronounced. Although Forest Service estimates for 1945 showed the character of cutting to be good or better on only 8 per cent of the land in private ownership and fair on an additional 28 per cent, corresponding figures for lands held by large owners were 29 per cent and 39 per cent. In the South, cutting was characterized as good or better on 52 per cent of the large forest holdings. The greater part of this area was owned by pulp and paper companies, which throughout the country have taken the leadership in the adoption of improved practices.

Progress was even more rapid after the close of the Second World War. Estimates obtained by the American Forestry Association on 235 million acres of private commercial forest land in twenty-five states showed the following comparison between the status of forest management in 1944 and 1949:

	Per cent of area under	
	Intensive management	Extensive management
Industrial holdings		
1944	11	32
1949	22	40
Small, nonfarm holdings		
1944	2	16
1949	3	22
Farm holdings		
1944	2	16
1949	5	22
All ownerships		
1944	4	20
1949	9	27

Quite evidently both extensive management and intensive management were being practiced much more widely by industrial owners than by other owners, whether the latter are farmers or nonfarmers. The difference is particularly striking in the case of intensive management, which is, of course, the goal. All classes of owners stepped up the adoption of improved practices at about the same rate during the five years from 1944 to 1949, but the area involved was much greater in the case of the industrial owners.

Although the picture as a whole is encouraging, there is still a long way to go before the situation can be regarded as satisfactory. This is also true with respect to commercial forest lands in Federal ownership. The American Forestry Association study showed that, in 1949, 14 per cent of these lands were under intensive management, 73 per cent under extensive management, and 13 per cent without management or non-operating.

Tree Farms. An important factor in the development of private forestry has been the tree-farm program sponsored by American Forest Products Industries. As officially defined, "a tree farm is an area of privately owned forest land devoted primarily to the continuous growth of merchantable forest products under good forest practices." To qualify for recognition as a tree farm, a tract must be approved by a certifying agency as meeting specified standards with respect to protection from fire, insects, disease, and excessive grazing, and the use of harvesting methods that will assure future crops of timber.

The program was inaugurated in 1941 by the Weyerhaeuser Timber Company with the establishment of the Clemons Tree Farm in Grays Harbor County in western Washington. First official certification of a tree farm under the national program, however, did not take place until the next year in Alabama. As of October 1, 1955, there were 7,152 tree farms in 38 states with an area of 37 million acres, or more than 10 per cent of the area of commercial forest land in private ownership. Certification takes place only after field inspection by a responsible agency (private in some states, public in others), which establishes and enforces standards of practice. Decertification is possible if an owner fails to maintain these standards, and it has been applied in a few cases.

The purpose of the program is to encourage forest owners to adopt improved forest practices by public recognition of their activities. Its success is demonstrated both by the rapid growth in the number and area of tree farms and by the common, although technically inaccurate, use of the term to designate any area under satisfactory management. Actually there is a substantial area of well-managed forest properties not registered as tree farms both in states having tree-farm programs and in other states.

Enlightened self-interest, rather than public compulsion or private altruism, is basically responsible for the upsurge of interest in forest management on the part of private owners who have previously regarded it as impracticable. A more favorable economic climate resulting from higher stumpage prices and improved technology in the use of wood are primarily responsible for the change.

Industry Provides Education and Service. Industry is active in promoting the more widespread practice of forestry by private owners at both the regional and the national level. Before presenting the national picture, it may be well to describe briefly the activities of three typical regional organizations.

The Southern Pulpwood and Conservation Association was organized in 1939, and fifteen years later represented about 85 per cent of the pulpwood consumption in the South. Its primary purpose is to secure an adequate, continuing supply of pulpwood for the Southern pulp and paper industry by helping forest owners to grow more wood at a greater profit. To this end it conducts group demonstrations, offers technical service to individual owners, distributes planting stock, and operates forestry training camps for farm youth. In 1953 the association itself employed area foresters in North Carolina, Georgia, Alabama, and Louisiana, while members of the association employed 126 consulting foresters who devoted full time to assisting small landowners.

Trees for Tomorrow was organized in 1944 by the paper mills and power companies in northern Wisconsin to assist private owners in the adoption of improved forest practices. It does this through a general educational program and by distributing planting stock, providing tree-planting machines, and preparing forest-management plans. In 1953 it reached 1,000 students and 2,000 adults through its conservation workshop at Eagle River.

The Industrial Forestry Association evolved in 1953 from the Forest Conservation Committee of the Pacific Northwest Forest Industries. Like its predecessor, it aims to promote private forestry in that region by stimulating better protection, timber-harvesting methods, and forest-management planning. It sponsors the West Coast Tree Farms and operates the forest industries' tree nursery at Nisqually, Washington, from which 7 million trees were distributed in 1953. It also represents its membership in all matters relating to forest policy, legislation, taxation, and public relations.

American Forest Products Industries. In the national field, the outstanding educational agency of the wood-using industries is American Forest Products Industries (A.F.P.I.). It was established in 1941 as "a non-profit, non-political, public-service organization sponsored and supported by forest land owners and wood processors," in the belief that "the industry had to present its case at the bar of public opinion, forcefully and effectively." Two conditions it regarded as necessary to attain this objective: "First, industry's forestry programs must deserve the public faith that was sought; second, the industry must have the means of presenting forestry facts and persuading people to listen to them."

A.F.P.I.'s campaign to get forestry widely practiced in the woods is furthered by three specific programs. The Tree Farm program, already described, gives public recognition to private owners who practice good management and harvest regular timber crops. The Keep Green program, also inaugurated in Washington in 1941, aims to strengthen forest-fire control in the participating states, particularly in the field of fire prevention. The More Trees for America program sponsors intensive state-wide campaigns to interest small owners in timber as a cash crop.

Public education is accomplished by a wide variety of means. These include informative publications, motion-picture films, radio talks, advertising mats, news services, and special editorial services. Particular attention is paid to the nation's schools, to about half of which teaching aids were distributed in 1952. The dual program of educating both timberland owners and the general public was furthered by a national conference held in Chicago, June 25 to 26, 1953, to discuss ways and means of improving the management of farm woodlands and other small forest properties, where such improvement is most urgently needed.

National Lumber Manufacturers Association. The National Lumber Manufacturers Association has long been interested in policies and programs affecting the forest lands of the country. Following the Second World War its board of directors approved a much fuller statement of policy than that adopted in 1937, when the board had urged its regional affiliates to continue the development of suitable forest-practice rules and methods.

The new statement opposed Federal regulation of private forest-management practices and Federal control of state forest policies, subjects which at the time were arousing considerable discussion. It also expressed belief that "only those lands which private ownership cannot keep productive should be taken into public ownership, primarily by the states and secondarily by the Federal government"; and it favored "the transfer of public forest lands to private ownership and of Federal land to State ownership where the public interest will be served." At the same time it recognized industry's obligation to maintain its forest lands in productive condition and invited all organizations and individuals to join in furthering the continuous production of forest crops through the Tree Farm, Keep America Green, and More Trees for America programs.

A revised and somewhat expanded statement of forest policy was approved by the National Lumber Manufacturers Association on September 14, 1953. In addition to the subjects covered by the previous statement, it advocated equitable taxation of forest properties; maximum utilization of each tree that is cut and of all forests that should be harvested or salvaged; promotion of industrial research; construction of access roads to open up currently inaccessible timber on Federal lands; prompt completion and maintenance of the basic inventory of forest resources; and promotion of forest credit and forest insurance by private financial institutions. An unusual item for inclusion in a forest policy was that urging "recognition of the paramount rights and interests of the states over their water resources."

Important additions were made by the association at its annual meeting in 1954.

The lumber industry [it stated] has long recognized the multiple use principle in the management of its forest lands. . . . Our industry emphasizes that good forest management includes the maintenance of forest soils, the production of wildlife, and the development of recreational values compatible with forest production. Private forest landowners should seek to develop further use of their lands for hunting and fishing which would not be detrimental to the major purposes of management.

With respect to the relations between forests and water, the association declared that

in the management of forested watersheds it is recognized that the harvesting of timber, according to tree farm principles, provides substantially greater supplies of usable water than will a program of management which eliminates timber harvesting.

Forest Industries Council. The Forest Industries Council, first suggested in 1940, was formally organized on May 10, 1943, by the American Paper and Pulp Association, the American Pulpwood Association, and the National Lumber Manufacturers Association. These associations are said by the council to represent industries using about four-fifths of the wood harvested annually from American forests. Its objectives are to make recommendations on matters where joint action of the forest industries is deemed advisable and to execute any program or policy approved by the participating associations.

In 1945 the council adopted a statement of forest policy expressing "faith that private enterprise and initiative can provide the most effective management, use, and renewal of our nation's forests." Emphasis was placed on protection from fire, insects, and disease, adoption of improved forest practices, more complete utilization of forest products, equalization of state and local taxes, and adequate support of state forestry organizations.

On the controversial subjects of regulation and ownership, the council agreed to support "public regulation when necessary or desirable, to be administered under state law"; to encourage "the sale and exchange of public lands in order to restore desirable lands to private ownership as well as to consolidate public holdings"; and to encourage "public ownership and practical management of forest lands which are incapable of producing sufficient wood to maintain profitable private ownership."

"Growing Trees in a Free Country." In the fall of 1953 the Forest Industries Council adopted a new statement of policy, accompanied by considerable explanatory material, which was published in a brochure entitled *Growing Trees in a Free Country*. In this statement the council pledged itself

to provide united leadership in improving America's woodlands and in maintaining a continuous and adequate flow of products from all forest lands. In the fulfillment of this leadership, the Forest Industries Council and its members agree to:

1. Promote the extension, to all forest lands, of adequate, dependable and economic protection against fire.

2. Cooperate with all agencies in preventing and reducing losses caused by forest insects and diseases.

3. Urge continuance and expansion of forest practices that will yield perpetual crops of forest products, protect our soil and watershed values, and insure full use of all resources on our forest lands.

4. Encourage private ownership and management of forest lands for the continuous and profitable production of tree crops, in preference to public ownership and management of such lands.

5. Support measures that will provide for management of public forest lands on the same high level as that practiced on the better managed private forest lands.

6. Advocate equitable taxation of timber and forest lands.

7. Support in each forested state a competent, adequately staffed and financed state forestry organization.

8. Encourage maximum utilization of each tree that is cut and of all forests which should be harvested or salvaged.

9. Promote industrial research and cooperate in effective public research, where appropriate, in the utilization of forest products and in commercial forest management.

On the subject of voluntary versus mandatory cutting practices, the statement had this to say:

The forest industries are committed to promoting cutting methods that will not only maintain but also increase the productivity of woodlands. . . . Some states have enacted codes for forest practice. Since neither the forest nor the economy is static, progressive changes in the provisions and interpretation of these laws will be necessary.

Chamber of Commerce of the United States. The Chamber of Commerce of the United States, representing more than 2,700 local chambers of commerce and 500 trade associations, has long taken an interest in forest policies and programs. In 1922 it appointed a special committee on forestry, and a few years later in a referendum vote it approved the principles that it believed should underlie a national forestry policy. The points most strongly stressed were cooperative action between the Federal government and the states to assure efficient fire protection and the adoption by the states of the principle of the yield tax.

Acting through its Natural Resources Department, the chamber has cooperated with several state chambers of commerce in arranging conferences to encourage citizens of the states concerned "to review their present state forest policies, and explore ways of applying sound forestry practices to forest lands." In 1949 it published a "Forestry Manual" showing how a local chamber of commerce "may organize to safeguard the future of its forest industries, and protect and preserve one of its most valuable natural resources, within the pattern of private enterprise."

Each year the chamber issues a set of "Policy Declarations on Natural Resources." Its 1955 declarations emphasized forest protection, forest management, tree planting, reduction of waste, research in both management and utilization, and retention of forest land in taxpaying private ownership wherever practicable. "Only where lands now in such ownership are submarginal and cannot be economically maintained and satisfactorily managed in private ownership, should they be taken into government ownership—primarily by the states or by local governmental units." No reference was made to either state or Federal control of cutting on private lands.

With respect to organization,

The Chamber regards as of primary importance the consolidation of the Forest Service and the forest and range functions of the Bureau of Land Management in a combined Forest and Range Service.

We believe it is of the utmost importance that the Congress, in conjunction with this consolidation, should legislate statutory codes providing the basic administrative principles and policies related to all agencies dealing with forest and range resources and thereby reduce to a minimum departmental regulation.

Pending the enactment of such legislation, the chamber made no recommendation as to the Department in which the new unit should be located.

The chamber expressed the belief that Federal lands on which agricultural and grazing uses are the predominant values should be made available for private ownership unless it is clear that such ownership is not in the public interest. It also recommended a time limit on temporary withdrawals of public land, doubtless having in mind the still-continuing "temporary" withdrawals made by President Roosevelt in 1934 and 1935.

National Association of Manufacturers. "Industry recognizes the paramount importance of conservation, development and wise utilization of soil, water and forest resources to the economic and social well-being of all our people." This attitude was expressed by the National Association of Manufacturers in a booklet entitled "Industry Believes," published April 21, 1954. In attaining the proposed objective it recommended the establishment of a comprehensive and unified program of action which should provide for increased state and local participation.

The association viewed with deep concern the continuing trend toward increasing government landownership (both Federal and state) and voiced the belief that "the successful operation of our national economic system depends largely on the private ownership of land and the development of the land's resources by individual enterprise." It urged the Federal government, in cooperation with private industry and state agencies, to make accurate periodic inventories of each of the nation's renewable natural-resource materials and recommended the development of detailed plans for the integrated utilization of public and private resources in times of national emergency. In opposition to Federal Valley Authorities, it proposed the development of watershed conservation on a basis of watersheds that people recognize and identify themselves with.

Congress of Industrial Organizations. The Congress of Industrial Organizations (C.I.O.) has taken a keen interest in forest policies and programs, particularly through its National Committee on Regional Development and Conservation and through the International Wood Workers in the Pacific Northwest. The Committee on Regional Development and Conservation on May 3, 1953, unanimously adopted a comprehensive national forestry program embodying the basic principles which the C.I.O. has long supported.

On the subject of regulation of forest practices, the committee stated:

We call for direct federal regulation of cutting practices on lands owned by large interests. . . . Much ballyhoo is issued through slick-paper publications about the magnificent conservation work these big corporations are doing; but official figures show that the talk runs far ahead of the practice. Such groups are far too powerful to be regulated by state legislatures, which in most cases the companies can control as they desire. The only answer is direct federal regulation of a vigorous kind, and we are prepared to support suitable national legislation to that end.

[The committee regarded] the rapid expansion of state and federal timberland holdings [as a second essential]. In general the best record of management thus far in America is held by the federal government as distinguished from the states and we would give priority to the enlargement of our national forests. . . . Greatly expanded federal appropriations will be essential for such purposes, and we shall fight for such public investment.

C.I.O. on Forest Management. With respect to management, the committee believed that

the principles on which our timberlands should be managed are those of sociological and ecological forestry. . . . In general we call for operation on the basis of sustained yield, selective cutting, multiple-use, and intensive management. . . . We stand for large-log forestry, which is high-wage forestry and which also guarantees the perpetuation of the big structural timber and plywood industries. . . . It is apparent that any program for light selective cutting on a short cycle basis must depend for its effectiveness on the development of an adequate system of access roads. . . . All timberlands should produce large logs for lumber, and smaller materials for poles, posts, ties, and pulp. . . . The need is for integrated sawmill, pulp and paper, wood chemistry plants of moderate size in each timber area. . . . Federal government pilot plants and federal government stimulation of integrated industrial development by credit, technical assistance, and otherwise is greatly needed.

The methods which must be followed to obtain good forestry on small holdings were considered to differ greatly from those applicable to the holdings of the big interests.

Where woodlots are an integral part of farming operations, regulation would intrude on the privacy of personal holdings. Such small owners and operators are in great need of technical assistance, the provision of machinery on rent or loan, the provision of seedlings and growing stock, and the like; and such aid should be given by the federal government in cooperation with the states or directly. Such owners are also in great need of financial assistance, and credit should be made available to them, suited to their needs. We also believe, however, that a system of voluntary long-term tree-farm contracts between farmers and small holders and soil conservation or watershed management districts would be of great assistance.

Other Policies. The committee favored the readjustment of state and local taxation policies in the direction of yield taxes instead of ad valorem taxes on forest property, and a great enlargement of public credit facilities. It opposed the establishment of cooperative sustained-yield units authorized by the act of 1944

on an inequitable and monopolistic basis. . . . This means generally speaking, that we oppose the use of such procedures entirely, because in practice they have almost always been inequitable and monopolistic. . . . The true answer to getting better forestry on the privately owned lands of the big interests lies in direct Federal regulation of such interests and the extension of National Forest holdings.

With respect to proposals "to have the National Forests turned over to the states or to private ownership" and "to vest ownership in the grazing-lands in private hands," the committee stated that

at the present time both of these threats are intensified in the national political picture. The off-shore oil steal may well be followed by attempts to turn the National Forests over to state or private management or to grant rights amounting to ownership to monopolistic grazing interests. All such attempts must be fought by the labor movement and by forward looking groups throughout the country with every resource available.

American Federation of Labor. The American Federation of Labor has consistently recognized that workers have a large stake in the management and conservation of forest resources. Its position was clearly indicated when its seventy-second annual convention in 1953 unanimously urged the Federation to "continue its active interest for the protection of the rights of the American citizen in matters affecting conservation and development of natural resources in the best public interests."

Julius C. Viancour, an official of the Northwestern Council of Lumber and Sawmill Workers, who spoke for the Federation at the Fourth American Forest Congress in October, 1953, emphasized the fact that "the more than two million workers and their families whose livelihood depends on forest-based industries and activities have a very direct interest in what happens to our forests." Additional millions of workers, he pointed out, have a vital interest in the nation's forests as consumers of forest products and as users of the recreational opportunities which they afford. Specifically, he recommended the building of sufficient public access roads "to make timber available to all potential purchasers, both large and small" and "to permit full use of the national forests for recreational purposes"; urged that national-forest timber be sold in small enough tracts to permit small purchasers to bid on them; supported the timber resource review of the Forest Service; and opposed opening the national parks to logging.

American Farm Bureau Federation. The keen interest that the American Farm Bureau Federation has long taken in policies affecting the conservation of natural resources was well illustrated by the resolutions adopted at its thirty-sixth annual meeting on December 16, 1954. The general attitude of the federation found expression in the affirmation that "we believe that private ownership and operation of the nation's land resources is in the national interest in most instances." With respect to forestry, it had this to say:

We favor the development of a privately-owned, sustained-yield forest industry with state and federal governments assisting with essential supplemental services, including fire protection, insect and disease control, and forestry research and education. We support state and federal tax programs which will provide incentive for the long-range investment required. We recommend a more aggressive program of research on the problems of forest and land management and utilization. Research on hybrid tree varieties holds great promise for an expanded development of commercial forestry and economic farm woodlot production.

We recommend increased emphasis on good management of public lands suitable for timber production, including insect control, access roads, and other necessary management practices to produce a maximum amount of timber on a sustained-yield basis.

The Forest Service has frequently exchanged timber for land, thus expanding its land holdings without specific Congressional authorization and avoiding payments to local governments. We oppose this policy.

The controversy then current with respect to the desirability of legislation dealing with range lands in Federal ownership was reflected in the following resolution:

> Where the federal government is a landlord, it should promote the development of good landlord-tenant relationships which will encourage the tenant to improve the property. We recommend the enactment of legislation which will stabilize and clarify the status of private users of federal lands suitable for grazing. Grazing users should be provided as much security of tenure as is compatible with public interest in the management of the land. Every effort should be made to increase the carrying capacity of these lands in preference to reducing the grazing permits. Where reductions in grazing units are essential for the protection of the resource, the users should have a sufficient period of time to adjust their operations.

International Conferences. The government of the United States has cooperated actively with the Food and Agriculture Organization of the United Nations in the organization and administration of its extensive activities dealing with forests and forest products. Forestry has also played an important part in the Point Four program, the Technical Cooperation program, and the Organization of American States.

Some of the international conferences in which the United States has participated include the following:

World Forestry Congress
International Congress of Forest Research Organizations
Inter-American Conference on the Conservation of Renewable Natural Resources
Latin-American Forestry Commission
United Nations Scientific Conference on the Conservation and Utilization of Resources
Technical Conference for the Protection of Nature
International Grasslands Congress
International Botanical Congress
Pacific Science Congress

Although it is difficult to identify any direct influence of these international connections on the development of American forest policy, that they have had an indirect and salutary effect cannot be doubted.

Summary. The report of the Joint Congressional Committee of Inquiry intensified the controversy over public control of cutting on privately owned forest lands. It also paved the way for the Sustained-Yield Forest

Management Act of 1944, the Forest Pest Control Act of 1947, and the Cooperative Forest Management Act of 1950, and for strengthening Federal cooperation with the states in fire control, reforestation, and educational activities under the Clarke-McNary Act of 1924.

Beginning in 1941, several states enacted legislation aimed at compelling or encouraging the adoption of improved cutting practices by private owners. Both Federal and state forests increased in area, and there was marked improvement in their management, particularly in the case of state forests. Research in forest management and wood technology went forward at an accelerated pace. The nationwide forest survey and other studies provided valuable information concerning the area, volume, growth, and use of the forest resources of the country.

Particularly significant was the improvement in the management of the larger forest properties in industrial ownership. Increased stumpage prices, coupled with new and better methods of wood utilization, combined to bring about a change of attitude on the part of many leaders in the wood-using industries that was little short of revolutionary. Favorable economic and technological conditions were also effectively supplemented by the policies and programs adopted by several national and regional associations. For the first time, timberland owners and operators throughout the country were cooperating actively with state and Federal agencies in a concerted effort to bring about the general practice of satisfactory management on all privately owned forest lands.

REFERENCES

American Forestry Association: "The Progress of Forestry, 1945 to 1950," Washington, 1951.

Greeley, William B.: "Forests and Men," chaps. 10, 13, Doubleday & Company, Inc., New York, 1951.

————: "Forest Policy," chap. 17, McGraw-Hill Book Company, Inc., New York, 1953.

Peffer, E. Louise: "The Closing of the Public Domain," chaps. 15–16, 18, Stanford University Press, Stanford, Calif., 1951.

Proc. Amer. Forest Cong., 1946, American Forestry Association, Washington, 1947.

Proc. Amer. Forest Cong., 1953, American Forestry Association, Washington, 1954.

CHAPTER 12

Summary and Analysis

The preceding chapters have given a roughly chronological account of the development of Federal, state, and private forest policies in the United States. The present chapter attempts to summarize the present status of those policies and the steps by which they came into being and to analyze the major issues to which they give rise. Emphasis will be placed on ownership, cooperation, regulation, research, and education as the principal subjects with which policy has been concerned.

Public Domain. Between 1781 and 1853 the United States by cession, treaty, and purchase acquired about a billion and a half acres, or approximately three-fourths of its continental land area. This land constituted the public domain, complete control over the disposal and management of which was placed by the Constitution in the hands of Congress. Many public-land laws have sold or donated large areas to states, corporations, and individuals. Other laws have permanently reserved or temporarily withdrawn from entry less extensive but still enormous areas.

Today the public domain comprises about 411 million acres, or slightly less than one-fifth of the continental land area. Since 1935, when President Roosevelt withdrew the remaining unreserved public lands from entry, they have been available for private acquisition only under the mineral laws or if classified by the Secretary of the Interior as suitable for sale or entry under applicable legislation. Most of the present public domain lies in the eleven Far Western states, where the percentage of total land area in Federal ownership runs from 35 per cent in Washington to 85 per cent in Nevada.

The policy followed by Congress in the handling of the public domain evolved gradually from one of virtually complete disposal, first by sale and then by extensive grants, to one of virtually complete reservation. Beginning with the establishment of Yellowstone National Park in 1872 and the creation of forest reserves under the enabling acts of 1891 and 1897, the national-park and national-forest systems were well established by the beginning of the present century. Then came legislation authorizing the reservation of national monuments (1906), the temporary withdrawal of public land from sale or entry (1910), the retention of revested O. and C. grant lands (1916), the leasing instead of the sale of certain mineral lands (1920), and the establishment of grazing districts (1934).

That the reservation policy as a whole is here to stay seems certain. Proposals relating to the disposal of certain lands in national forests and grazing districts will be discussed later in connection with those classes of reservations, but that they will be abolished or even greatly reduced in area seems unlikely. More probable is legislation limiting the period for which "temporary" withdrawals can be made under the act of 1910.

In addition to the public domain, the United States owns some 44 million acres which have been acquired, chiefly by purchase, for specific purposes such as national forests, wildlife refuges, military reservations, hospitals, post offices, and customhouses. As a trustee, it also manages about 56 million acres of Indian lands.

Federal Timberlands Prior to 1891. In the early days when forests were apparently "inexhaustible," the policy was to encourage, or at least not to interfere with, their unrestrained exploitation by private owners. Large areas were transferred from public to private ownership. Cutting was done with no thought of the future and was usually followed by fire.

Until 1878 the only laws dealing specifically with the disposal of public timberlands were those of 1817 and 1827 authorizing the creation of reserves for the purpose of assuring an adequate supply of live oak and red cedar to meet the needs of the Navy. Less than 300,000 acres of reserves were actually established, mostly in Florida, and widespread trespass virtually nullified their value. With the passing of wooden ships, the naval-timber reserves were gradually abolished, until in 1923 the last 3,000 acres were restored to entry.

The most important result of the premature and abortive attempt to establish forest reserves was an act passed in 1831 forbidding timber trespass both on the reserves and on the unreserved public domain. A similar act in 1859 forbade the unlawful cutting of timber on public lands reserved or purchased for military or other purposes.

Despite this legislation, timber stealing was for many years the order of the day, perhaps inevitably so in the absence of any satisfactory provision for acquisition by private owners and operators of the timber so urgently needed in the development of the country. In 1877 Secretary of the Interior Carl Schurz started a vigorous campaign of law enforcement which led the next year to enactment by Congress of the Free Timber Act and the Timber and Stone Act. The former made timber on mineral lands in the public domain in the Rocky Mountain states available without charge for building, agricultural, mining, or other domestic purposes. The privilege was later extended to include land of any character in any of the public-domain states and to permit the use of timber for manufacturing purposes.

The Timber and Stone Act provided for the sale at not less than $2.50 per acre of nonmineral public land in the Pacific Coast states (later all public-domain states) primarily valuable for timber or stone and unfit for cultivation. Both acts proved so effective in hastening the dissipation of the country's forest resources and in promoting fraud that they lent strength to the proposal that the government retain title to the lands and sell the timber under supervision.

National Forests. Warnings that continued forest destruction would inevitably result in timber shortages, accelerated erosion, and injury to the water supply led in 1876 to an appropriation of $2,000 for the employment of a man of "approved attainments" to study the situation. Then in 1891 a rider to an omnibus act revising the general land laws authorized the President to set aside timber and brushlands in the public domain as forest reserves. Another rider to the Sundry Civil Appropriations Act of June 4, 1897, provided for administration of the reserves. Among other things it instructed the Secretary of the Interior to protect the reserves from destruction by fire and depredations and authorized him to make rules and regulations for their occupancy and use. Jurisdiction over the forest reserves was transferred to the Secretary of Agriculture in 1905, and in 1907 their name was changed to national forests.

The constitutionality of the basic acts was upheld in 1911 by the Supreme Court, which ruled that Congress has complete power to handle the public domain in any way it sees fit. Particularly noteworthy was the decision that the broad authority to make rules and regulations granted to the Secretary by the act of 1897 did not constitute a delegation of legislative power and that the Secretary was entirely justified in using that authority to control and to make a charge for grazing, which is not mentioned in the act.

Under an act passed in 1907 and subsequent amendments, the President no longer has authority to establish national forests from the public domain by proclamation in any of the Far Western states except Montana, Utah, and Nevada. On the other hand, national forests can be created by purchase under the Weeks Act of 1911 and the Clarke-McNary Act of 1924 and by exchange with states or private owners under the General Exchange Act of 1922. As of June 30, 1953, there were 153 national forests with a net area of about 181 million acres, of which 21 million acres were in Alaska. The figures include 19 million acres of acquired land, mostly in the Eastern states.

National forests can legally be created only for the protection of the water supply and the production of timber. They can, however, be used for any other purposes not inconsistent with these primary objectives. Grazing, recreation, and mining constitute the other major uses. They are thus multipurpose reservations, with strong but not exclusive emphasis on the use of their resources for the support of economic activities. Co-ordination of these diverse and often conflicting uses, with due regard to biological, physical, economic, social, and aesthetic considerations, offers one of the most difficult problems in the administration of the forests.

Probable Permanence. That national forests are an integral and per-manent part of the forest policy of the United States is now generally taken for granted. There is, however, far from universal agreement as to whether their present area (about 19 per cent of the commercial forest area of the country plus a large additional acreage chiefly valuable for watershed protection, livestock production, and recreation) should re-main approximately stationary or should be materially expanded or con-tracted.

Some feel strongly that Federal holdings should be greatly expanded by the acquisition of private land within existing national forests and purchase units and by the establishment of new purchase units. Others regard private and state ownership as being more in accord with the underlying principles of the free-enterprise system and of states' rights. They would therefore transfer to private ownership or to state ownership (and probably eventually to private ownership) much land now in na-tional forests. Excepted land would be that on which it is unlikely that private or state owners could make a financial profit and that which is more valuable for the protection of the water supply than for the produc-tion of timber and forage. With respect to these latter resources, advo-cates of the proposal believe that private and state owners are now cap-able of as efficient management as the Federal government, irrespective of what the case may have been in the past.

Most people familiar with the present situation do not go to either of these two extremes. They favor a careful and impartial study of the entire pattern of forest ownership, in which they believe Federal forests have an important part to play. They neither anticipate nor favor any drastic or abrupt change in the present pattern but feel that some adjustment in existing boundaries might be to the advantage of all concerned. Con-siderable additions to some purchase units may be highly desirable, while it is probable that others should be abandoned because the need for them no longer exists.

Progress in the adoption of improved forest practices by private owners and the states has been sufficient to make any large additions to the area of the national forests unlikely. On the other hand, any large reduction of their area seems equally unlikely whatever may be the future development of private forestry. This is true not only because of general public recognition of their value as stabilizers in the field of timber production, but perhaps even more because of the services they render in the fields of watershed protection and recreation.

There is strong feeling in some quarters that more intensive management of the national forests and the utilization of their timber resources to the full extent of their sustained-yield capacity would be highly desirable. The former will require increased appropriations, and the latter the construction of many miles of access roads to permit the harvesting of currently inaccessible stands. Access roads are also urgently needed, both on national forests and on revested lands in the O. and C. grant, to permit the salvaging of immense quantities of wind-thrown and insect-killed timber.

General agreement exists as to the desirability of more intensive management of national forests and other Federal timberlands, although there are those who fear that greater accessibility will decrease recreational values and will threaten the very existence of wilderness areas. This situation is part and parcel of the problem created by the fact that simultaneous maximum use and enjoyment of all resources is a biological and physical impossibility. Policy and practice must be so framed as to bring about such use of resources in timber, water, forage, wildlife, scenery, and other values as will best serve the interests of the general public.

Oregon and California Railroad Lands. Violation of the terms of the land grant made to the Oregon and California Railroad in 1866 led Congress in 1916 to revest in the United States title to the lands in the grant not already patented to the company. Provision was made for the homesteading or sale of all land except power sites. Destruction of timber resources under this arrangement led to such vigorous criticism that Congress in 1937 authorized the Secretary of the Interior to classify the lands and to handle under sustained-yield management those classified as timberlands.

The revested lands in the original grant contain some 2.1 million acres of high-quality Douglas fir and associated species. In addition there are slightly less than half a million acres in the indemnity limits of the grant, jurisdiction over which was claimed both by the Department of the Interior and the Department of Agriculture. Congress settled the controversy in 1954 by declaring them to be O. and C. lands but placing them under the administration of the Forest Service. The act also provided for the blocking up by exchange of intermingled O. and C. and national-forest lands.

Another major question of policy is whether all of the revested lands should not be incorporated in the national forests, which they resemble in character. This question is complicated by the fact that the counties receive 75 per cent of the gross receipts (minus costs of access-road construction) from the revested lands but only 25 per cent of the receipts from national forests. Eventual consolidation of the two classes of holdings under one administration seems probable.

Grazing Districts. Until twenty years ago the range lands in the public domain were virtually a commons. No measures were taken to prevent or control grazing on them, and the attempt to get them into private ownership by means of the Stockraising Homestead Act of 1916 proved a failure. Nothing came of the 1931 recommendation of the Commission on Conservation and Administration of the Public Domain that they be turned over to the states, largely because the states displayed conspicuous reluctance to accept the proposed gift. By this time, depletion had progressed to the point where the lands were generally regarded as a liability rather than an asset.

Constructive action to improve the situation came with the Taylor Grazing Act of 1934, which authorized the Secretary of the Interior to establish grazing districts in the unreserved public domain and to make rules and regulations for their occupancy and use. This authority was, however, hedged about by many more restrictions than was the case in the act of 1897 providing for administration of the national forests. In 1953 grazing districts comprised some 160 million acres of land administered by the Bureau of Land Management, all in the Western states.

The future of the grazing districts is much less certain than that of the national parks, monuments, and forests. Technically, they are regarded as "withdrawn" rather than "reserved" from entry and appropriation. The language of the act, which authorizes their withdrawal and management pending "final disposal," implies that the arrangement may be a temporary one. Contradictory proposals that they be permanently retained in Federal ownership and that they be turned over to the states or to private owners both have their vigorous proponents. A policy issue of outstanding importance still awaits decision.

Differences between Grazing Districts and National Forests. Grazing districts differ from national forests in many respects. In addition to being classified as withdrawals rather than as reservations, they are established by the Secretary of the Interior instead of by the President or by act of Congress. They are administered by the Secretary of the Interior and not by the Secretary of Agriculture, although similar lands in the national forests are administered by the latter. This matter of jurisdiction was one of the factors that delayed legislation on the subject. It is quite possible that the coolness which developed between the stockmen and the Forest Service during the debates over the administration of grazing on the national forests may have played an important part in swinging the decision in favor of the Department of the Interior.

The penalty for violation of the law and of the Secretary's regulations is less severe in the case of grazing districts than in the case of national forests and can be imposed only for "willful" violation "after actual notice." Contrary to the usual procedure, ignorance of the law can be advanced as an excuse for its violation. Land exchanges with states can be made on the basis of equal area as well as equal value.

Fees average much lower. One-eighth of the receipts from grazing fees, instead of one-fourth, goes to the counties. Perhaps the most important difference of all is the inclusion in the law of the detailed conditions under which permits for the use of grazing districts must be issued by the Secretary of the Interior, as compared with the much greater freedom allowed the Secretary of Agriculture in determining the conditions under which the national forests may be used for any purpose, and particularly for grazing.

Considerable differences exist in the principles and practices followed in the administration of grazing on the grazing districts and the national forests. These are apt to be irritating to the stockmen, many of whom utilize both classes of land. Although great improvements have been made in the condition of the grazing districts, there is general agreement that their management has not been as effective as that of range lands in the national forests.

National Parks and Monuments. In 1832 Hot Springs Reservation in Arkansas became the first unit in today's extensive system of reservations. Yellowstone National Park, established in 1872, was the first true national park in the modern sense of the term. Four more national parks were established prior to 1900—Yosemite (1890), Sequoia (1890), General Grant (1890), and Rainier (1899). Since then their number has grown to 28 (including one each in Hawaii and Alaska), with a total area of some 13 million acres.

National parks are created only by Congress. They are administered under an act passed in 1916 which established the National Park Service and directed it to "conserve the scenery and the natural and historic objects and the wildlife therein and to provide for the same in such manner and by such means as will leave them unimpaired for the enjoyment of future generations." The act prohibited commercial utilization of the resources in national parks, with the single exception that the Secretary of the Interior was authorized to permit grazing when in his judgment it would not be detrimental to the primary purpose. Within these limitations, the Secretary was given general authority to make rules and regulations for the proper use and management of the parks.

The American Antiquities Act of 1906 authorized the President by proclamation to establish national monuments for the preservation of features of historic, prehistoric, and scientific interest. Since 1933 they have been under the jurisdiction of the Secretary of the Interior. Congress as well as the President can, of course, create national monuments and in a few instances it has done so. In 1953 there were 86 monuments with a total area of some 9 million acres.

National parks and national monuments share one common and major characteristic. They are both set aside and maintained for noneconomic purposes. Parks are intended to preserve for all time outstanding scenery and natural wonders; they are museum pieces of the best that nature has to offer. Monuments are also museum pieces, but with the emphasis on features of historic and scientific, rather than primarily scenic, interest. The line between the two is often hard to draw. Mesa Verde National Park, for example, is preeminently of prehistoric interest, while Yellowstone, Mammoth Cave, and Everglades National Parks contain certain features of great scientific interest. The Grand Canyon, which is scenically outstanding, was for many years a national monument before it was "elevated" by Congress to national-park status.

Problems and Threats. Although the national-park system, of which the parks and monuments are the chief constituents, is now almost universally accepted as a permanent institution, controversy still exists with respect to certain matters of policy connected with its establishment and administration. For example, question is often raised as to the wisdom of completely reserving its resources from utilization for such purposes as grazing, logging, mining, and reservoir construction.

A considerable area was actually opened to grazing in the First World War. During the Second World War strong pressure was exerted on the Secretary of the Interior to recommend that the Olympic National Park be opened for limited logging operations in order to provide spruce for airplane construction. In 1951 and again in 1953, the Secretary approved a proposal, not acted on by Congress up to 1955, for the construction of a reservoir in the Dinosaur National Monument for power and irrigation purposes. Should the United States become engaged in a war in which there appears to be need for minerals found in the parks or monuments, there can be no doubt that opening them to mining would be strongly urged.

Any activities of these sorts are vigorously opposed by the National Park Service, the National Parks Association, and many other organizations and individuals on the ground that they would violate the very purpose for which the reservations were established—to preserve natural conditions. In their judgment this should be done only in case of a national emergency which cannot be met by any other means. To identify such an emergency to the satisfaction of all concerned will obviously be difficult.

A real threat to the integrity of natural conditions comes from another form of utilization which cannot be prevented. Congress has specifically directed that national parks shall be so administered as to provide for their "enjoyment," which of course means that they must be open to visitors. How to accommodate the ever-increasing number of visitors and at the same time leave the natural resources "unimpaired for the enjoyment of future generations" is a problem to which there is probably no completely satisfactory solution. The choice appears to lie between concentrating visitors so far as practicable in a few places where the interference with natural conditions will be considerable and dispersing them as widely as practicable with less drastic but more widespread interference with natural conditions and the probable eventual disappearance of real wilderness areas. In either case much larger funds for protection and maintenance than are now available will be essential to prevent serious deterioration of irreplaceable values.

Another problem in the realm of policy has to do with the character and size of areas that should be included in the national-park system. Great concern has at times been expressed as to the danger of lowering standards by the inclusion of areas with no particularly outstanding qualifications which local communities desire to have set aside as national parks because of the advertising value of the name. The Sullys Hill Reservation, created as a national park in 1904, was changed in 1914 to a national game preserve, as a more appropriate classification. A bill to create the Ouachita National Park in Arkansas was pocket-vetoed by President Coolidge in 1929 because of his belief that the area was of substandard quality. What, if any, areas now outside of the system truly deserve addition to it is a matter on which there is wide diversity of opinion.

Opinions also differ as to whether some national parks now in the system, such as Platt, Wind Cave, and Isle Royale, really belong there. Many believe that the Olympic National Park is larger than necessary to serve the purpose adequately and that a considerable area of forest land at the lower elevations should be eliminated in order to make the timber available for the support of economic activities.

The question of size arises even more forcibly in connection with national monuments because of the legal proviso that they must be as small as is compatible with the proper care and management of the objects to be preserved. In the case of the Grand Canyon, prior to its creation as a national park, the Supreme Court ruled that the area was not unduly large to accomplish the intended purpose. More recently Jackson Hole National Monument was severely criticized on many grounds, including its size, and was saved from abolition prior to its creation as a national park only by a Presidential veto. The controversy over this particular monument also brought forth recommendations, on which Congress has taken no action, that the President's power to establish national monuments be revoked and that the consent of the state concerned be made a prerequisite to their establishment.

Indian Lands. Long-continued conflict between the Indians and the white man over the ownership and use of lands occupied by the Indians finally resulted in limiting the equities of the Indians to reservations which now total some 55 million acres in the continental United States and 4 million acres in Alaska. The resources on these reservations are managed by the Federal government as trustee for the Indians through the Office of Indian Affairs in the Department of the Interior.

Some 6 million acres of commercial forest land and 44 million acres of range land are handled by two branches of the Bureau. Only one Indian forest has been formally established, on the Red Lake Indian Reservation in Minnesota in 1916, but Congress in 1934 directed that all Indian forest units should be handled on the principle of sustained yield.

Little is now heard of a former suggestion that the 6 million acres of forest land might advantageously be managed by the Forest Service as agent for the Department of the Interior. An important item of policy involves the broad question as to whether Indian affairs should be handled by the Department of the Interior because they involve primarily the management of natural resources or by the Department of Health, Education, and Welfare because of the many social problems involved in attaining the basic objective of promoting the welfare of the Indians. Still more important from the standpoint of resource management is the question as to whether the Federal government should terminate its trusteeship and let the Indians handle their own lands, as is apparently contemplated in two acts passed in 1954 providing for the orderly liquidation of Federal supervision over the members and property of the Menominee Indian Tribe in Wisconsin and the Klamath Indian Tribe in Oregon.

Departmental Jurisdiction. The Department of the Interior now has jurisdiction over Federal lands in national parks, national monuments, grazing districts, wildlife refuges, and the unreserved public domain. The Department of Agriculture has jurisdiction over the national forests and the submarginal lands acquired during the depression of the 1930's. Never-ending discussion as to whether or not this is a desirable arrangement was intensified by the report of the Hoover Commission in 1949.

That Commission recommended that the activities of the Bureau of Land Management in the Department of the Interior be transferred to the Department of Agriculture and combined with the activities of the Forest Service in a new Forest and Range Service. It also recommended combining the civilian activities of the Army Corps of Engineers with those of the Bureau of Reclamation in the Department of the Interior in a new Water Utilization and Development Service. Three members of the Commission in a vigorous minority report agreed with both of these proposed combinations but urged strongly that the new bureaus be located in a new Department of Natural Resources and that the Department of the Interior be abolished. Two other members, in another minority report, dissented violently from the proposal to strip the Army Corps of Engineers of its civilian activities.

The most striking fact about these diverse recommendations is that they agree in proposing consolidation of the activities relating to national forests, O. and C. revested lands, and grazing districts in a single bureau. The underlying logic is so obvious that the proposal has aroused relatively little open opposition. National-forest lands and O. and C. lands are intermingled or adjacent and are substantially alike in character. National forests and grazing districts are also often adjacent to each other; more important, their range resources are similar in character and to a large extent are used by the same permittees at different seasons of the year. Both efficiency and economy of management would seem clearly to be promoted by placing all three classes of reservations under the same administrative direction.

Controversy centers chiefly around the question whether the new bureau should be located in the Department of Agriculture, the Department of the Interior, or a new Department of Natural Resources. Those who would place it in the Department of Agriculture argue that forests and forage are crops and should therefore be under the jurisdiction of the department that handles other crops and that is bound in any event to be concerned with forests in farm wood lots. They emphasize the advantages accruing to the agency in charge of forest and range management from close cooperation with the other research and service agencies in the Department of Agriculture, and they point to the excellent record of achievement by the Forest Service as a unit in that Department.

Opponents of this view argue that the government's responsibilities in these fields are centered largely in the management of the public lands, where the problems are very different from those on farms; that cooperation with states and private owners in the management of natural resources will be carried out primarily through state departments of forestry and conservation rather than state departments of agriculture; that the relations between land and water resources are so close that they should be handled together; and that it would be unwise to place this tremendous responsibility in the Department of Agriculture, the major purpose of which is to render service to the farmer.

The consolidation of forest and range activities, on the desirability of which there is fairly general agreement, is made more difficult because of the fact that they are now divided between two different departments, with strong disagreement about the department in which they should be located. Conflicting considerations make any completely satisfactory answer practically impossible.

At one time there were many who thought it highly important that the national forests and the national parks should be in the same department. Apparently this feeling is still held by those who believe that the proposed Forest and Range Service should be in the Department of the Interior or a new Department of Natural Resources, but not by those who believe that it should be in the Department of Agriculture. The present situation indicates that the conflicts which at one time existed between the Forest Service and the National Park Service have largely disappeared.

Federal activities dealing with natural resources (soil, water, minerals, fish and wildlife, and crops of all kinds) ramify so widely and at the same time are so closely interrelated that it is difficult to conceive of any departmental organization for handling them which would be entirely satisfactory. Apparently the best that can be done is to adopt whatever setup promises the most advantages and the fewest disadvantages.

Contributions in Lieu of Taxes. It is well settled in law that property owned by the Federal government is not subject to state or local taxation without the consent of Congress. Voluntary contributions in lieu of taxes have, however, been authorized by Congress for most classes of real property owned by the government. The usual procedure with respect to land-utilization and conservation projects is to pay a specified percentage of the receipts from such projects to state or local governmental units, with or without limiting the purpose for which the payment may be used. The great diversity that exists between different classes of property is indicated by the following tabulation of contributions in lieu of taxes as of 1954:

National forests: 25 per cent of the gross receipts to the counties for schools and roads

Land-utilization projects: 25 per cent of the gross receipts to the counties for schools and roads

O. and C. revested lands: 75 per cent of the gross receipts (less cost of access roads) to the counties

Coos Bay Wagon Road reconveyed lands: equivalent of taxes that would be paid by private owners to the counties

Grazing districts (excluding Indian lands): 12½ per cent of the gross receipts from grazing fees to the counties

"Section 15" range lands: 50 per cent of the gross receipts to the counties

National parks and monuments: none

Wildlife refuges: 25 per cent of the receipts to the counties for schools and roads

Mineral leasing: 37½ per cent of the gross receipts to the counties for schools and roads

Water-power licenses on national-forest and public lands: 37½ per cent of the gross receipts to the states

Sites for reclamation structures: none

Unreserved public lands: 5 per cent of the net receipts from the sale of either the land or materials therefrom to the states for schools and roads.

With the exception of the Coos Bay Wagon Road reconveyed lands, contributions are based on receipts, but with widely varying percentages. While 25 per cent is the most common figure, the extremes range from zero in the case of national parks and monuments to 75 per cent (less cost of access roads) in the case of O. and C. revested lands. The generous treatment accorded the communities affected by the reconveyed and revested lands in Oregon is due to the fact that these lands were once in private ownership, but no difference is made in the contributions to local communities from national forests created from the public domain and those formerly in private ownership which were acquired under the Weeks Act and the Clarke-McNary Act.

Principles and Practices. A Federal Real Estate Board appointed by President Roosevelt in 1939 came to the following conclusions in 1943:

(1) Each class of real estate should be considered separately . . . ; (2) the amount of the Federal contribution should take into consideration the extent of actual tax loss, the local benefits from Federal ownership and its effect on requirements for services of State and local governments; (3) where determination of these factors is difficult or impracticable, contributions may be made on a receipts-sharing basis so determined as to approximate the desired results; (4) Federal contributions ought not to be made in such a way as to encourage perpetuation of uneconomic units of government or to impede reforms in the organization and functioning of local government.

In the light of these principles the Board recommended certain changes in current practices, the most important of which was the suggested payment to local communities of 0.75 per cent of the value of acquired land in national forests. Numerous bills incorporating this and other proposals failed to obtain congressional approval in the decade preceding 1955. The only exception was the drastic change in the allocation of fees from grazing districts effected in 1947.

In general, the payment to local communities of a specified percentage of the gross receipts from the classes of Federal land considered in this book seems to work reasonably well. Some means should, however, be devised, such as use of a five-year moving average, to avoid the extremes in contributions that are now possible as between years of very large and very small receipts. It is doubtful whether any restriction should be placed on local communities in their use of Federal contributions. Greater uniformity in the percentage of receipts to be paid from similar classes of land is also desirable. Adoption of an arrangement satisfactory alike to Federal, state, and local governments would do much to meet the common complaint that Federal ownership takes land off the tax rolls, with the implication that there is no satisfactory substitute for the payment of taxes by private owners.

Another aspect of the disposition of receipts from Federal lands concerns their earmarking for specific purposes. Examples are the earmarking of 10 per cent of the receipts from national forests for the construction of roads and trails; of all of the range-improvement fee in grazing districts for improvement of the range; and of 52½ per cent of the receipts from mineral leases for the reclamation fund. Whether, and to what extent, this practice should be extended is a moot point. Advocates of a particular activity, such, for example, as the improvement of recreational facilities on national forests, feel that adequate funds can be obtained only in this way. Others maintain that the control over appropriations should at all times remain in the hands of Congress and that fixed allocations are almost certain to make too much or too little available for the purpose in view. Theoretically, the latter group have the best of the argument; practically, the earmarking process has so many advantages from the standpoint of special-interest groups that it may be difficult to stop.

State Activities. During the 1860's several states appointed commissions of inquiry into their forest situation and attempted to encourage tree planting by offering bounties and tax exemptions. The first permanent state forestry organization was established in New York in 1885. By 1900, the number of states with some form of administrative organization for forestry had increased to 9, by 1910 to 25, and by 1953 to 44. State interest in forestry matters was greatly stimulated by the Governors' Conference of 1908.

Early state activities were confined largely to fire control and public education except in New York, Pennsylvania, and Wisconsin, where there were considerable areas of state-owned forests. A few states undertook to assist private owners by providing nursery stock for planting and by offering advice on the management of their woodlands. Later these activities increased greatly in scope and effectiveness, with Federal cooperation, particularly after passage of the Clarke-McNary Act of 1924, the Cooperative Farm Forestry Act of 1937, and the Cooperative Forest Management Act of 1950. Progress in forest-fire control, to which the Federal government contributes about 25 per cent of the cost, has been especially noteworthy.

Management of state forests has also become a major activity in a considerable number of states, and the regulation of cutting practices is assuming increasing importance.

State Forests and Parks. The states in 1954 owned nearly 30 million acres of forest land, or about 6 per cent of the commercial forest area of the country. Most of this land had been acquired by grants from the Federal government, by tax delinquency, and by purchase. Some 12 million acres were in legally constituted state forests, with nearly two-thirds of this area in the three states of Michigan, Minnesota, and Pennsylvania. New York's "forest preserve" of 2.5 million acres constituted by far the largest area of state-owned land set aside for park purposes.

State forests have three main objectives: demonstration of satisfactory management practices, rehabilitation of depleted areas, and provision of opportunities for recreation. In the few states with large areas they will gradually become an increasingly important source of wood for commercial purposes. Watershed protection, although often substantial, is less significant than in the national forests, the bulk of which are located in the mountains of the West where most of the water of the region originates. State parks are commonly well forested and usually have attractive scenic features, but the scenery is seldom as outstanding as in national parks.

Private Ownership and Management. Largely as a result of the disposal of the public lands, about three-fourths of the commercial forest area of the country passed into private ownership. Management of these lands for continuous forest production developed slowly. Although a few owners took steps in this direction early in the present century, the great majority felt that the relation between costs and probable future returns did not justify expenditures beyond those necessary for fire control. Many felt that the practice of forestry was, and always would be, a public function. Difficult times during the 1920's, followed by the severe depression of the early 1930's, did little to change prevailing attitudes and practices.

The turning point was perhaps reached in 1934, when the Code of Fair Competition for the Lumber and Timber Products Industries focused attention on the problem of private forest management and forced the formulation of specific rules of forest practice. For the first time representatives of private owners of all classes sat down with representatives of public agencies to determine just what measures were required to make continuous forest production on their lands a reality.

The resulting familiarity with the problem aroused an interest that continued after the National Industrial Recovery Act under which the Code had been prepared was invalidated by the Supreme Court. This interest was enhanced by increased stumpage prices and technological improvements in wood utilization which gave reasonable assurance that investments in forest management would yield satisfactory dividends. The establishment of American Forest Products Industries and the initiation of the Tree Farm program in 1941 were also decidedly helpful influences.

During the last decade or two progress in private forestry has proceeded at an impressive rate. The advance has been especially striking among the larger owners, among pulp and paper companies, and in the South and the Pacific Northwest. A significant aspect of the situation is the constructive leadership being exercised by leaders in the wood-using industries to bring about improved practices not only on their own lands but on the lands of the smaller owners.

Changing Patterns of Ownership. The pattern of forest landownership has been, and doubtless will continue to be, a continually changing one. During the past 150 years the change has, indeed, been almost revolutionary. As a result of the disposal of the public domain, the relative positions of the Federal government and of individuals and corporations as owners of forest land have been practically reversed. In 1954 approximately 75 per cent of the commercial forest area was in private ownership, 19 per cent in Federal ownership, and 6 per cent in state and local government ownership.

Although the most extensive transfers between public and private owners have probably already taken place, there is no reason to believe that the process is at an end. Neither the advocates of a marked expansion of public ownership nor the advocates of a marked reduction seem likely to convince the great body of voters that there is anything seriously wrong with the status quo; but it would be surprising indeed if some changes in ownership in both directions did not prove to be in the public interest.

Moreover, within the bounds of either private or public ownership, permanence of the existing pattern is far from assured. In the Federal field, for example, shifts between national forests, national parks, grazing districts, wildlife refuges, and other reservations may well prove to be desirable. Landownership, like land use, must be constantly adjusted to meet the requirements of changing economic and social conditions and goals. Continuing study of the problem, perhaps by a permanent Federal lands commission as well as by other agencies, offers the best means of assuring its intelligent solution under a policy that is dynamic rather than static.

State forests are now generally accepted as a permanent and valuable feature of a state's forestry program. In spite of their relatively small area, they serve several highly useful purposes, especially in the fields of education and recreation. Question occasionally arises as to the relative value of different possible products, as for example timber and wildlife. In New York there is continual pressure to open the timber resources in the "forest preserve" to logging under conservative methods of management, or in effect to change its status from that of a park to that of a state forest in the generally accepted sense of the latter term. Proposals to transfer certain state-owned lands to private ownership may be advanced, particularly in states with the larger holdings, as increased productivity of the land makes it an attractive object of private investment.

"Community forests," which is the generic term usually applied to forests owned by counties, townships, school districts, and other local units of government, have been increasing steadily in number and area. Although they comprise only about 4,500,000 acres, their influence as an educational factor is much increased because they are usually located near centers of population. The major problem which they present is that of assuring continuity of competent management.

Public Cooperation. The steady growth of public cooperation with private owners during the last fifty years has been one of the most striking and most promising developments in the field of forest policy. One of Pinchot's first moves as Chief of the Division of Forestry (1898) was to offer cooperation with landowners in the preparation of forest working plans and planting plans. The offer was primarily of interest as breaking new ground in an important field, since the failure of most owners to put the plans into practical operation was disappointing.

During the first ten years of the century several states, notably in the Northwest and the Northeast, joined hands with private owners in the protection of forests from fire. Progress in this field was furthered in 1911 by passage of the Weeks Act, which appropriated $200,000 of Federal funds to be spent through the states in the control of fire on state and private lands. Another major advance came in 1924 when the Clarke-McNary Act authorized a large increase in the financial contribution of the Federal government and at the same time conditioned its cooperation on the adoption by the states of satisfactory standards of protection.

The Clarke-McNary Act also authorized Federal financial participation in the programs of assistance to private owners in forest management and forest planting which had already been undertaken by several of the states. These programs were aimed primarily at the small owner, and especially the farmer, whose need for help was felt to be greater than that of the larger owner. Greatly increased Federal participation was authorized by the Cooperative Farm Forestry Act (Norris-Doxey Act) of 1937. Additional help was extended by the Soil Conservation Service through the soil-conservation districts.

Amendments to the Clarke-McNary Act in 1944 and 1949 and passage of the Forest-Pest Control Act in 1947 and the Cooperative Forest Management Act in 1950 strengthened still further the Federal government's part in the cooperative program. The last-named act authorized aid to nonfarm as well as farm owners of small forests in the management of their woodlands and the marketing of the harvested products and also aid to the processors of forest products. Meanwhile the states have continued to expand the scope of their cooperative activities and to increase their expenditures, which in the aggregate greatly exceed those of the Federal government.

The cooperative principle is now thoroughly established and generally supported as a desirable form of state activity. Nearly everywhere the states have taken over major, and in some cases practically exclusive, responsibility for the control of forest fires. The trend with respect to the control of insects and diseases is similar but less pronounced.

In the fields of planting, management, marketing, and processing, some question is now being raised as to the extent to which public assistance should be carried. This is particularly true with respect to Federal participation, which will be discussed in the next section. It is, however, appropriate to point out here that Federal expenditures, except those made under the Agricultural Conservation Program, are all handled through state agencies. Some contraction of Federal aid is favored by those who regard it as unnecessary on its present scale or as a form of "creeping socialism" and by some consulting foresters who fear the loss of potential clients. What limited evidence is available on this latter point indicates that the very limited amount of service rendered by the state is apt to arouse a desire for further professional advice which tends to throw business to the consulting forester rather than to take it away from him.

Some disagreement existed at one time as to the respective roles of state forestry organizations and state extension services in the program of cooperative assistance. This situation was largely cleared up in 1948 by the acceptance of an agreement between the Association of Land Grant Colleges and the Association of State Foresters that the state extension services would normally conduct all educational activities of a general nature and that the state forestry organizations would conduct all service activities involving specific on-the-ground assistance in management, marketing, or processing. Under this arrangement, which seems to be working out satisfactorily in practice, the Federal Extension Service cooperates with forty-five state extension services, while the Federal Forest Service cooperates with thirty-eight state forestry departments.

Grants-in-Aid. A grant-in-aid is a device by which one level of government provides funds to a lower level of government for the conduct of certain activities under specified conditions. These conditions usually give the grantor considerable control over the expenditure of the funds and require at least a partial matching of the funds by the grantee. The grant commonly continues from year to year but may be discontinued by the grantor at any time.

The appropriation of $200,000 made by the Weeks Act in 1911 for cooperation with the states in control of forest fires is often regarded as the first of the modern grants-in-aid. Subsequent grants in other fields were made by the Clarke-McNary Act of 1924 and its amendments of 1944 and 1949, by the Norris-Doxey Act of 1937 (repealed in 1950), and by the Cooperative Forest Management Act of 1950. Total grants under these various acts during the fiscal year 1953 amounted to approximately 11 million dollars—a mere bagatelle in comparison with the total for all groups of about 3 billion dollars, in which grants for road construction and welfare rank high. These figures do not include subsidies to private owners, such as benefit payments for the adoption of recommended forestry practices, in forestry as well as in other fields.

Among the arguments in favor of grants-in-aid by the Federal government to the states are that they stimulate the undertaking of desirable activities; that by redistributing the national income they make it possible for the poorer states to meet standards of accomplishment otherwise beyond their means; that they give the states the benefit of the wider experience, technical competence, and leadership of the Federal government; and that joint supervision over the expenditure of funds results in their more effective use.

Among the arguments against grants-in-aid are that they stimulate (or "bribe") states to undertake undesirable or relatively undesirable activities; that the redistribution of income which they effect is unfair to the richer states and tends to pauperize the poorer ones; that Federal control may be arbitrary and uninformed with respect to local conditions and needs; that administration is unduly expensive, since a dollar loses weight in traveling to Washington and back; and that division of responsibility between Federal and state governments is likely to result in less rather than more effective supervision over the selection of projects and the expenditure of funds.

Whatever one may think of the relative merits of the opposing arguments, the system is apparently too well entrenched to make probable any material modification either in forestry or other fields. Fortunately the record in forestry is an excellent one, and administration of the grant for cooperation in fire control is often cited as a model of efficiency. Much hard work, many conferences, and above all a determination on the part of Federal and state officials to adjust their differences for the attainment of a common objective have combined to bring about this result.

Interstate Compacts. The interstate compact is a device to promote cooperation among the states which is attracting increasing attention. Until 1954 the Northeastern States Forest Fire Protection Compact constituted the sole example of its use in the field of forestry. Although it has been in existence only since 1949, it has already proved of value in strengthening individual state organizations through the exchange of information, in developing their personnel through joint training programs, and in preparing plans for concerted attack on any conflagration that may occur. Use has also been made of the Southern Region Educational Compact to develop plans for the coordination of professional instruction and research in forestry at the schools of forestry in the Southern states.

These developments suggest the strong possibility that the interstate compact may prove to be a useful tool in other fields than fire control and forestry education. Some of those that might be considered include control of insects and diseases, production of nursery stock, and regulation of cutting practices. The argument against the latter that it discriminates against residents of a particular state might be overcome by joint action of the states in a given region.

Private Cooperation. In addition to improving the management of their own timberlands, private owners and operators have been doing much in recent years, both individually and through their organizations, to bring about better practices by the smaller owners. The Tree Farm and More Trees for America programs, the adoption of rules of forest practice, and the publication of bulletins and leaflets describing and illustrating effective methods of growing and harvesting forest products have had an influence that is hard to measure but that is certainly substantial.

Even more important is the giving of on-the-ground advice by company or association foresters to owners who are not themselves in a position to employ professional assistance. This practice, which is growing, particularly in the South, resembles closely that used by state forestry departments in their cooperative programs. Another educational measure is for the company or the contractor to favor the owner who adopts approved methods of cutting. Some companies have recently organized "tree farm families," consisting of owners of officially certified tree farms, from whom they agree to buy all the wood that can properly be harvested under sustained-yield management.

The larger private owners, because of their intimate contacts with smaller owners from whom they obtain a large part of their timber supply, can if they will exert a potent influence in improving forest practices throughout the country. They can most effectively supplement, but probably never entirely supplant, the educational and service activities of public agencies. For many years to come the coordinated efforts of Federal, state, and private agencies will certainly be needed to raise present cutting practices to a satisfactory standard. How best to promote such coordination at both the national and the local level is a problem that deserves the earnest attention of all concerned.

Consulting foresters will play an increasingly important role in bringing about the practice of forestry as the value of their services is more generally recognized.

Public Regulation. In 1919 both the Forest Service and the Society of American Foresters started campaigns to bring about public regulation of cutting on privately owned forest lands. The basic arguments were that an ample supply of forest products is essential to assure the prosperity of the country in time of peace and its safety in time of war and that since private forest owners had failed voluntarily to adopt practices that would produce such a supply they must be forced to do so.

Two schools of thought soon developed. One favored Federal regulation on the ground that the problem was a national one and that the Federal government alone was big enough and strong enough to enforce effective control. The other favored state regulation on the ground that it was more democratic and more in accord with the American way of doing things; that in the long run it was likely to prove more effective; and that it was not open to the charge of unconstitutionality. Still a third school of thought, less vocal at the time, did not favor any regulation on the ground that it was intrinsically objectionable and that in the fullness of time economic conditions would lead private owners voluntarily to practice more intensive forestry than could be forced by public compulsion.

The result of the wordy, and occasionally acrimonious, argument was a stalemate so far as regulation was concerned. It did, however, result in the passage of four important pieces of Federal legislation—the General Exchange Act of 1922, the Clarke-McNary Act of 1924, the McNary-Woodruff Act of 1928, and the McSweeney-McNary Act of 1928.

The subject of public regulation, either Federal or state, remained in the background until after 1933, when the advent of F. A. Silcox as Chief of the Forest Service again brought it into prominence. His annual report for 1937 announced "A Three-point Program" of public ownership and management, public cooperation with private owners, and public regulation. Silcox, Clapp, and Watts all advocated state regulation with Federal cooperation, but also urged direct Federal regulation if the states failed to pass and enforce satisfactory legislation within a reasonable time. Since 1952 the subject has again disappeared from the limelight.

Meanwhile, between 1941 and 1949, several states passed laws dealing with forest practices. In five states, acceptable cutting practices are incorporated in the statutes; in two states, they are promulgated by regional boards; and in three states, private owners are encouraged to comply voluntarily with cutting standards established by state or regional agencies.

These laws have met with relatively little opposition from timberland owners, who have in some cases taken the initiative in obtaining their passage. The general position of the forest industries is one of outright opposition to Federal regulation and of watchful waiting with respect to state regulation. They wish to see how the laws already enacted work out in actual practice before opposing or supporting further developments along these lines.

Constitutionality and Trends. There now seems to be no question as to the constitutionality of public regulation. So far as the states are concerned, the matter was settled in 1949 by the action of the state Supreme Court and the United States Supreme Court in upholding the Washington statute as a valid exercise of the state's police power. So far as the Federal government is concerned, Supreme Court decisions over the last twenty years apparently leave no doubt that properly drawn Federal legislation would be upheld as valid under the constitutional powers of Congress to control interstate commerce and to provide for the public welfare. Decision as to the best course to pursue can therefore be based on considered judgment as to what type of regulatory legislation, if any, will in the long run be most effective in strengthening the national economy.

Recent state legislation, the steady improvement in cutting practices on private lands, and the present temper of the people with respect to any extension of Federal activities in the economic sphere make highly unlikely the passage of Federal regulatory legislation in the near future. The long-time prospect depends on future developments in the management of private lands and in political philosophies.

On the other hand, the passage of additional state legislation and the amendment of existing laws in the light of experience with them appear highly probable. Although no one believes that legislation can force unwilling owners to adopt intensive forest practices, there is widespread belief that it can force them to stop destructive practices. Such legislation is also proving to be an effective educational device, since it is bringing the subject of forest management to the attention of indifferent owners in a way to make them do some thinking about it. The results can hardly fail to be beneficial.

Two trends in regulatory state legislation seem to be both clear-cut and desirable. The first, almost universal, trend is to permit an owner, with the approval of the appropriate public agency, to substitute his own plan of management for the cutting practices prescribed by the statute or by a state or regional board. The other trend is to place the formulation and enforcement of standards of cutting practice in the hands of local boards containing representatives both of the general public and of timberland owners. In both cases the effect is to promote better management by making it possible to adopt the method of cutting best suited to specific properties, each of which has its own individual requirements. Flexibility is an indispensable prerequisite for the application of optimum practices of management.

Few, if any, regard public regulation (Federal or state) as a panacea. An increasing number, however, apparently believe that state regulation is a desirable supplement to, but not in any way a substitute for, state cooperation. Properly framed and administered, it can perform a useful function in the United States, as it has in other countries.

Forest Taxation. The ad valorem general property tax has long been recognized as imposing a hardship on owners of forests from which a current income is not being regularly received. It tends to encourage the premature cutting of old-growth timber and to discourage the holding of cutover lands until another crop is ready for harvesting. Timber owners have sometimes averred that it constitutes one of the chief obstacles to the holding and management of forest lands for sustained production.

In order to meet this situation, twenty-six states now have thirty-one special forest tax laws. The method most commonly used to grant relief is to replace the ad valorem general property tax, usually on an optional basis, by an annual tax on the land and a yield tax on the value of the timber at the time of cutting. Owners who elect to take advantage of the opportunity are commonly required to comply with conditions not imposed on other owners, such as following certain forest practices or opening their lands to the general public for hunting and fishing.

The total area taxed under special laws is relatively small. It does not appear likely to increase rapidly unless other states should follow the example of New Hampshire, which applies its recent annual land tax and yield tax to all forest land irrespective of the wishes of the owner. The need to replace or modify the general property tax becomes less urgent as forest properties are placed under sustained-yield management which produces an income annually or at short intervals. With equitable assessments and a mathematically correct relation between the property tax and the yield tax, the burden on a forest with a sustained annual income would be identical under the two methods of taxation. As more and more forest properties approach this condition, both the necessity and the pressure for special forest-tax legislation may be expected to decrease.

Estate and inheritance taxes may have an adverse effect on forest management in cases where the beneficiary is forced to sell a considerable part of the inherited land or timber in order to meet tax payments. The threat is particularly serious in the case of a forest under sustained yield, the management of which may be seriously affected by heavy overcutting or by breaking it up into smaller parcels. Even corporation-owned properties may be in danger if the forced sale of stock by the heirs of an individual with a controlling interest in the company transfers control to persons who prefer liquidation to sustained-yield management. The often-urged amendment of Federal and state laws to permit the payment of death taxes in installments during a reasonable time would materially improve the present situation.

Previous discriminations against forest owners under the income tax were removed in 1944 by amendments liberalizing the provisions of the law with respect to capital gains and the classification of current expenses. A further amendment that might encourage the prompt restocking of cutover areas would be to permit the owner to report as a current expense the cost of planting immediately following cutting. Such cost would then be treated in the same way as expenditures incurred to obtain natural reproduction.

Insurance and Credit. For various reasons, forest insurance has made little progress in the United States. Sporadic interest in the subject by insurance companies and timberland owners has not resulted in any comprehensive coverage either by private companies or public agencies.

Felled timber and stacked logs are commonly insured, but the coverage on standing timber and young growth is negligible. In 1944 the Federal Crop Insurance Corporation was authorized to insure timber and forests but took no steps in that direction. Several states have given the subject consideration but have taken no action.

Insurance is such an integral part of the entire system of private enterprise that it seems unlikely that the present situation will continue indefinitely. As investments in timber production increase, more and more owners will feel the need for protection at reasonable cost against the losses that are inevitable no matter how greatly management may improve. Whether that need will be met by business or by government is still uncertain.

Credit has so far been adequate for the purchase and harvesting of merchantable timber, but not for the practice of forestry. Commercial banking institutions have been hesitant about making loans on terms that would really encourage the building up and continuous management of stable forest properties, particularly by small owners. Until 1953, when Congress authorized loans up to ten years, national banks were not allowed to make loans secured by forest lands, which were classified as unimproved property. Limited credit has been extended by several of the Federal Reserve Banks and by the Farmers Home Administration.

Many owners would doubtless be interested in more intensive management practices if credit were available at a low rate of interest and for a long enough period (perhaps up to fifty years) to enable them to establish and build up a forest property as a going concern. A committee appointed by the Governor of the Farm Credit Administration in 1945 recommended the creation of a national Forest Credit Bank as the best means of meeting this need. Under the proposed plan, borrowers would be required to follow approved forest practices and to take out insurance, which would be provided by a government corporation. Insurance would also be made available through the corporation to other forest owners. It seems not unlikely that provision of better facilities in the credit and insurance fields will proceed simultaneously, whether under private or public auspices.

Research. Forest research has long been an important activity of the Forest Service and other Federal agencies. Work in that field was greatly strengthened by the establishment of the regional forest experiment stations (from 1908 on) and the Forest Products Laboratory (1910), the organization of the Branch of Research (1915), and the passage of the McSweeney-McNary Act (1928). Schools of forestry have always regarded research as a major part of their educational function. State agricultural experiment stations and state departments of forestry have made limited contributions, and in recent years industry has entered the field in a substantial way, particularly in the field of forest products.

There is complete agreement that further progress in forest production and utilization is dependent on research and that present activities by all agencies should be materially strengthened. There is also agreement that increased emphasis should be placed on fundamental as contrasted with applied research. The main problems are to determine the general fields and the specific projects most in need of attention and to obtain effective coordination of effort among the many agencies concerned. Much help in finding the answers should be provided by a comprehensive study of forest research conducted during 1953 and 1954 by the Society of American Foresters and the National Research Council, with the financial support of the Rockefeller Foundation.

Education. Education, both professional and popular, is universally recognized as another indispensable element in continued progress. Adequate facilities for professional instruction are now available in the twenty-four schools of forestry which were accredited by the Society of American Foresters in 1954. Increased emphasis on breadth of training and the encouragement of graduate work are desirable trends.

Education of a nonprofessional character is being steadily extended to forest owners and wood processors, with increased effectiveness. The cooperative approach being used by public agencies and private industry is gradually bringing about changes in forest production and utilization that only a few years ago would have been regarded as revolutionary. Its success to date augurs well for much greater future accomplishments.

A third type of education which has always been stressed by public agencies and recently by industry aims at giving the man on the street accurate information as to the contributions made by forest products and forest services to the national well-being and as to the measures that must be taken to perpetuate these contributions. Constructive public policies, both national and state, can be adopted and maintained only with the support of an informed electorate, including school children who will be the citizens of tomorrow. The increasing interest and knowledge being shown by teachers, editors, legislators, and businessmen in every walk of life gives hope that forest policies can be formulated and forest programs carried forward far more effectively than has been the case in the past.

Professional Leadership. Much of the progress that has taken place in forestry has been due to the leadership of professional foresters, both individually and collectively through the Society of American Foresters.

Organized in 1900, that society has been a powerful force in developing constructive policies and in maintaining high standards of technical competence and ethical conduct. Through the *Journal of Forestry* it has provided an effective medium for the presentation of information and the interchange of ideas, and through its geographical sections and its subject-matter divisions it has promoted friendly personal relationships and informal discussion of matters of common interest. It has been an integrating and stimulating agency with widespread influence on the country's forest policies and practices.

Conclusion. The forest and range policies of the United States are not single entities but are made up of the policies of governments, associations, companies, and individuals. Each of these policies in turn has its own characteristic components. All have evolved gradually in response to stimuli exerted by ever-changing conditions.

That evolution has now reached the stage where the activities of public and private agencies form a pattern that gives promise of relative stability. The task of managing the country's forest and range lands so as to assure an adequate and permanent supply of products and services has become largely a cooperative enterprise. There will inevitably be some adjustments in the areas in different classes of ownership; in the scope and character of the cooperative and regulatory activities of public agencies; in the programs adopted by industry to promote better management; and in the division of responsibility for education and research. The main problem today is not to invent new methods of attaining the goals which all agree to be desirable, but rather to sharpen the tools already in existence.

National policy will always be a mosaic in a country characterized by private enterprise and a Federal system of government, with ample opportunity for the free expression of diverse philosophies. Continual rearrangement and improvement of the parts of the mosaic can give the picture which it presents a steadily increasing unity and strength that bodes well for the future.

REFERENCES

Cameron, Jenks: "The Development of Governmental Forest Control in the United States," chap. 12, Johns Hopkins Press, Baltimore, 1928.
Clawson, Marion: "Uncle Sam's Acres," chap. 10, Dodd, Mead & Company, Inc., New York, 1951.
Dana, Samuel T.: "Forest Policy in the United States," University of Britis Columbia, Vancouver, B.C., 1953.
Greeley, William B.: "Forests and Men," chap. 15, Doubleday & Company, Inc., New York, 1951.

————: "Forest Policy," chaps. 23–24, McGraw-Hill Book Company, Inc., New York, 1953.

Gulick, Luther Halsey: "American Forest Policy," chaps. 5–7, Duell, Sloan & Pearce, New York, 1951.

Hibbard, Benjamin H.: "A History of the Public Land Policies," chap. 28, The Macmillan Company, New York, 1924.

Ise, John: "The United States Forest Policy," chap. 13, Yale University Press, New Haven, Conn., 1920.

Peffer, E. Louise: "The Closing of the Public Domain," chap. 19, Stanford University Press, Stanford, Calif., 1951.

Pinchot, Gifford: "Breaking New Ground," chap. 90, Harcourt, Brace and Company, Inc., New York, 1947.

Winters, Robert K. (ed.): "Fifty Years of Forestry in the U.S.A.," chap. 1, Society of American Foresters, Washington, 1950.

Survey of Federal Policy on Wildlife, Soil, Water, and Minerals

As a help in obtaining a broader view of the picture as a whole, this appendix presents briefly the chief events in the development of Federal policy with respect to four other important resources with which forest and range lands are often associated. All natural resources are so closely related that policies relating to one resource are likely also to affect other resources, particularly where multipurpose land management is or should be practiced.

WILDLIFE

Early Abundance. Wildlife, regarded today as primarily a recreational asset, was valued chiefly in colonial times as a source of pelts, hides, and food. Present in extraordinary variety and abundance, it constituted a source of wealth to be had for the taking. By far the most important from the economic standpoint were the fur-bearing animals such as mink, otter, weasel, marten, fisher, beaver, fox, and bear. These formed the basis for a lucrative international trade that put cash in the pockets not only of hunters, trappers, and traders but of the frontier farmer. Many of the early explorations in the North and the West were motivated by the search for new sources of supply.

Millions of migratory waterfowl furnished a cheap and plentiful source of food. Canvasback duck, for example, were fed to slaves along Chesapeake Bay until they rebelled at the monotony of the fare. Wild turkeys were a delicacy available to any skillful hunter. Fresh-water fish abounded in the many lakes and streams and provided a welcome variety in food.

Far more important from the financial point of view were the offshore and ocean fisheries. These gave rise to another important and profitable form of international commerce, primarily with Western Europe and the West Indies. Although fishing was particularly intensive on the Newfoundland banks, American ships were soon making heavy catches of cod, haddock, halibut, and mackerel all up and down the coast and were scouring the seven seas in search of whales. Apparently unlimited supplies caused no concern for the future.

Although Plymouth Colony in 1627 decreed that "fowling, fishing, and hunting be free," legislation protecting deer by a closed season was enacted by Massachusetts as early as 1693. By 1776 all of the colonies except Georgia had enacted some type of game legislation, and in 1776 the Continental Congress ordered a closed season on deer in all of the colonies except Georgia. That such legislation had any material effect in restraining the hunter or in reducing the kill is highly improbable.

As settlement moved westward, the pioneers encountered vast herds of antelope and buffalo, countless millions of passenger pigeons, large numbers of coyotes and buffaloes, and innumerable prairie chickens, sage grouse, and sharp-tails. Still farther west deer, elk, bear, mountain sheep, and mountain goats were abundant. Gradually trapping, hunting, and change of environment made it clear that the supply of wildlife was no more inexhaustible than the supply of timber. State Legislatures attempted to meet the situation in various ways, and by 1880 all of the states and territories had some sort of game legislation, but this was for the most part poorly drawn, weakly enforced, and did little to remedy the situation.

Federal Research. The Federal government's first recognition that it had any responsibility in connection with wildlife came in the field of research. In 1871 Congress authorized the President to appoint a Commissioner of Fish and Fisheries to study the causes of the decrease in the supply of food fish of the coast and lakes of the United States. The Commissioner reported directly to the President and Congress—a form of organization that the American Association for the Advancement of Science attempted unsuccessfully to have followed in the inauguration of Federal work in forestry. In 1903 the Commission of Fish and Fisheries became the Bureau of Fisheries in the Department of Commerce and Labor, which was established that year.

In 1887 an Entomological Commission was appointed in the Department of the Interior to study one of the greatest outbreaks of migratory locusts in the history of the country. This Commission was transferred to the Department of Agriculture in 1880 and its activities combined with other entomological work of the Department in the Division of Entomology. During the 1890's and later, this Division undertook economic investigations of the bark and timber beetles of North America to determine the character and extent of damage caused by these insects and to find methods of preventing losses from their attacks.

Efforts by the American Ornithological Union to bring about the creation of a division of economic ornithology in the Department of Agriculture resulted in 1885 in an appropriation to initiate work in this field in the Division of Entomology. The next year (1886) the work was expanded and placed in a new Division of Economic Ornithology and Mammalogy. In 1896 this Division in turn became the Division of Biological Survey, and in 1905 the Bureau of Biological Survey. Much of the research by the Division and the Bureau dealt with birds and mammals of interest in forest management.

Protection of Wildlife. The responsibility of the Biological Survey for research relating to wildlife was extended to the administrative field by the Lacey Act of May 25, 1900. As a means of helping the states to suppress illicit traffic in wildlife, this act forbade the interstate transportation of wild animals or birds taken or possessed in violation of the laws of the state from which or to which they were shipped. First attention was paid to stopping illegal interstate traffic in quail, grouse, and deer. The act also prohibited the importation into the United States of the mongoose, fruit bat, English sparrow, starling, and such other birds and mammals as the Secretary of Agriculture might declare to be injurious to the interests of agriculture or horticulture.

The first wildfowl sanctuary in the country was established by California in 1870 in what is now the heart of the city of Oakland. No further action of this sort was taken until 1903, when President Roosevelt set aside Pelican Island in Florida as the first Federal wildlife refuge. During the next ten years, refuges were established by Indiana (1903), Pennsylvania (1905), Alabama (1907), Massachusetts (1908), Idaho (1909), and Louisiana (1911). Federal refuges also increased in number, thirty-six being set aside in 1918 alone, but these were mostly small and intended primarily for the protection of nesting birds. Proclamation of the Wichita Forest Reserve in Oklahoma as a game refuge in 1905 and acquisition of the National Bison Range in Montana in 1908 initiated the Federal program of refuges intended specifically for the preservation of big-game animals. National parks and national monuments also served as wildlife refuges.

Protection of Fur Seals. By the beginning of the present century the supply of fur seals had been so far reduced as to jeopardize an important industry and even to threaten extinction of the species. To meet this situation the United States, Great Britain, Japan, and Russia in 1911 entered into a convention to prohibit pelagic sealing in the Pacific Ocean north of the 31st parallel of north latitude. Each government thereupon became responsible for the taking of seals within specific areas and for distributing the skins in accordance with specified ratios. The arrangement proved so successful that in the Pribilof Islands, where the United States controls the take, the population of fur seals increased from 130,000 to about 1,500,000 in 1953.[1]

The convention continued in effect until the Second World War, when Japan and Russia withdrew. In 1942 the United States entered into a provisional agreement with Canada, under which 20 per cent of the sealskins harvested on the Pribilof Islands are turned over to Canada.

Protection of Other Marine Animals. The policy of international cooperation in the protection of marine biological resources, first established by the fur seal treaty, has been continued through a number of other international agreements. These include the Pacific Halibut Convention of 1923 with Great Britain, the Sockeye Salmon Convention of 1930 with Canada, the multilateral Whaling Convention of 1931, the Conventions with Mexico (1949) and Costa Rica (1949) for the scientific investigation of tuna, the International Convention of 1949 for the Northwest Atlantic Fisheries to which eleven countries are signatories, and the North Pacific Fishery Convention of 1952.

[1] The Fish and Wildlife Service now believes that previous estimates of more than three million fur seals were much too high. That figure was based on a computation which considered largely factors dealing with mortality, at a time when the herd was small. In recent years, population estimates have been based on (1) aerial photographic census of rookery areas, (2) land counts of pups and breeding bulls, and (3) returns from extensive tagging and banding experiments.

A convention with Canada for the protection and improvement of the Great Lakes fisheries was signed by the United States in 1946 but has not yet been ratified by the Senate.

Protection of Migratory Birds. Continuing efforts to provide national protection for migratory birds, the supply of which was rapidly diminishing under the wholly inadequate protection afforded by the individual states, finally led in 1913 to passage of the Weeks-McLean Migratory Bird Act. This act declared migratory birds to be under "the custody and protection of the United States" by virtue of the control of Congress over interstate commerce and authorized the Secretary of Agriculture to establish regulations to prevent their destruction. Attempts to enforce these regulations led to decisions by both state and Federal courts that the act was unconstitutional because it did not in fact deal with interstate commerce but only with the taking of wild animals, which were universally recognized to be under the custody of the state.

The fact of state custody had been definitely established as far back as 1896, when the United States Supreme Court, in the case of *Geer v. Connecticut*, upheld the right of Connecticut and other states to preserve their wildlife. The decision was based on the principle that the state as a sovereign entity owns all the wildlife within its borders and that what it owns it can dispose of as it sees fit. The immediate effect of the decision was to stimulate the enactment by the states of additional game legislation.

Even while the Weeks-McLean bill was under consideration, resolutions were introduced in the Senate requesting the President to negotiate treaties with other countries for the protection of migratory birds. This suggestion led to the Migratory Bird Convention of 1916 between the United States and Great Britain. Among other things, the convention provided closed seasons of various lengths for certain specified migratory game birds, migratory insectivorous birds, and other migratory nongame birds and arranged for the supplementary detailed legislation necessary to effectuate its general provisions.

In 1918, Congress passed the Migratory Bird Treaty Act, which was even more drastic in some respects than the 1913 act. In addition to including the specific provisions incorporated in the convention, it banned interstate or international shipment of birds illegally taken; authorized and directed the Secretary of Agriculture to adopt suitable regulations for the control of such activities as the taking, sale, or transportation of any birds covered by the act; authorized enforcing officers to arrest without warrant; and made violation of the convention, the act, or the Secretary's regulations an offense punishable by a fine of not more than $500 and imprisonment for not more than six months. The act also permitted any state to give such further protection to migratory birds as it might desire, and several states have taken advantage of this provision in such matters as closed seasons, bag limits, and methods of taking.

Constitutionality of Migratory Bird Treaty Act. Several Federal district courts upheld the constitutionality of the act, including that for the District of Eastern Arkansas, which had held the 1913 act unconstitutional. The state of

Missouri appealed one of these decisions to the United States Supreme Court in the case of *Missouri v. Holland*, on the ground that ownership of game resides in the state and that the law violated the Tenth Amendment. In 1920, the Supreme Court, with two justices dissenting, upheld the constitutionality of the act on such broad grounds as to make the case one of interest in other fields than the protection of migratory birds.

"If the treaty is valid," said Justice Holmes, who wrote the decision of the Court, "there can be no dispute about the validity of the statute. . . . It is said that a treaty cannot be valid if it infringes the Constitution, that there are limits, therefore, to the treaty-making power, and that one such limit is that what an act of Congress could not do unaided, in derogation of the powers reserved to the States, a treaty cannot do. . . . It is obvious that there may be matters of the sharpest exigency for the national well being that an act of Congress could not deal with but that a treaty followed by such an act could, and it is not lightly to be assumed that, in matters requiring national action, 'a power which must belong to and somewhere reside in every civilized government' is not to be found. . . . The treaty in question does not contravene any prohibitory words to be found in the Constitution. The only question is whether it is forbidden by some invisible radiation from the general terms of the Tenth Amendment. We must consider what this country has become in deciding what that Amendment has reserved.

"The State founds . . . its claim of exclusive authority upon the assertion of title to migratory birds, an assertion that is embodied in statute. No doubt it is true that as between the State and its inhabitants the State may regulate the killing and sale of such birds, but it does not follow that its authority is exclusive of paramount powers. To put the claim of the State upon title is to lean upon a slender reed. Wild birds are not in the possession of anyone; and possession is the beginning of ownership. The whole foundation of the States' rights is the presence within their jurisdiction of birds that yesterday had not arrived, tomorrow may be in another State, and in a week a thousand miles away. . . .

"Here a national interest of very nearly the first magnitude is involved. It can be protected only by national action in concert with that of another power. The subject matter is only transitorily within the State and has no permanent habitat therein. But for the treaty and the statute there soon might be no birds for any powers to deal with. We see nothing in the Constitution that compels the Government to sit by while a food supply is cut off and the protectors of our forests and our crops are destroyed. It is not sufficient to rely upon the States. The reliance is vain, and were it otherwise, the question is whether the United States is forbidden to act. We are of the opinion that the treaty and the statute must be upheld."

Other Migratory Bird Legislation. Other protection to migratory birds was given by acts passed in 1924 and 1928, authorizing respectively appropriations of $1,500,000 and $350,000 to enable the Secretary of Agriculture to purchase lands for the establishment of the Upper Mississippi River Wild Life and Fish Refuge and the Bear River (Utah) Migratory Bird Refuge. The more general Migratory Bird Conservation Act of 1929 authorized an appropriation of nearly

$8,000,000 for a ten-year program of land examination, acquisition, and development to provide an extensive system of Federal refuges for migratory birds. The latter act also created a Migratory Bird Conservation Commission, consisting of the Secretary of Agriculture, the Secretary of Commerce, the Secretary of the Interior, and two members each from the Senate and House of Representatives, to approve of purchases and otherwise to direct the development of the program. The similarity between this Commission and the National Forest Reservation Commission is clear.

Additional funds for the acquisition of migratory-bird refuges were provided by an act passed in 1934 requiring hunters of migratory birds to buy a $1 Federal stamp. Receipts from this fee, which in 1949 was increased to $2, have been substantial.

Migratory Bird Treaty with Mexico. Further protection to migratory birds was afforded in 1936 by the conclusion of a migratory-bird treaty with Mexico. The treaty was effectuated the same year by passage of an act amending the Migratory Bird Treaty Act of 1918 so as to extend its provisions to cover the migratory birds included in the treaty with Mexico. In addition, the act made it unlawful to export or import "any game mammals, dead or alive, or parts or products thereof, except under permit or authorization of the Secretary of Agriculture in accordance with such regulations as he may prescribe."

Expansion of Refuge Program. The establishment and development of Federal wildlife refuges proceeded at a greatly accelerated pace during the period from 1933 to 1953. In the thirties, some $8,500,000 of emergency funds were made available for the advancement of the refuge program, and in addition extensive improvements were effected by the C.C.C. From less than a million acres in 1933, the area administered by the Fish and Wildlife Service had risen by 1953 to approximately 17½ million acres, of which some 8 million acres were in Alaska, and some 6 million acres were on lands used also for other purposes such as grazing districts and reclamation projects. Much the largest number of refuges (202 out of a total of 272) is for migratory water fowl, but the largest area (11 million acres) is for big game. In addition, wildlife is protected in all of the national parks and national monuments, and there are substantial areas of state game refuges in the national forests. There are also other extensive areas of state wildlife refuges, and a much smaller area in private ownership.

Coordination Act. The Coordination Act of 1934 authorized the Secretary of Agriculture and the Secretary of Commerce to cooperate with Federal, state, and other agencies in developing a nationwide program of wildlife conservation and rehabilitation and to determine the effects of pollution on wildlife, with recommendations for remedial measures. The act also provided that whenever the Federal government impounds water for any use, the Bureau of Fisheries and/or the Bureau of Biological Survey (now combined in the Fish and Wildlife Service) shall be given an opportunity to make such uses of the impounded waters as are not inconsistent with their primary use; and that in connection with any future dam construction, either by the Federal government or by any agency under government permit, provision shall be made, if economically practicable, for the migration of fish life between the upper and lower waters by means of fish lifts, ladders, or other devices.

In addition, the act provided for the better protection of wildlife resources of all kinds on Indian reservations and unallotted Indian lands; for the preparation of a program for the maintenance of an adequate supply of wildlife on any lands owned or leased by the Federal government; for the establishment thereon of game farms and fish cultural stations; and for the acceptance of donations of land, funds, and other aids to the development of the program authorized by the act. No financial authorization was included. The act was amended and strengthened in 1946.

Federal Aid in Wildlife and Fishery Management. Federal aid to the states in wildlife management was afforded by the Wildlife Restoration Act of 1937, commonly known as the Pittman-Robertson Act. This act authorized distribution to the states of the Federal tax on firearms, shells, and cartridges, on the basis of their respective areas and number of licensed hunters. The funds were made available for research and for the purchase and development of game refuges and public hunting grounds. An amendment in 1946 authorized the use of not more than 25 per cent of the Federal contribution for the maintenance of completed wildlife-restoration projects. States accepting grants are required to provide by law that all receipts from the sale of hunting licenses will be used for the State Fish and Game Department, to submit all proposed projects to the Fish and Wildlife Service for approval, and to contribute from state funds at least 25 per cent of the total cost of all approved projects.

Similar aid in fishery management was provided by the Dingell-Johnson Act of 1950, which authorized use of the Federal tax on fishing rods, creels, reels, and artificial lures, baits, and flies for cooperative effort in this field. The act closely resembles the Pittman-Robertson Act in purpose and requirements, but differs slightly in the formula for the apportionment of funds to the states.

The acts have greatly stimulated state activity, and extensive programs in both fields have been conducted under them in all parts of the country.

An act of 1948 provided that upon request certain property controlled by Federal agencies but no longer needed by them might be transferred (1) to the jurisdiction of a state wildlife agency, without compensation, for wildlife conservation purposes other than for migratory birds or (2) to the Secretary of the Interior if the property has particular value in carrying out the national migratory-bird management program.

Interstate Fishery Compacts. Within recent years all the states bordering on the Atlantic Ocean, the Gulf of Mexico, and the Pacific Ocean have used the interstate compact as a means of protecting the marine, shell, and anadromous fisheries in which they have a common interest. The Atlantic States Marine Fisheries Compact was approved in 1942, the Pacific Marine Fisheries Compact in 1947, and the Gulf States Marine Fisheries Compact in 1949. In the case of the Atlantic states and Gulf states compacts, the Fish and Wildlife Service acts as the primary research agency of the respective commissions; while in the case of the Pacific compact, the fisheries research agencies of the signatory states act in collaboration as the official research agency of the Commission.

Summary. Federal participation in wildlife activities was limited to research until 1900, when Congress forbade the interstate shipment of wild animals or birds taken or possessed in violation of the laws of the states concerned. Protection of migratory birds was materially strengthened by legislation enacted under authority of treaties negotiated with Great Britain (acting in behalf of Canada) in 1916 and with Mexico in 1936. The Federal program of wildlife refuges, first started in 1903, expanded until by 1954 it included nearly 9½ million acres in the states and 8 million acres in Alaska. Federal aid to the states, granted under the Pittman-Robertson Act of 1937 and the Dingell-Johnson Act of 1950, has greatly stimulated state activity in the fields of wildlife and fishery management.

SOIL

Americans have used soil as prodigally as they have all other natural resources. As usual, a few wise men, and only a few, foresaw the dangers of continued exploitation, and tried to do something about it. Even in colonial days a handful of the more progressive farmers, including Washington and Jefferson, recommended the use of marl and lime, fertilizers (particularly animal manure), soil-building crops, cover crops, crop rotations, contour plowing, terracing, and strip cropping as means of building soil fertility and preventing erosion. For various reasons these measures were never widely or successfully employed. Until recently there was always unplowed soil, just as there were uncut forests, beyond the horizon.

Soil Conservation Service. Toward the latter part of the 1800's soil surveys and soil research were initiated by the state agricultural experiment stations and the United States Department of Agriculture. Soil erosion, which by 1935 was estimated to have affected half the land area of the United States, to have destroyed or seriously damaged some 100 million acres of crop land, and to have materially reduced the fertility on at least an equal area, attracted gradually increasing attention. The dust storms of May, 1934, when soil from the Great Plains was blown to the East coast and out to sea, finally convinced the public that the problem was one in the solution of which the Federal government must take an active part.

The first steps were taken under a provision of the National Industrial Recovery Act of 1933 specifically authorizing the construction of works for the prevention of soil and coastal erosion. Under this authorization Public Works Administrator Harold L. Ickes made an allotment of 5 million dollars to himself as Secretary of the Interior for work of this character. The Secretary indicated his recognition of the importance of the work in the statement, "It will be impossible to maintain permanent prosperity over large areas of the United States if the present rapid destruction and impoverishment of our most valuable agricultural lands by accelerated erosion is permitted to continue."

The well-known erosion-control project in Coon Valley, Wisconsin, was the first to be undertaken. In October, 1933, the Soil Erosion Service was established in the Department of the Interior, and in March, 1934, an additional 5 million dollars was made available for its activities. The work was transferred to the Department of Agriculture by Executive order in March, 1935, and the next month was put on a permanent basis by act of Congress.

The Soil Conservation Act of April 27, 1935, "recognized that the wastage of soil and moisture resources on farm, grazing, and forest lands of the Nation, resulting from soil erosion, is a menace to the national welfare and that it is hereby declared to be the policy of Congress to provide permanently for the control and prevention of soil erosion and thereby to preserve natural resources, control floods, prevent impairment of reservoirs, and maintain the navigability of rivers and harbors, protect public health, public lands and relieve unemployment, and the Secretary of Agriculture, from now on, shall coordinate and direct all activities with relation to soil erosion." In order to effectuate this policy the Secretary was authorized to conduct research and demonstrational projects and "to carry out preventive measures, including, but not limited to, engineering operations, methods of cultivation, the growing of vegetation, and changes in use of land" both on Federal and other lands. In the case of other lands, he was to require reasonable safeguards for the permanence of soil-conservation measures, together with such cooperative contributions as he might deem wise.

The Secretary of Agriculture was instructed to establish a Soil Conservation Service, which replaced the former Soil Erosion Service, to carry out the provisions of the act. Under the dynamic leadership of Hugh H. Bennett, a veteran in the soil-conservation field and an evangelist of the first water, the new Service expanded by leaps and bounds. Almost overnight, people everywhere were made conscious that soil conservation is one of the country's most imperative needs; and Hugh Bennett was its prophet.

Soil-conservation Districts. The outstanding accomplishment of the Soil Conservation Service has been the establishment of a nationwide system of soil-conservation districts. One of its first activities was to draft a model law for such districts which has now been adopted by every state in the Union, with appropriate modifications to meet local conditions. Each district is an autonomous body which is controlled by the landowners who comprise its membership and which receives technical assistance from the Soil Conservation Service. More and more, this assistance has taken the form of preparing farm plans that provide for devoting each part of the farm to the purpose for which it is best suited —cultivated crops, pasture, woods, ponds—with special reference to soil improvement and erosion control. In 1953 there were about 2,500 districts, which included about 80 per cent of all the agricultural land and 85 per cent of all the farms and ranches. Not all of the farms within the districts, however, participate in the program.

Various "land capability" classes have been developed that help greatly in interpreting soil and topography in terms of the particular form of land use to which they are best adapted. Forestry, grazing, and wildlife management nearly always occupy an important place in the farm plan. Such measures as contour plowing, terracing, strip cropping, and rotation of cultivated crops are supplemented by proposals for tree planting and the management of existing woodlands in places where trees clearly constitute the best and safest crop. The whole program aims at better land use, in which forestry normally plays a prominent part. The great strength of this approach lies in its treatment of woodland management, which has been generally neglected as something foreign to the main interests of the farmer, as an integral part of the farm enterprise deserving of the same careful attention as crop management or pasture management.

Soil conservation on Federal lands is handled by the agency in charge of their administration, such as the Forest Service and the Bureau of Land Management, with cooperation in technical matters from the Soil Conservation Service.

Soil-conservation Payments. Soil conservation also became the justification of the acreage allotments and benefit payments included in the Agricultural Adjustment Act of 1933 and its many amendments. After the original act was declared unconstitutional by the Supreme Court in 1936, on the ground that its primary purpose was to regulate agricultural production, a power not delegated to the United States and therefore reserved to the individual states, Congress promptly attempted to attain its objectives by making them part of a national program of soil conservation. The Soil Conservation and Domestic Allotment Act of February 29, 1936, amended the Soil Conservation Act of 1935 by declaring it to be the policy of that act also to secure "(1) preservation and improvement of soil fertility; (2) promotion of the economic use and conservation of land; (3) diminution of exploitation and wasteful and unscientific use of national soil resources; (4) the protection of rivers and harbors against the results of soil erosion in aid of maintaining the navigability of waters and water courses and in aid of flood control; and (5) reestablishment . . . of the ratio between the purchasing power of the net income per person on farms and that of the net income per person not on farms that prevailed during the five-year period August 1909–July 1914, inclusive. . . ."

In furtherance of this policy, the Secretary of Agriculture was authorized to make benefit payments or grants of other aid to agricultural producers for the purpose of encouraging soil restoration, soil conservation, prevention of erosion, and desirable changes in land use, and also the control of production in accordance with national requirements. He "shall in every practical way encourage and provide for soil conserving and soil rebuilding practices rather than the growing of soil depleting commercial crops."

The Agricultural Adjustment Act of 1938 provided in great detail for benefit and parity payments to farmers, stressed the idea of the "ever-normal granary," and inaugurated Federal crop insurance for wheat through the creation of the Federal Crop Insurance Corporation under the control of the Secretary of Agriculture. Federal insurance was later extended to certain other specified crops, but not to timber. The act also provided for the establishment by the Department of Agriculture of four regional laboratories for the conduct of investigations relating to the industrial use of farm products. These laboratories were located at Philadelphia, Pa., Peoria, Ill., New Orleans, La., and Albany, Calif.

Retirement of Submarginal Land. Another aspect of the widely ramifying and highly complicated program for agricultural adjustment in which soil was concerned provided for the retirement of submarginal lands from cultivation and resettlement of their occupants. These and related activities were handled under the Federal Emergency Relief Act of 1933, the Bankhead-Jones Farm Tenant Act of 1937, and other legislation. Several emergency-relief agencies were involved, prominent among which were the Resettlement Administration and its successor the Farm Security Administration. Careful plans were prepared, but did not materialize, for the resettlement of farm families in communities on the national forests, where they could make a living by a combination of subsistence farming and work in the forests.

Large expenditures were made for the purchase of submarginal land, which was used for grazing, forest production, recreation, and other wild-land purposes. Some tracts were turned over to the states, but the bulk of the area was retained by the Federal government. Administration of the purchased units, which became known as "land-utilization projects," was handled at first by the Farm Security Administration and later by the Soil Conservation Service. On November 2, 1953, their administration was transferred to the Forest Service under a reorganization of the Department of Agriculture.

Summary. Although the maintenance of soil fertility and the prevention of soil erosion were early recognized as serious problems, progress toward their solution was slow. The first really effective governmental action to control erosion came with the establishment of the Soil Erosion Service in 1933 and the Soil Conservation Service in 1935. Since then much has been accomplished, largely through the 2,500 soil-conservation districts that have been organized throughout the country. Reliance has been placed on research, education, and cooperation rather than on regulatory measures.

WATER

Navigation. Under its constitutional power to regulate foreign and interstate commerce, Congress has always been concerned with the use of both coastal and interior waters for navigation. Since 1824, when Congress definitely assigned responsibility for the handling of internal improvements to the Army Corps of Engineers, improvement of rivers and harbors has been the major civilian activity of the Corps. Well over a billion dollars have been spent for this purpose.

Inland Waterways Commission. A broader approach to the subject of navigation was taken by President Roosevelt in March, 1907, when he appointed an Inland Waterways Commission with instructions to prepare a comprehensive plan for the coordinated use of the water resources of the country, which would take into consideration "the relations of the streams to the use of all the great permanent natural resources and their conservation for the making and maintenance of prosperous homes." The following February (1908) the Commission, in a brief report with a lengthy appendix, pointed out that every river system is a unit from its source to its mouth and that it should be treated as such. Local interests should be considered in relation to regional and national interests. "Hereafter plans for the improvement of navigation in inland waterways, or for any use of these waterways in connection with interstate commerce, shall take account of the purification of the waters, the development of power, the control of floods, the reclamation of lands by irrigation and drainage, and all other uses of the waters or benefits to be derived from their control."

The Commission saw big problems in a big way. It struck the characteristic Roosevelt-Pinchot note of warning against monopoly in the control of natural resources, an excessive share of which had been "diverted to the enrichment of the few rather than preserved for the equitable benefit of the many." Its activities helped much to round out and to advance the conservation movement.

National Waterways Commission. In 1909, Congress created a National Waterways Commission to continue the work of its predecessor on questions pertaining to water transportation and the improvement of waterways. The executive branch of the government was not officially represented on the Commission but was undoubtedly helpful in an advisory capacity.

The Commission submitted a preliminary report in January, 1910, and a final report in March, 1912. Among other things, the latter dealt at length with the subjects of impounding reservoirs, forest influences, and control of water power. The practicability of storage reservoirs for the prevention or mitigation of floods, for the improvement of navigation, and for the development of water power was recognized, as was also the fallacy of expecting the same reservoir to be equally effective for all three purposes. An interesting conclusion in the light of subsequent developments was that "the Federal Government has no constitutional authority to engage in works intended primarily for flood prevention or power development. Its activities are limited to the control and promotion of navigation and works incident thereto . . . if navigation is not concerned, the Federal Government should have nothing to do with flood prevention. A method is provided in the Constitution by which the States may cooperate for this purpose."

After careful consideration of the many factors involved, the Commission decided that "whatever influence forests may have upon precipitation, run-off, and erosion, it is evidently greatest in the mountainous regions where the rainfall is heaviest, slopes steepest, and run-off most rapid." It therefore favored strongly the protection and perpetuation of forests in such regions both for the influence they might have on runoff and more particularly on erosion and for the production of timber on lands not suited to other purposes. Since the Federal government "has no power to regulate the methods of farming or the cutting of timber on private lands," the Commission felt that any responsibility which the public may have in this direction must be assumed by the states under their police power. Federal cooperation with states in the prevention of forest fires and the introduction of scientific methods was endorsed.

With respect to the development and control of water, the Commission believed that Supreme Court decisions had left no doubt that "the authority of Congress reaches to the remotest sources in the mountains of every navigable stream. . . . There should be no misunderstanding as to the extent of the authority of Congress if it should become necessary or expedient to exercise its full powers. With the increasing unity of our national life and the growing necessity of securing for human needs the maximum beneficial use of the waters of every stream it will become increasingly necessary to treat every stream with all its tributaries as a unit. In the nature of the case so comprehensive a policy could be successfully administered only by the Federal Government, and consequently the eventual desirability of Federal control is easy to predict."

As to the development of water power on the public domain, the Commission pointed out that the government had already undertaken to control the situation through its proprietorship of the public lands in two ways: (1) by withdrawing water-power sites from entry and (2) by authorizing the Department of the Interior and the Department of Agriculture to make rules for the use of public lands. The Commission made numerous suggestions for improving the current situation which closely resembled many of the provisions later embodied in the Federal Power Act of 1920.

Both preliminary and final reports were accompanied by substantial reports by experts on a wide variety of subjects relating to the work of the Commission. One of the most valuable of these appendices was an exhaustive statement by Raphael Zon of the Forest Service on "Forests and Water in the Light of Scientific Investigation."

Flood Control. In spite of the belief of the National Waterways Commission that the Federal government has no constitutional authority to engage in works intended primarily for flood prevention, Congress in 1927 took action to provide Federal help for the control of floods on the Mississippi River. The Flood Control Survey Act of 1936, however, constituted the first recognition for the nation as a whole that flood control on navigable waters or their tributaries is a proper activity of the Federal government, in cooperation with the states and their political subdivisions. It also gave explicit recognition to the fact, implied but not stated in the Weeks Act and the Clarke-McNary Act, that forests and other vegetative cover have something to do with flood control.

The act provided that thereafter Federal investigations and improvements of rivers and other waterways for flood control and allied purposes should be handled by the War Department and that Federal investigations and measures for water-flow retardation and soil-erosion prevention on watersheds should be handled by the Department of Agriculture. It also authorized interstate compacts for flood control and a long list of specific control projects for prosecution by the Army Engineers.

Amendments of the 1936 act in 1937 and 1938 authorized additional surveys and examinations at specified localities and provided that the Secretary of Agriculture should make runoff and erosion surveys on all watersheds previously specified for flood-control surveys by the Secretary of War. The comprehensive flood-control act of 1944, in addition to continuing the duties of the Secretary of Agriculture in this field, provided that any electric power generated at multi-purpose dams constructed by the Army Engineers should be distributed and sold by the Department of the Interior and that any irrigation water made available at such dams should be handled under the reclamation laws.

Substantial appropriations have been made to the Department of Agriculture under these various acts both for research and for actual upstream control of runoff and erosion. Another significant move was the appropriation by Congress in 1953 of 5 million dollars for the specific purpose of trying out practices for the control of runoff and erosion on several small experimental watersheds. The next year the Watershed Protection and Flood Prevention Act went still further by authorizing the Secretary of Agriculture to cooperate with states and local organizations for the purpose of preventing erosion, floodwater, and sediment damages and of furthering the conservation, development, utilization, and disposal of water.

Irrigation. In recognition of the growing interest in the irrigation of arid lands, Congress in 1866 granted water owners rights of way over public lands for the construction of ditches and canals. A decade later came the passage of the Desert Land Act of 1877, the provisions and results of which are discussed in Chapter 2. The recommendation of the Public Land Commission, in 1880, that irrigable lands be sold in unlimited quantities, instead of the 640 acres specified by the Desert Land Act, was not acted on by Congress.

The increasingly widespread practice of irrigation by private owners led Congress in 1888 to instruct the Geological Survey (established in 1879) to determine the extent to which the arid lands could be reclaimed by irrigation and authorized it to reserve all lands that might be selected as sites for reservoirs and irrigation canals and all other lands capable of being irrigated as a result of the construction of such reservoirs and canals. J. W. Powell, director of the Geological Survey, took advantage of this authority to withdraw 127 reservoir sites and more than 30 million acres of irrigable land. His action aroused so much protest that Congress in 1890 restricted withdrawals to reservoir sites and in 1891 to land actually needed for the construction and maintenance of reservoirs.

The first Interstate Irrigation Congress, held at Salt Lake City in 1891, urged the Federal government to cede all of the irrigable public lands to the states and territories in which they were situated. Although this proposal was never approved, the Carey Act of 1894 did offer to donate to certain states not more than 1,000,000 acres of public land which they should cause to be settled, irrigated, and in part cultivated. Of the seven states which applied for land, four succeeded in reclaiming about 600,000 acres.

By the end of the nineteenth century it had become clear that most irrigation projects which could be handled by individuals, corporations, or even the states had already been developed and that if much further progress were to be made the Federal government would have to take a hand. Pressure from the Interstate Irrigation Congress, the National Irrigation Association (organized by George W. Maxwell in 1897), and the White House led to passage of the Reclamation Act of 1902, often known as the Newlands Act, which provided the basis of all future government irrigation activities. The Reclamation Service was first organized in 1903 under a chief engineer as a division of the Geological Survey, but in 1907 was promoted to the status of a separate Bureau with a director reporting directly to the Secretary of the Interior. Frederick H. Newell was in charge of the work in both capacities. In 1924 its name was changed to Bureau of Reclamation.

The Reclamation Act has been the subject of innumerable amendments, court decisions, and administrative regulations. Expenditures have increased until they now run into hundreds of millions of dollars a year. The program has had its successes and its failures, its friends and its enemies. It originated as part of the broad conservation movement fathered by Roosevelt and Pinchot and has always recognized the beneficial influence of well-managed forests and range lands on the water supply.

Numerous interstate compacts have dealt with the rights of the states concerned to the waters of interstate streams for irrigation and other purposes. Notable among these are the Colorado River Compact of 1928 for equitable distribution of the waters of the Colorado River, and the Upper Colorado River Compact of 1949 to determine the rights and obligations of the five signatory states.

Water Facilities. The Water Facilities Act of August 28, 1937, authorized the Secretary of Agriculture to develop facilities for water storage and utilization in the arid and semiarid regions. Although public lands were not excluded from the operation of the act, in practice it was limited almost entirely to lands owned by farmers and ranchers. In 1954 the Secretary's authority was extended to all parts of the country.

An act of August 11, 1939, authorized the Secretary of the Interior to undertake the construction of water and conservation projects in the Great Plains and in the arid and semiarid areas of the United States. It was intended to facilitate the development of water supplies on lands not reached by the large irrigation projects.

Drainage. Drainage of swamp and overflowed lands to make them fit for cultivation started early in the 1800's and was encouraged in various ways by many of the states. The Federal government first concerned itself with the problem through passage of the swampland acts of 1849, 1850, and 1860, which are discussed in Chapter 2.

Aside from research by the Department of Agriculture, the Federal government did not itself participate in drainage projects until the depression of the 1930's, when funds for the purpose were made available through emergency appropriations for the Public Works Administration and the Works Progress Administration. Because of criticism that some of these projects dealt with land more valuable for use by wildlife than for crop production, cooperative arrangements were effected in 1935 between the Works Progress Administration and the Bureau of Biological Survey for investigation by the latter of all proposed drainage projects that might materially affect wildlife environment. This procedure prevented the undertaking of many ill-advised projects.

Hydroelectric Power. Prior to 1920, the Federal government concerned itself with hydroelectric power only to the extent of granting permits for power development by private enterprise on national forests and occasionally on other public lands. The situation was unsatisfactory because the permits could not be issued for a long enough period to justify the heavy investment ordinarily required of the permittee. There was also strong feeling on the part of many, including Pinchot, that monopoly of water power had proceeded to a point justifying Federal intervention.

In 1915, the Senate passed a resolution asking the Secretary of Agriculture for detailed information concerning the ownership and control of water-power sites in the United States, including any facts bearing on the existence of a monopoly of hydroelectric power. The resulting report, prepared by the Forest Service under the direction of its chief engineer, O. C. Merrill, showed a marked concentration of control of a large percentage of developed water power by a very few companies. It also showed the national forests to contain almost a third of the potential water-power resources of the Western states. Existing developments which utilized national-forest land had 42 per cent of the total developed water power in the United States, and an additional 14 per cent either occupied public land or were dependent on reservoirs on such land.

In view of this situation, a decision by the United States Supreme Court in 1917 in the case of the *Utah Power & Light Co. v. the United States* was of major importance. The company was occupying without permission certain lands within the national forests for the purpose of generating and distributing electric power. It claimed that it was entitled to do so without obtaining any license from the Secretary of Agriculture; that in using the land it had acquired vested rights; and that, except for land used for strictly governmental purposes, lands owned by the government are subject to the laws of the state in the same manner as other lands. The Supreme Court ruled against the company on all of these claims. The decision clearly established the jurisdiction of the government over its own lands and emphasized the desirability of a definite leasing law.

Such a law was the Federal Water Power Act of June 10, 1920, which aimed to safeguard the rights of both the government and hydroelectric companies. It created a Federal Power Commission with authority to issue licenses for a period not exceeding fifty years "for the development and improvement of navigation, and for the development, transmission, and utilization of power across, along, from or in any part of the navigable waters of the United States, or upon any part of the public lands and reservations of the United States (including the Territories), or for the purpose of utilizing the surplus water or water power from any Government dam." At the expiration of the license the United States might take over the properties of the licensee for its own use, permit them to be taken by another, or issue a new license to the old licensee. Regulation of the services rendered by the licensee (including rates and profits) may be exercised by the Commission until such control is assumed by the state concerned.

Of the receipts from public lands (other than Indian reservations), 50 per cent is paid into the reclamation fund, 37½ per cent to the states, and 12½ per cent to the Treasury of the United States. In 1921, existing national parks and national monuments were excluded from the provisions of the act, and in 1935 this exclusion was extended to all national parks and monuments.

The scope of the act is extremely broad in that it gives the government control over the development of water power on any navigable stream or on land owned by the United States, including the national forests. The significance of this control is emphasized by the definition of navigable streams incorporated in the act: "Navigable waters means those parts of streams or other bodies of water over which Congress has jurisdiction under its authority to regulate commerce with foreign nations and among the several States, and which either in their natural or improved condition, notwithstanding interruptions between the navigable parts of such streams or waters by falls, shallows, or rapids compelling land carriage, are used or suitable for use for the transportation of persons or property in interstate or foreign commerce including therein all such interrupting falls, shallows, or rapids; together with such other parts of streams as shall have been authorized by Congress for improvement by the United States or shall have been recommended to Congress for such improvement after investigation under its authority."

Passage of the act resulted in the submission during the next eight months of applications involving the development of 13 million horsepower, as compared with applications totaling only 2,500,000 horsepower during the preceding twenty years. Evidently Congress had at last succeeded in finding a formula that would encourage industrial development and at the same time adequately protect the public interest.

More recently the government has itself gone into the development and sale of hydroelectric power in a big way. Conspicuous examples of such development exist at the Hoover Dam, the Bonneville Dam, the Grand Coulee Dam, and throughout the Tennessee Valley. There has been an increasing tendency to make all water-development projects multipurpose projects and to include the production of hydroelectric power in Bureau of Reclamation projects concerned primarily with irrigation and in Army Engineer projects concerned primarily with navigation and flood control. Power developed by these two agencies, but not by the Tennessee Valley Authority, is marketed by the Bonneville Power Administration, the Southwestern Power Administration, and the Southeastern Power Administration in the Department of the Interior. How far the government should get into the hydroelectric business and at what rates and under what conditions it should sell the power are among the controversial questions of the day.

Water Pollution. From the 1860's on, the states have shown awareness of the need to check the steadily increasing pollution of harbors, rivers, and lakes by municipal and industrial waste. Interstate compacts attempting joint remedial action include the Tri-State Pollution Compact of 1935 to deal with pollution in New York Harbor; the Potomac Valley Pollution and Conservation Compact of 1940 to abate and control pollution in the Potomac Valley; the

Ohio River Valley Sanitation Compact of 1940 for the control and reduction of pollution of the streams of the Ohio River drainage basin; and the New England Pollution Compact of 1947 for the control and reduction of pollution of the streams and watersheds of the New England states.

Federal authorities have also recognized the problem, but except for research and education were able to do little to improve the situation until passage of the Taft-Barkley Water Pollution Control Act of June 30, 1948. That act declared it to be the policy of Congress to recognize, preserve, and protect the primary responsibilities and rights of the states in controlling water pollution; "to conduct research on the treatment of industrial wastes; and to provide technical service and financial aid to State and interstate agencies and to municipalities, in the formulation and execution of their stream pollution abatement programs."

The act further declared the pollution of interstate waters which endangers the health or welfare of persons in a state other than that in which the discharge originates to be a public nuisance. Whenever the Surgeon General finds such a public nuisance to exist, he is authorized to take certain steps to secure its abatement, including recommendation to the appropriate state or interstate agency that it initiate suit to abate the pollution. If that agency fails to respond, the Federal Security Administrator (now the Secretary of Health, Education, and Welfare), with the consent of the agency, may request the Attorney General to bring suit on behalf of the United States to secure abatement of the pollution.

No use has been made of this roundabout process, and the Public Health Service has confined itself chiefly to research, surveys, and educational activities. The main purpose of the act was to enable the Federal government to cooperate with the states in solving a problem that it regards as primarily their responsibility. In 1952 the financial authorizations approved by the act for a five-year period were extended to cover an eight-year period ending June 30, 1956.

Summary. Congress used the public domain to promote reclamation of swamp and overflowed lands through the swampland acts of 1849, 1850, and 1860, and the irrigation of arid lands through the desert-land act of 1877 and the Carey Act of 1894. It involved the Federal government itself in large-scale reclamation projects with passage of the Reclamation Act of 1902.

Promotion of navigation by the improvement of rivers and harbors has always been a function of the Federal government. The need to develop each river system as a unit from its source to its mouth was recognized by the Inland Waterways Commission in 1908 and the National Waterways Commission in 1910 and 1912. Participation in flood control was recognized as a Federal responsibility in 1927 and more fully in the Flood Control Acts of 1936 and 1944.

Generation of hydroelectric power has come to be an important part of practically all multipurpose water-development projects. Cooperation in the control of pollution is extended to state and interstate agencies under the Water Pollution Control Act of 1948.

MINERALS

Early Legislation. Congress always recognized the need for handling mineral lands differently from agricultural lands, but was slow to provide clear-cut legislation for their disposal and use. As was noted in Chapter 2, the Ordinance of 1785 reserved "one-third part of gold, silver, lead, and copper mines, to be sold or otherwise disposed of as Congress shall hereafter direct"; and the act of 1796 reserved all salt springs and licks and the section in which they were located. The saline lands, aggregating a few hundred thousand acres, were later turned over to the several states at the time of their admission to the Union.

Anything even approaching a definite mineral-land policy was slow in developing. Legislation passed between 1807 and 1846, providing first for the lease and later for the sale of certain mineral lands, proved unsatisfactory. A step forward was taken in 1847 when lands in the Lake Superior District in Michigan and the Chippewa District in Wisconsin containing copper, lead, or other valuable ores were opened to sale at a minimum cash price of $5 per acre. This arrangement did not last long, however, and in 1850 such lands were opened to sale and preemption in the same manner as other public lands.

The discovery of gold in California in 1848 and the subsequent rush of immigrants to that state and later to the Rocky Mountain states created new problems in the mineral field. The invasion was as impossible to control as was the invasion of the timberlands in the Lake states and the South. "Possessory rights" were universally respected. Hundreds of local associations were organized along lines similar to the older claim associations and like them promulgated regulations which became in effect the law of the land. The miner was in the saddle; all attempts to control his activities other than by his own regulations were vigorously resisted.

In 1864 the Governor of Idaho called the attention of the Legislature to the need for "better protection of that hardy class of our population to whom danger is not a sentiment—who, amid trackless wastes . . . pursue their enterprises with no capital but their rough hands, with no defender but their revolvers." The next year Chief Justice Chase of the United States Supreme Court stated that "a sort of common law of the miners, the offspring of a nation's irrepressible march . . . has sprung up on the Pacific Coast, and presents in the value of a 'Mining Right' a novel and peculiar question for this court."

General Sales System. Congress finally took action in 1866, when it made the mineral lands of the public domain, both surveyed and unsurveyed, "free and open to exploration by all citizens of the United States, and those declaring their intention to become citizens, subject to such regulations as may be prescribed by law" and "subject also to the local customs or rules of miners in the several mining districts, so far as the same may not be in conflict with the laws of the United States." Patents to lode mines could be obtained at a price of $5 per acre if the claimant had occupied them according to local rules and had expended as much as $1,000 in labor and improvements.

In 1870, Congress provided for the survey and sale of placer mining lands at $2.50 per acre. Two years later (1872) it constituted mineral lands a distinct class subject to sale under prices and requirements differing materially from those applying to other lands. The established prices of $5 per acre for lode claims and $2.50 per acre for placer claims were retained. Claimants were required to spend not less than $100 a year on labor or improvements, and the total expenditure required before patent could be obtained was reduced to $500. Mineral claims are legally transferrable, and there is no specified time within which application for patent must be submitted. In practice, the claims system has proved to be subject to many abuses. As pointed out in Chapter 11, these abuses have interfered seriously with the administration of the national forests.

In 1873 iron lands were removed from the operation of the act of 1872 and were made available for purchase at auction at not less than the standard minimum price of $1.25 per acre. Coal lands were not included in any of these laws. At first they were disposed of under the laws applying to public lands in general, but after 1864 they were sold under special legislation.

Mineral Leasing. No radical changes in these methods of disposing of mineral lands took place until the passage of the Mineral-Leasing Act of February 25, 1920. That act put a stop to sales and provided in detail for the leasing by the Secretary of the Interior of deposits of coal, phosphate, sodium, oil, and oil shale on public lands. It was later (1926 and 1932) extended to deposits of sulfur, and similar legislation applicable to potash was enacted in 1917 and 1927. The Secretary was authorized to reserve the right to sell, lease, or otherwise dispose of the surface of leased lands if not necessary for the use of the lessees. Lessees pay both an annual rental and a royalty per unit of the mineral removed. Of the amount received, 52½ per cent goes to the reclamation fund, 37½ per cent to the states for the construction of roads or the support of education, and 10 per cent to the Treasury of the United States.

The act applied to national forests created from the public domain, but not to acquired national forests (in which mineral lands were leased under the act of August 11, 1916), to national parks, or to military reservations. General authority for the leasing of minerals on acquired lands was granted to the Secretary of the Interior in 1947.

Use was sometimes made of legislation aimed to promote the settlement of agricultural lands, and particularly of the Preemption Act and the commutation provision of the Homestead Act, for the fraudulent acquisition of mineral lands. The practice was not, however, nearly so extensive as in the case of timberlands.

A few nonmetalliferous minerals have not yet been included in the mineral-leasing laws; and all metalliferous minerals continue to be subject to location, entry, and patenting under the general laws applying to that class of minerals. Receipts from mineral leases and permits on public lands and acquired lands totaled more than 49 million dollars for the fiscal year 1953.

Petroleum. The existence of lands chiefly valuable for petroleum was not specifically recognized by Congress until 1897. They were then made subject to entry under the placer mining law of 1870, under which they had previously been sold in spite of considerable question as to its applicability. Sizable areas of oil land were withdrawn from entry by the President both before and after passage of the Withdrawal Act of 1910. Then in 1920, all oil lands were withdrawn from sale and made available for use by leasing under the Mineral-Leasing Act of that year. The sale of petroleum from certain lands reserved for the benefit of the Navy led to the "Teapot Dome" scandals of the 1920's.

The increasing demand for and decreasing supply of petroleum led during the latter part of the 1940's to vigorous controversy as to the ownership of the petroleum underlying the waters over the continental shelf, commonly but mistakenly referred to as "tidelands oil." The United States Supreme Court in 1947 and 1950, in three separate suits relating to the situation in California, Texas, and Louisiana, without definitely settling the question of ownership, ruled that the United States had "paramount rights" over all resources in the continental shelf beyond the low-water limit. Both decisions, however, implied that ownership of the petroleum could by law be vested in the states if Congress so desired. Two bills taking this action were vetoed by President Truman.

The question was finally settled by the Submerged Lands Act of May 22, 1953, which confirmed and established the titles of the states to lands beneath navigable waters within state boundaries and to the natural resources within such lands and waters. At the same time it confirmed the jurisdiction of the United States over the natural resources of the sea bed of the continental shelf seaward of state boundaries. Authority for the Secretary of the Interior to lease such lands for certain purposes, including, of course, the utilization of petroleum, was provided by the act of August 7, 1953.

Efforts by several states to conserve their supplies of petroleum were supplemented by Congress in 1935 by passage of the Connally "hot oil" act. This act follows the precedent set by the Lacey Act of 1900 dealing with wildlife by prohibiting the transportation in interstate commerce of petroleum produced, transported, or withdrawn from storage in violation of the laws of the state in which it is produced.

Summary. No general policy for the disposal of mineral lands existed until after the Civil War, when acts passed in 1866, 1870, and 1872 provided for the sale of lode and placer claims. Since 1920, many nonmetalliferous minerals have been withdrawn from sale and can be utilized only under lease from the government. Congress in 1953 vested in the states title to land and natural resources beneath navigable waters within state boundaries and confirmed the jurisdiction of the United States over the natural resources in the continental shelf seaward of state boundaries.

Chronological Summary of Important Events in the Development of Colonial and Federal Policies Relating to Natural Resources

This summary includes only the major features of the events listed. Items of particular interest from the standpoint of forest and range policy are treated at greater length in the main text.

Dates regarded as especially significant for the purposes of this book are printed in bold-faced type.

When two page numbers are given in a citation, they refer respectively to the first page of the act cited and to the page on which the item in question appears. "Stat." refers to United States Statutes at Large, "U.S." to United States Supreme Court decisions, "P.L." to Public Law, and "T.I.A.S." to Treaties and Other International Agreements Series.

1609. First shipment of masts from the Colonies was sent from Virginia to England.

1626. Plymouth Colony forbade the selling or transportation of timber out of the Colony without the approval of the governor and council.

1631. First commercial sawmill in the Colonies was probably established at Berwick, Maine. (Claim of a sawmill at York, Maine, in 1623 is poorly substantiated.)

1631. Massachusetts Bay Colony forbade the burning of any ground prior to March 1. Subsequent legislation in Massachusetts and other colonies forbade burning at other times and specifically recognized damage by fire not only to timber but also to young growth, soil, and domestic stock.

1651. First of the Navigation Acts attempting to limit English and colonial trade to ships of English registry and to channel raw materials from the Colonies to England to pay for British manufactures was passed.

1668. Massachusetts reserved for the public all white pine trees fit for masts in certain parts of the town of Exeter.

1681. William Penn provided that for every 5 acres of forest cleared 1 acre should be kept in trees.

1691. William and Mary in a new charter creating the Province of Massachusetts Bay forbade the cutting, without permission of the British government, of all trees 24 inches or more in diameter at 12 inches from the ground growing on land not theretofore granted to a private person, under penalty of £ 100. This became known as the Broad Arrow policy because of the practice of marking trees reserved under it for the use of the Crown with the broad arrow of sovereignty.

1704. Bounties were offered for naval stores and masts shipped to England, and these items were put on the enumerated list under the Navigation Acts. The law also placed a penalty on injuring pitch pine through fire or cutting.

1705. The British Parliament prohibited the felling of all "Pitch Pine and Tar Trees" less than 12 inches in diameter and not growing on private property in the various colonies.

1708. New Hampshire enacted legislation embodying the British Broad Arrow policy of 1691.

1711. Broad Arrow policy was extended to include all white pine trees fit for masts 24 inches or more in diameter at 12 inches from the ground, and not private property, anywhere in New England, New York, and New Jersey. A penalty of £5 was provided for unlawfully marking any tree with the broad arrow.

1721. Broad Arrow policy was broadened to forbid the cutting of any white pine trees not growing within a township, from Nova Scotia to New Jersey, under penalties ranging from £5 to £50.

1729. Broad Arrow policy was reenacted with somewhat stricter provisions as to what constituted private land and with better machinery for enforcement and was extended to every part of America which belonged to Great Britain or should thereafter be acquired. It remained in effect in this form until the Revolution. The act also reduced somewhat the bounties on naval stores provided by the act of 1704.

1739. Massachusetts undertook to check the encroachment of sand dunes at Truro and on Plumb Island in Ipswich Bay by regulating timber cutting, grazing, and burning. Later acts applied to other parts of Cape Cod.

1743. New York authorized anyone to call for help in fighting forest fires in certain counties.

1744. Massachusetts authorized groups of five or more owners in the town of Ipswich to apply for the establishment of a common woods. If two-thirds of the proprietors within the proposed limits approved, all of the lands involved became subject to the joint control and management of the proprietors.

1752. Connecticut forbade appropriation by others than their owners of logs and other forest products being floated down the Connecticut River. This action was followed by similar legislation in other colonies and states which led to abandonment of the English common law that only tidal streams are navigable and substitution therefor of the doctrine that any stream which will float a log or boat is navigable and consequently a public highway.

1760. New York authorized the election of "firemen," or wardens, in certain counties, and authorized them to call for help.

1772. New York forbade the bringing to Albany for fuel of more than six pieces of wood per load under 6 inches in diameter at the large end for pine and under 4 inches for other species.

1776. Continental Congress offered land bounties to deserters from the enemy army and to soldiers who should serve throughout the war.

1780. Continental Congress resolved that lands ceded to the United States should be used for the common benefit of all the states.

1781. New York ceded its western lands to the Federal government. Virginia, Massachusetts, Connecticut, South Carolina, North Carolina, and Georgia presently followed suit, the last cession being made by Georgia in 1802. The 233 million acres included in these cessions started the public domain.

1783. Massachusetts passed an act substantially equivalent to the Broad Arrow policy of the British.

1785. Ordinance of May 20 provided for the rectangular system of survey of the public lands. After survey, the lands were to be sold at auction for cash to the highest bidder at not less than $1 per acre. Sections 8, 11, 26, and 29 were reserved for later disposal by the government, and Section 16 for common-school purposes. Reservation was also made of one-third of all gold, silver, lead, and copper mines to be sold or otherwise disposed of as Congress should direct.

1787. Ordinance of April 21 provided that one-third of the sale price of public lands should be paid immediately and the balance in three months.

1788. First patent to public land was issued by the government on March 4.

1789. Constitution provided (Art. 4, Sec. 3, Par. 2) that "the Congress shall have Power to dispose of and make all needful Rules and Regulations respecting the Territory or other Property belonging to the United States." This provision has been repeatedly interpreted by the Supreme Court (14 Peters 526, 13 Wallace 92, and other cases) as giving Congress complete control over the public domain.

1796. Act of May 18 (1 Stat. 464) provided for a Surveyor General and gave directions for applying the rectangular system of survey adopted in 1785. Sales were to be made at auction to the highest bidder, with a minimum price of $2 per acre, payable in full within one year. All navigable streams within the territory covered by the act were declared to be public highways.

1799. Act of February 25 (1 Stat. 622) appropriated $200,000 for the purchase of timber or of lands on which timber suitable for purposes of naval construction was growing and for the preservation of such timber for future uses.

1800. Act of May 10 (2 Stat. 73) provided that the townships west of the Muskingum River which, under the act of May 18, 1796, were to be sold as quarter townships should be subdivided and sold as half sections (320 acres). One-twentieth of the purchase price (minimum of $2 per acre) had to be paid at the time of sale, one-fourth within forty days, another fourth within two years, a third fourth within three years, and the final fourth within four years. Lands remaining unsold after the public sales could be disposed of by the local land offices at private sale at not less than the minimum price.

1802–1803. Acts of April 30, 1802 (2 Stat. 173, 175) and March 3, 1803 (2 Stat. 225, 226), providing for organization of the state of Ohio, granted Section 16 in each township to the state for school purposes. Ohio was also granted 3 per cent of the net proceeds from all sales of public lands within the state for the construction of roads, and 5 per cent of the net proceeds was to be used by Congress for the construction of roads leading to and through the state. In return, Ohio agreed to exempt from taxation all land sold by the government for five years from the date of sale.

1803. Act of October 31 (2 Stat. 245) authorized the President to take possession of the Louisiana Purchase, which added some 523 million acres to the public domain.

1804. Act of March 26 (2 Stat. 277) reduced the minimum area of public lands offered for sale to a quarter section (160 acres). The minimum price of $2 per acre was retained, but no interest was to be charged on deferred payments unless they became delinquent. Section 16 in each township and all salt springs were reserved for educational purposes, but the other four sections near the center of each township previously reserved were now to be sold.

1807. Act of March 3 (2 Stat. 445) forbade anyone to settle on or occupy the public lands until authorized by law. The President was authorized to direct the marshal to remove trespassers and to take such other measures and to use such military force as necessary for the purpose.

1807. Act of March 3 (2 Stat. 448, 449) reserved lead mines in Indiana Territory for future disposal and authorized the President to lease such mines for terms not exceeding five years.

1811–1812. Acts of February 20, 1811 (2 Stat. 641) and April 8, 1812 (2 Stat. 701) admitting Louisiana to the Union granted the state 5 per cent of the net proceeds from the sale of public lands for the construction of roads and levees. Lands sold by the government were to be exempt from taxation for five years, and lands belonging to nonresident citizens were never to be taxed higher than those belonging to residents.

1812. Act of April 25 (2 Stat. 716) established the General Land Office in the Treasury Department.

1816. Act of April 19 (3 Stat. 289) admitting Indiana to the Union, in addition to the usual provisions concerning school lands and exemption from taxation, reserved 5 per cent of the net proceeds from the sale of public lands for the construction of public roads and canals; of this amount, three-fifths was to be spent by the state and two-fifths by Congress for the construction of roads leading to the state. An additional township, to be designated by the President, was reserved for a seminary of learning.

1817. Act of March 1 (3 Stat. 347) authorized the Secretary of the Navy, with the approval of the President, to reserve from sale public lands containing live oak and red cedar for "the sole purpose of supplying timber for the navy of the United States." Unauthorized removal of any timber from such reservations or of any live oak or red cedar from any other public lands was punishable by a fine of not more than $500 and imprisonment for not more than six months. Administration of the reserves was under the Navy Department.

1819. Act of March 3 (3 Stat. 523) authorized the President to take possession of the territories included in the Florida purchase, which added some 43 million acres to the public domain.

1820. Act of April 24 (3 Stat. 566) provided that public lands should thereafter be offered at public sale to the highest bidder in half-quarter sections (80 acres); reduced the minimum price to $1.25 per acre; and required full payment at the time of sale. Private sale (at not less than the minimum price) of lands unsold at public auction was authorized.

1821. Attorney General ruled that under the act of 1807 timber trespassers on public lands could be removed by military force and subjected to fine and imprisonment.

1822. Act of February 23 (3 Stat. 651) authorized the President to employ so much of the land and naval forces of the United States as might be necessary effectually to prevent the cutting and carrying away of the timber of the United States in Florida and to take such other measures as he might deem advisable for the protection of such timber.

1823. Act of February 28 (3 Stat. 727) granted to Ohio a right of way 120 feet wide, together with a strip of land 1 mile in width on each side thereof, to aid in the construction of a proposed wagon road from the lower rapids of the Miami River to the western boundary of the Connecticut Western Reserve. The road was to be completed in four years, and none of the land was to be sold for less than $1.25 per acre.

1824. Act of April 30 (4 Stat. 22) assigned responsibility for the handling of internal improvements, including improvement of rivers and harbors, to the Army Corps of Engineers. The first act dealing only with rivers and harbors was passed in 1826 (4 Stat. 175).

1824. Act of May 26 (4 Stat. 47) granted to Indiana a strip of land 90 feet wide on each side of a proposed canal to connect the Wabash and Miami Rivers. The law was never utilized.

1827. Two acts of March 2 (4 Stat. 234, 236) granted to Illinois and Indiana, respectively, for canal construction a quantity of land equal to one-half of five sections in width on each side of proposed canals in the two states, reserving each alternate section to the United States. The canals were to be free public highways for the use of the United States.

1827. Act of March 3 (4 Stat. 242) authorized the President to take proper measures to preserve the live oak timber growing on the lands of the United States and to reserve from sale such public lands as might be found to contain live oak or other timber in sufficient quantity to render it valuable for naval purposes.

1828. Naval Appropriations Act of March 19 (4 Stat. 254, 256) appropriated not more than $10,000 for the purchase of lands necessary and proper to provide a supply of live oak and other timber for the Navy. This was for the Santa Rosa naval timber reserve and experiment station.

1828. Henry M. Brackenridge, in a letter to Secretary Southard of the Navy Department, discussed the culture of live oak in one of the first American papers on silviculture.

1828. Act of May 23 (4 Stat. 290) granted Alabama 400,000 acres for improvement of navigation on the Tennessee River to be sold at not less than the minimum price charged for public lands. Subsequent grants for river improvement were made to Wisconsin and Iowa in 1846.

1830. Act of May 29 (4 Stat. 420) granted preemption rights for one year to settlers on the public lands. Such settlers, on proof of settlement or improvement, might purchase not more than 160 acres on payment of the minimum price of $1.25 per acre; but the act was not to delay the regular sale of any of the public lands, and preemption rights were not transferable. Temporary preemption laws were also enacted in 1832, 1833, 1838, and 1840. Numerous laws relating to preemption in restricted localities and for specified purposes had been passed prior to 1830.

1831. Act of March 2 (4 Stat. 472) imposed a fine of not less than three times the value of the timber and imprisonment for not more than twelve months (1) on anyone who should unlawfully cut or remove any live oak, red cedar, or other timber from lands reserved or purchased for the use of the Navy and (2) on anyone who should cut or remove any live oak, red cedar, or other timber from any other lands of the United States without written authorization or with intent to export it or use it for any other purpose than for the Navy of the United States. The commissioners of the Navy pension fund were authorized to mitigate in whole or in part any fine, penalty, or forfeiture incurred under the act.

1832. Secretary of the Navy Woodbury, in an estimate of the total area which should be reserved to provide adequate supplies of live oak for the Navy, became a pioneer in the field of forest regulation.

1832. Act of April 5 (4 Stat. 503) reduced to 40 acres the minimum size of tracts offered at private sale.

1832. Act of April 20 (4 Stat. 505) reserved the hot springs in Arkansas from entry, together with four sections of land surrounding them, for future disposal by the United States.

1832. Act of July 9 (4 Stat. 567) constituted the last of the laws providing relief for settlers who had purchased lands under the credit system.

1833. Act of March 2 (4 Stat. 646) provided that ships transporting live oak were to be granted clearance only on proof of its lawful acquisition.

1833. Act of March 2 (4 Stat. 662) modified a former canal grant to Illinois to permit its use in building a railroad, but no advantage was ever taken of the privilege.

1835. From 1835 on, Congress frequently granted to railroads a free right of way through public lands. The privilege was made general in 1852 (10 Stat. 28).

1836. Act of July 4 (5 Stat. 107) reorganized the General Land Office.

1836. Specie Circular of July 11 required local land officials to accept only gold and silver in payment for public land, except for actual settlers buying not more than 320 acres.

1841. Act of September 4 (5 Stat. 453) covered three important points:

    1. It granted 10 per cent of the net proceeds from the sale of public lands in Ohio, Indiana, Illinois, Alabama, Missouri, Mississippi, Louisiana, Arkansas, and Michigan to the state concerned. The balance (after deduction of certain expenses) was to be distributed quarterly to the states, the District of Columbia, and the territories of Wisconsin, Iowa, and Florida for such use as the Legislatures might direct. Distribution was, however, to cease in case of war, if the minimum price of public lands was increased above $1.25 per acre (except in alternate sections), or if the duties fixed by the act of March 2, 1833, were raised above 20 per cent.

2. For purposes of internal improvement 500,000 acres was granted to each of the nine states named above and to such new states as might later be admitted to the Union. Grants already received from the Federal government were to be deducted from this figure. The lands were to be disposed of by the states for not less than $1.25 per acre and the proceeds used only for internal improvements.

3. The preemption privilege was made general by providing that every head of a family, widow, or single man over twenty-one years of age who was a citizen of the United States or had declared his intention to become a citizen could settle upon and purchase at the minimum price of $1.25 per acre not more than 160 acres of surveyed, nonmineral, unoccupied, and unreserved public lands, subject to certain restrictions. Preemptors must inhabit and improve the land and must swear that the land was being taken up for their own exclusive use and benefit. Final proof of settlement and habitation had to be made within one year of the date of settlement, but preemption was not to delay the regular sale of any of the public lands of the United States.

1842. Act of August 4 (5 Stat. 502), a forerunner of the Homestead Act, offered to donate lands in Florida to actual settlers in lots of 160 acres up to a grand total of 200,000 acres.

1842. Act of August 30 (5 Stat. 548, 567), dealing primarily with the tariff, suspended the distribution of proceeds from the sale of public lands. Distribution was never resumed.

1843. The Senate asked the Secretary of the Navy for any evidence "that depredations of a most ruinous kind are being daily committed on the navy timber."

1843. Act of March 3 (5 Stat. 611) opened to settlement certain reservations of live oak lands in Louisiana. The last disposal of these reservations was made in 1923.

1843. Act of March 3 (5 Stat. 619, 620) provided that a person could not, after once exercising the right of preemption, file on a second tract.

1844. Act of May 23 (5 Stat. 657) provided for the disposal of town sites on the public lands in tracts not exceeding 320 acres.

1845. Joint Resolution of March 1 (5 Stat. 797) consented to the annexation of Texas, which had been an independent republic since 1836, with the provision that it should retain control of its public lands.

1846. Treaty of June 15 with Great Britain (9 Stat. 869) confirmed the claim of the United States to some 181 million acres of territory embracing the present states of Oregon, Washington, and Idaho, and parts of Montana and Wyoming and defined the northern boundary of the territory.

1846. Act of July 11 (9 Stat. 37) authorized the public sale of reserved lead mines in Illinois, Arkansas, Wisconsin, and Iowa at not less than $2.50 per acre.

1846. Act of August 3 (9 Stat. 51) authorized the Commissioner of the General Land Office to sell isolated or disconnected tracts of unoffered lands without the formality of a proclamation by the President.

1846. Act of August 8 (9 Stat. 83), granting certain public lands to Wisconsin for the construction of the Wisconsin-Fox River Canal, inaugurated the policy of charging the "double minimum" price of $2.50 per acre for land in the alternate sections retained by the government.

1847. Act of January 26 (9 Stat. 118) permitted all states admitted to the Union prior to 1820 to tax all public lands from and after the day of their sale, provided that lands belonging to citizens living outside the states were never to be taxed higher than those belonging to persons residing therein.

1847. Acts of March 1 (9 Stat. 146) and March 3 (9 Stat. 179) established the Lake Superior district in Michigan and the Chippewa land district in Wisconsin for purposes of mineral survey and authorized sale in quarter sections, after six months' notice, of lands containing copper, lead, or other valuable ores at a minimum price of $5 per acre.

1848. Treaty of February 2 with Mexico (9 Stat. 922, 926) added some 335 million acres to the public domain.

1848. Beginning with the act of August 14 (9 Stat. 323, 330) establishing the territorial government of Oregon, Sections 16 and 36 in each township were reserved for school purposes.

1849. Act of March 2 (9 Stat. 352) granted to Louisiana all of the swamp and overflowed lands in that state unfit for cultivation, with the proviso that the proceeds should be used exclusively, as far as necessary, for the construction of levees and drains.

1849. Act of March 3 (9 Stat. 395) created the Department of the Interior and gave the Secretary responsibility for administering the public lands (except reservations administered by another department, such as the naval-timber reserves). The new Department acquired the General Land Office, Indian Affairs, and the Patent Office (in which the government's agricultural services were then carried on) by transfer respectively from the Treasury Department, War Department, and State Department.

1850. Act of September 9 (9 Stat. 446) added some 79 million acres to the public domain by purchase from Texas.

1850. Act of September 20 (9 Stat. 466), the first of the railroad land grants, granted to the states of Illinois, Alabama, and Mississippi, to aid in the construction of the Illinois Central Railroad (which was privately built): (1) a right of way not over 200 feet wide; (2) free use of construction material, such as earth, stone, and timber; and (3) every alternate section of land designated by even numbers for 6 sections in width on each side of the road, with the right to make lieu selections in place of alienated lands to a distance of not more than 15 miles from the road. The alternate sections retained by the government were to be sold at not less than $2.50 per acre. Property and troops of the United States were at all times to be transported over the railroad free of charge, and the mails at such rates as Congress might fix.

1850. United States Supreme Court (9 Howard 351), in the case of *United States v. Briggs,* declared constitutional that part of the act of March 2, 1831, forbidding the removal of any timber on any public lands, in spite of the fact that the title referred only to timber on naval reservations.

1850. Act of September 26 (9 Stat. 472) amended the acts of March 1 and March 3, 1847, to provide for the public sale of mineral lands in the Lake Superior district in Michigan and the Chippewa district in Wisconsin in the same manner, and subject to the same minimum price and the same rights of preemption, as other public lands.

1850. Act of September 28 (9 Stat. 519) granted to Alabama, Arkansas, California, Florida, Illinois, Indiana, Iowa, Michigan, Mississippi, Missouri, Ohio, and Wisconsin all of the swamp and overflowed lands in those states, under conditions similar to those contained in the grant to Louisiana.

1850. Special agents were appointed, probably for the first time, to suppress timber trespass on the public lands generally.

1852. Act of March 22 (10 Stat. 3) made all warrants for military bounty lands assignable.

1852. Act of August 4 (10 Stat. 28) granted free right of way through public lands to all railroads already chartered or to be chartered within ten years, together with free use of timber and other construction materials.

1853. Act of March 3 (10 Stat. 259) authorized the sale of certain red cedar lands in Alabama that had been reserved for naval purposes under the act of March 1, 1817.

1853. Gadsden Purchase of December 30 (10 Stat. 1031) added some 19 million acres to the public domain.

1853–1854. Several acts permitted preemption on unsurveyed lands in California, Oregon, Washington, Kansas, Nebraska, and Minnesota.

1854. On January 19 responsibility for the protection of the public lands was transferred from the Secretary of the Interior to the Commissioner of the General Land Office.

1854. Graduation Act of August 4 (10 Stat. 574) reduced the price of land according to the time it had been on the market, with a minimum of 12.5 cents per acre after thirty years. The act did not apply to alternate sections in grants made for internal improvements or to mineral lands held at more than $1.25 per acre. Purchasers had to swear that they were acquiring the land for their own use and for the purpose of actual settlement and cultivation.

1855. Act of March 3 (10 Stat. 701) liberalized military bounties so as to grant 160 acres to all participants in all wars from the Revolution to date.

1855. On December 24 prevention of timber trespass was transferred from special agents to the local land officers, and a circular of instructions was issued for the guidance of the latter.

1855. André François Michaux bequeathed $12,000 to the American Philosophical Society in Philadelphia for forestry instruction.

1859. Act of March 3 (11 Stat. 408) forbade the unlawful cutting of timber on lands of the United States reserved or purchased for military or other purposes. This proviso became Section 5388 of the Revised Statutes of 1878.

1860. Act of March 12 (12 Stat. 3) extended the provisions of the swamp-land grants to Minnesota and Oregon.

1862. Act of May 15 (12 Stat. 387) established a Department of Agriculture (not of Cabinet rank), headed by a Commissioner of Agriculture.

1862. Homestead Act of May 20 (12 Stat. 392) authorized any person who was head of a family or over twenty-one years of age, and who was a citizen of the United States or had declared his intention to become such, to enter upon not more than 160 acres of unappropriated land subject to preemption and sale at a minimum price of $1.25 per acre, or not more than 80 acres subject to sale at a minimum price of $2.50 per acre. Free patent could then be secured by the settler for his exclusive use and benefit on proof that he had resided upon or cultivated the land for five years, provided that if he should actually change his residence during that period or should abandon the land for more than six months at any one time, it was to revert to the government. Commutation, or purchase of the land at its regular price, was possible at any time after six months from the date of filing.

1862. Act of June 2 (12 Stat. 413) extended the Preemption Act to unsurveyed lands in all the public-land states and territories and repealed the Graduation Act of 1854.

1862. Act of July 1 (12 Stat. 489) granted the Union Pacific and Central Pacific railroads alternate, odd-numbered sections of land for 10 miles on each side of the road. This distance was increased to 20 miles in 1864 (13 Stat. 356). Mineral lands were not included, but the amendatory act of 1864 excluded coal and iron land from this category. Government mails, troops, and supplies were to be transported at fair and reasonable rates. Three years after completion of the roads any lands remaining were to be subject to preemption by settlers and sold to them for not more than $1.25 per acre. The price of the sections retained by the government was not increased.

1862. Morrill Act of July 2 (12 Stat. 503) granted to each state 30,000 acres of nonmineral public land (with minimum price of $1.25 per acre) for each senator and representative to which it was entitled under the census of 1860. The proceeds were to be invested in a permanent fund and the interest used for the establishment of colleges of agriculture and the mechanic arts. States without public lands were given an equivalent amount of scrip, purchasers of which were not to take up more than a million acres in any one state.

1864. Act of June 30 (13 Stat. 325) granted the Yosemite Valley and the Mariposa Big Tree Grove to California to be held forever "for public use, resort, and recreation."

1864. Act of July 1 (13 Stat. 343) repealed the act of May 23, 1844, relating to town sites, and provided in detail for the sale at auction of town sites established on public lands at not less than $10 per lot (not to exceed 4,200 square feet in size).

1864. Act of July 1 (13 Stat. 343) provided for the sale of coal lands in the public domain at auction at a minimum price of $20 per acre.

1864. Act of July 2 (13 Stat. 365) granted the Northern Pacific Railroad alternate, odd-numbered sections of nonmineral land for 40 miles on each side of the road in the territories traversed, and half that amount in the states. Lieu lands could be selected within 10 miles of the outer limit of the primary grant. The price of the alternate sections retained by the government was raised to $2.50 per acre. Five years after completion of the road, the railroad was required to sell all unmortgaged lands still in its possession for not more than $2.50 per acre. Transportation of government mails, troops, and supplies was to be furnished under regulations imposed by Congress.

1865. Rev. Frederick Starr of St. Louis, Missouri, in an article in the annual report of the Commissioner of Agriculture, pointed with alarm to the destruction of the forests and recommended various preventive measures, including the establishment of forest experiment stations.

1866. Act of June 21 (14 Stat. 66) withdrew all lands in Alabama, Mississippi, Louisiana, Arkansas, and Florida from disposal except under the Homestead Act.

1866. Act of July 25 (14 Stat. 239) granted the California and Oregon Railroad Company alternate, odd-numbered sections of nonmineral public land to a distance of 20 miles on each side of the road. Property and troops of the United States were to be transported without charge.

1866. Act of July 26 (14 Stat. 251) provided that the mineral lands of the public domain, both surveyed and unsurveyed, should be free and open to exploration and occupation by citizens of the United States or those who had declared their intention to become citizens. Lode mines could be purchased for $5 per acre if the claimant had occupied them according to local mining rules and had expended as much as $1,000 in labor and improvements.

1866. Act of July 26 (14 Stat. 251, 253) granted right of way for the construction of ditches and canals across public lands to persons having rights to the use of water for mining, agricultural, manufacturing, or other purposes.

1866. Act of July 27 (14 Stat. 292, 299) made grants to the Atlantic and Pacific Railroad (now the Atchison, Topeka, and Santa Fe) and to the Southern Pacific Railroad similar in amount to the grant to the Northern Pacific Railroad.

1867. By treaty of March 30 (15 Stat. 539) some 365 million acres in Alaska were purchased from Russia.

1869. Act of March 3 (15 Stat. 340) granted to Oregon, to aid in the construction of a military wagon road from Coos Bay to Roseburg, the odd-numbered sections of nonmineral land to 6 miles on each side of the road, which was to be a public highway for the free transportation of property, troops, and mails of the United States.

1870. Act of July 9 (16 Stat. 217) provided for the sale of placer mines at $2.50 per acre in tracts not exceeding 160 acres.

1871. Joint Resolution of February 9 (16 Stat. 593) authorized the President to appoint a civil officer of the government as Commissioner of Fish and Fisheries to study problems relating to the conservation of the food fishes of the coast and lakes of the United States. Deficiency Appropriations Act of May 18, 1872 (17 Stat. 122, 124) appropriated $3,500 for the prosecution of the work.

1871. Naval Appropriations Act of March 3 (16 Stat. 526, 527) appropriated $5,000 for the fiscal year 1872 for the protection of timberlands in naval-timber reservations. The last appropriation for this purpose was for the fiscal year 1876.

1871. Act of March 3 (16 Stat. 573) made the last railroad land grant, to the Texas Pacific Railroad Company.

1872. Act of March 1 (17 Stat. 32) reserved the Yellowstone National Park "as a public park or pleasuring-ground for the benefit and enjoyment of the people."

1872. Arbor Day was first celebrated in Nebraska on April 10 at the instance of J. Sterling Morton, later Secretary of Agriculture.

1872. Act of May 10 (17 Stat. 91) constituted mineral lands a distinct class and provided for their survey and sale at $2.50 per acre for placer mines and at $5 per acre for lode mines.

1872. Sundry Civil Appropriations Act of June 10 (17 Stat. 347, 359) appropriated $10,000 for the protection of public timberlands in general from trespass and fraud.

1873. Act of February 18 (17 Stat. 465) excluded lands containing iron, coal, or any other minerals in Michigan, Wisconsin, and Minnesota from the provisions of the mineral act of 1872 and opened them to exploration and purchase as before the passage of that act.

1873. Timber Culture Act of March 3 (17 Stat. 605) offered to donate 160 acres of public land to any person who would plant 40 acres to trees, not more than 12 feet apart each way (302 per acre), and keep them in a growing and healthy condition by cultivation for a period of ten years. Any homesteader who should, at the end of three years, submit satisfactory proof of having had under cultivation for two years 1 acre of trees for each 16 acres in his homestead claim was entitled to receive a patent at once, the planting and cultivation of the trees being accepted in lieu of the additional two years' residence required by the Homestead Act of 1862.

1873. Act of March 3 (17 Stat. 607) made more detailed provision than the act of 1864 for the sale of coal lands and reduced the price to a minimum of $10 per acre for lands more than 15 miles from a completed railroad.

1873. American Association for the Advancement of Science passed a resolution favoring the creation of Federal and state forestry commissions and appointed a committee with Franklin B. Hough as chairman to follow the matter up.

1873. John A. Warder represented the United States at the International Exposition in Vienna and prepared a report on European forestry.

1874. Act of March 13 (18 Stat. 21) amended the Timber Culture Act of 1873 by limiting it to heads of families or persons over twenty-one years of age who were citizens or had declared their intentions to become citizens and by reducing the period of cultivation to eight years. A person entering a quarter section had to plow 10 acres the first year, 10 acres the second year, and 20 acres the third year, and to plant 10 acres the second year, 10 acres the third year, and 20 acres the fourth year, with proportional areas for smaller claims.

1874. Act of June 22 (18 Stat. 194) allowed land-grant railroads to make lieu selections for land found to be in the possession of actual settlers.

1874. President Grant in a special message to Congress called attention to the urgent need for forest protection and transmitted a draft of proposed legislation on the subject.

1874. Report by William H. Brewer of Yale University on *The Woodlands and Forest Systems of the United States* was published in the "Statistical Atlas of the Ninth Census."

1875. Naval Appropriations Act of January 18 (18 Stat. 296, 300) made the last specific appropriation ($5,000) for protection of timber in naval-timber reserves for the fiscal year 1876.

1875. Act of March 3 (18 Stat. 482) granted railroads free rights of way through public lands and the use of timber and other materials for construction purposes.

1875. The American Forestry Association was organized in Chicago on September 10 under the leadership of Dr. Warder.

1876. Act of May 20 (19 Stat. 54) amended the Timber Culture Act by extending for one year the period during which cultivation and planting must be accomplished for each year that the trees were destroyed by grasshoppers or other inevitable causes. Planting of seeds, nuts, and cuttings was declared to constitute compliance with the law.

1876. Act of July 4 (19 Stat. 73) repealed the act of 1866 restricting disposal of public lands in the South to homesteaders and thereby opened them to sale.

1876. Appropriations Act of August 15 (19 Stat. 143, 167), through a rider, appropriated $2,000 for the employment by the Commissioner of Agriculture of an expert to study and report upon forest conditions. Franklin B. Hough was appointed.

1876. A forest-reserve bill was introduced by Representative Fort of Illinois but received no action.

1877. Act of January 12 (19 Stat. 221) provided that saline lands which had been reserved for granting to the states on their admission to the Union were to be examined and offered for sale at public auction at not less than $1.25 per acre if found to be actually saline.

1877. Act of March 3 (19 Stat. 344, 357) established an Entomological Commission in the Department of the Interior. It was transferred to the Department of Agriculture in 1881 (21 Stat. 259, 276).

1877. Desert Land Act of March 3 (19 Stat. 377) provided for the sale in eleven Western states and territories of 640 acres of nontimber, nonmineral land unfit for cultivation without irrigation to any settler who would irrigate it within three years after filing. A payment of $0.25 per acre was to be made at the time of filing and $1 per acre at time of final proof.

1877. The system of special agents to check timber trespass on the public domain was revived and expanded under Carl Schurz, Secretary of the Interior, and J. A. Williamson, Commissioner of the General Land Office. A revised and enlarged circular of instructions was issued for the guidance of timber agents.

1878. Free Timber Act of June 3 (20 Stat. 88) provided that residents of Colorado, Nevada, New Mexico, Arizona, Utah, Wyoming, Dakota, Idaho, or Montana might cut timber on public mineral lands for building, agricultural, mining, or other domestic purposes, subject to such regulations as the Secretary of the Interior might prescribe. This privilege was later extended in acts of 1891, 1893, and 1901.

1878. Timber and Stone Act of June 3 (20 Stat. 89) provided for the sale in Washington, Oregon, California, and Nevada of 160 acres of surveyed, nonmineral land, chiefly valuable for timber or stone and unfit for cultivation, which had not been offered at public sale, for not less than $2.50 per acre. The purchaser had to swear that the land was being acquired solely for his own use and benefit. In 1892 the provisions of this act were extended to all the public-land states.

In the states concerned, the act also forbade the unlawful cutting or wanton destruction of timber on any public lands or its removal for export or other disposal; granted permission to miners and farmers to clear land and to use such timber as necessary for improvements; relieved trespassers who had not exported the timber from the United States from further prosecution on payment of $2.50 per acre for the timber; and directed that all moneys collected should be covered into the Treasury of the United States.

1878. Act of June 14 (20 Stat. 113) reduced the area to be planted under the Timber Culture Act to not less than one-sixteenth of the amount entered for areas of not less than 40 acres. The number of trees to be planted was increased to 2,700 per acre, of which 675 had to be living at the time of final proof. An extension of one year in the cultivation and planting was allowed for each year the trees were destroyed by grasshoppers or drought.

1878. Senator Plumb of Kansas introduced a bill to withdraw all timbered lands from settlement.

1879. Sundry Civil Appropriations Act of March 3 (20 Stat. 377, 394) created both the Geological Survey and the Public Land Commission. A preliminary report by the Commission, submitted in 1880, offered a proposal for the classification of the public lands and recommended sale of timber without the land. Donaldson's comprehensive "Public Domain" followed later (1880–1884).

1879. United States Circuit Court (5 Dillon 405) ruled that the trespass acts of 1831 and 1859 did not apply to Indian reservations, since these are not "lands of the United States."

1880. Act of June 15 (21 Stat. 237) relieved timber trespassers on the public lands prior to March 1, 1879, from both civil and criminal prosecution on payment of $1.25 per acre.

1881. The forestry work started by Hough in 1876 was organized as a separate division under the Commissioner of Agriculture.

1882. American Forestry Congress met in Cincinnati in April, absorbed Dr. Warder's American Forestry Association in August, and in 1889 assumed that name.

1882. F. B. Hough started the *American Journal of Forestry,* which ran from October, 1882, to September, 1883.

1883. Nathaniel H. Egleston succeeded Hough in charge of the forestry work in the Department of Agriculture.

1884. Act of May 17 (23 Stat. 24, 26) made the laws of the United States relating to mining claims applicable in Alaska. This action was repeated, with certain provisos, in the act of June 6, 1900 (31 Stat. 321, 329) providing for a civil government for Alaska.

1885. Charles S. Sargent's *Report on the Forests of North America* was published as part of the Tenth Census.

1885. Act of February 25 (23 Stat. 321) forbade the making of enclosures on the public domain and authorized the destruction of unauthorized enclosures.

1885. Agricultural Appropriations Act of March 3 (23 Stat. 353, 354) included funds "for the promotion of economic ornithology" by the Entomological Division.

1886. Agricultural Appropriations Act of June 30 (24 Stat. 100, 101) established the Division of Economic Ornithology and Mammalogy in the Department of Agriculture.

1886. Act of June 30 (24 Stat. 100, 103) gave forestry statutory recognition as a distinct division of the Department of Agriculture. Bernhard E. Fernow took charge of it, succeeding Egleston. Edward A. Bowers entered the General Land Office as an inspector and later became assistant commissioner.

1887. Hatch Act of March 2 (24 Stat. 440) provided for financial assistance to states in the establishment of agricultural experiment stations, which included forestry in their activities.

1888. Act of January 20 (25 Stat. 1) removed the previous requirement that the Commissioner of Fish and Fisheries must be a civil officer of the government who would serve without additional compensation by authorizing the President to appoint "a person of scientific and practical acquaintance with the fish and fisheries" at a salary of $5,000.

1888. Act of June 4 (25 Stat. 166) amended the act of March 3, 1859 (Sec. 5388 of the Revised Statutes of 1878) to include timberlands in Indian reservations.

1888. Act of October 2 (25 Stat. 505, 526) provided for a survey of the public lands suitable for irrigation and directed that all lands selected as sites for reservoirs, canals, and ditches and all lands thereby made susceptible of irrigation should be withdrawn from entry.

1889. Act of January 14 (25 Stat. 642) provided for the cession to the government of certain lands in the Chippewa Indian Reservation in Minnesota and for the sale of land and timber thereon under government supervision, with the proceeds going into a permanent fund for the benefit of the tribe.

1889. Act of February 16 (25 Stat. 673) authorized the President to permit Indians to cut and sell dead timber on Indian reservations, provided the timber had not been intentionally killed.

1889. Act of March 2 (25 Stat. 835) gave cabinet rank to the Department of Agriculture, which was headed by the Secretary of Agriculture.

1889. Act of March 2 (25 Stat. 854) abolished private sale of timberlands in the South.

1889. Act of March 2 (25 Stat. 939, 961) authorized the President to reserve the land containing the Casa Grande ruin in Arizona, which thus became the first prehistoric-site reservation.

1890. Act of June 12 (26 Stat. 146) authorized the Secretary of the Interior, subject to the approval of the tribe, to employ Indians to cut not more than 20 million board feet of green timber per year on the Menominee Indian Reservation in Wisconsin.

1890. Act of August 30 (26 Stat. 371, 391) repealed that part of the act of October 2, 1888, relating to the withdrawal from entry of irrigable lands and of sites for canals and ditches, but permitted the continued withdrawal of sites for reservoirs. The act also forbade anyone to acquire title to a grand total of more than 320 acres of public land, thus cutting in half the amount previously available under the Desert Land Act of 1877.

1890. Second Morrill Act of August 30 (26 Stat. 417) provided for additional assistance to land-grant colleges out of proceeds from the sale of the public lands. An amendment in 1903 provided that in case these proceeds were insufficient to meet the amount appropriated, it should be paid from the Treasury.

1890. Act of September 25 (26 Stat. 478) set apart the Big Tree National Park as a public park, or pleasure ground.

1890. Act of September 29 (26 Stat. 496) forfeited and restored to the public domain all land in grants adjacent to the uncompleted sections of railroads to which grants had been made.

1890. Act of October 1 (26 Stat. 650) set apart the Yosemite National Park and General Grant National Park "as forest reservations."

1891. Act of March 3 (26 Stat. 1095) repealed the Timber Culture Act of 1878 and the Preemption Act of 1841; put a stop to auction sales of public lands except isolated tracts and abandoned military and other reservations; tightened up the requirements for improvement and cultivation under the Desert Land Act of 1877, and extended it to include Colorado; did not allow commutation under the Homestead Act of 1862 until fourteen months after filing; limited the time within which suit to annul patent might be brought; restricted withdrawals for reservoir sites to the area actually needed for that purpose; authorized rights of way for irrigation canals and drainage ditches through public lands and reservations; provided that in any criminal or civil prosecution for trespass on the public lands in any of the Rocky Mountain states or territories except Arizona and New Mexico and in the district of Alaska, it should be a defense if the timber had been cut for use in such state or territory by a resident thereof for agricultural, mining, manufacturing, or domestic purposes and had not been transported out of the same, and authorized the Secretary of the Interior to make rules and regulations for the carrying out of this provision; and empowered the President to set aside as forest reserves public lands covered with timber or undergrowth, whether of commercial value or not. The last provision (Sec. 24) was added by the conference committee and is often referred to as the Forest Reserve Act.

1891. Act of March 3 (26 Stat. 1093) amended the preceding act to provide that in order to avoid prosecution, cutting of timber on the public lands must have been done under rules and regulations prescribed by the Secretary of the Interior; authorized the Secretary to designate tracts where timber may be cut; and specified that the act did not repeal the act of June 3, 1878, relating to the cutting of timber on mineral lands (Free Timber Act).

1891. President Harrison on March 30 (26 Stat. 1565) proclaimed the Yellowstone Forest Reserve.

1892. Act of August 4 (27 Stat. 348) extended the Timber and Stone Act of 1878 to all of the public-land states.

1893. Act of February 13 (27 Stat. 444) extended the Free Timber Act of 1878, as amended in 1891, to New Mexico and Arizona.

1894. Act of May 7 (28 Stat. 73) prohibited the hunting of birds or wild animals in Yellowstone National Park, but permitted fishing with hook and line. It also authorized the Secretary of the Interior to make rules and regulations for the management and care of the park, "especially for the preservation from injury or spoliation of all timber, mineral deposits, natural curiosities, or wonderful objects"; for the protection of birds and animals from capture or destruction; and for the control of fishing.

1894. Section 4 (Carey Act) of the Sundry Civil Appropriations Act of August 18 (28 Stat. 372, 422) authorized the donation to states having desert lands (or to their assigns) of not more than 1 million acres each which they should cause to be settled, irrigated, and in part cultivated within ten years. Not more than 160 acres was to be sold or disposed of to any one person. Subsequent amendments made it possible for the states to obtain a total of 14 million acres, of which only about 600,000 acres has actually been granted.

1895. Act of February 26 (28 Stat. 687) amended the act of August 3, 1846, to authorize the sale at public auction of isolated tracts containing not more than 160 acres at not less than $1.25 per acre.

1895. Act of March 2 (28 Stat. 727, 735) established the Division of Agricultural Soils in the Department of Agriculture. Research relating to soils had been authorized by the act of August 8, 1894 (28 Stat. 264, 268, 274).

1896. At the request of Secretary of the Interior Hoke Smith, the National Academy of Sciences appointed a special committee, with C. S. Sargent as chairman, to investigate the forest-reserve situation and recommend a national forestry policy. Sundry Civil Appropriations Act of June 11 (29 Stat. 413, 432) appropriated $25,000 to meet the expenses of the committee.

1896. United States Supreme Court decision on March 2 in case of *Geer v. Connecticut* (161 U.S. 519) confirmed the right of the states to protect their wildlife.

1896. Agricultural Appropriations Act of April 25 (29 Stat. 99, 100) established the Division of Biological Survey in the Department of Agriculture.

1897. Act of February 11 (29 Stat. 526) made oil lands subject to the laws relating to placer mineral claims.

1897. Act of February 24 (29 Stat. 594) provided penalties for willfully or maliciously setting on fire any timber, underbrush, or grass on the public domain; carelessly or negligently leaving fire to burn unattended near any timber or other inflammable material; or failing to totally extinguish any campfire or other fire in or near any forest, timber, or other inflammable material before leaving it.

1897. Sundry Civil Appropriations Act of June 4 (30 Stat. 11, 34) specified the purposes for which forest reserves might be established and provided for their protection and administration.

1897. A Division of Geography and Forestry was established in the Geological Survey to handle surveying and mapping of the forest reserves and to collect data on their resources.

1898. Act of May 14 (30 Stat. 409, 414) extended the homestead laws to Alaska, with the proviso that no homestead should exceed 80 acres. It also authorized the Secretary of the Interior to sell timber on the public lands in Alaska at not less than its appraised value, for use in the district, and to grant free use of timber for specified purposes, under rules and regulations prescribed by him. Export of pulpwood and wood pulp was authorized in 1905 (33 Stat. 628) and of birch timber in 1920 (41 Stat. 874, 917).

1898. Act of May 18 (30 Stat. 418) abolished the distinction between offered and unoffered lands and provided that in disposing of public lands under the homestead laws and the timber and stone law all land should be treated as unoffered.

1898. Sundry Civil Appropriations Act of July 1 (30 Stat. 597, 618) made the first appropriation ($75,000) for protection and administration of the forest reserves.

1898. Gifford Pinchot succeeded Fernow as chief of the Division of Forestry, with the title of Forester.

1899. Act of February 28 (30 Stat. 908) authorized the Secretary of the Interior to lease ground near or adjacent to mineral, medicinal, or other springs in forest reserves for the erection of sanitariums or hotels, under such regulations as he might prescribe. All receipts were to be covered into a special fund to be expended in the care of public forest reservations.

1899. Act of March 2 (30 Stat. 993) established Mt. Rainier National Park.

1899. Act of March 3 (30 Stat. 1074, 1095) directed persons connected with the administration and protection of forest reserves to assist so far as practicable in the enforcement of state fish and game laws.

1899. Appalachian National Park Association was organized in North Carolina in November.

1900. Act of May 5 (31 Stat. 169) amended the act of February 24, 1897, by omitting the words "carelessly or negligently" in connection with leaving any fire to burn unattended.

1900. Lacey Act of May 25 (31 Stat. 187) prohibited the importation of any foreign wild animal or bird except under special permit from the Department of Agriculture and specifically prohibited the importation of the mongoose, fruit bat, English sparrow, starling, and such other birds or animals as the Secretary of Agriculture may declare injurious to the interest of agriculture or horticulture. It also prohibited the interstate transportation of wild animals or birds taken or possessed in violation of the laws of the state from which or to which they were shipped.

1900. Act of May 25 (31 Stat. 191, 197) appropriated $5,000 for investigating forest conditions in the Appalachians with a view to purchasing land for forest reserves.

1900. Sundry Civil Appropriations Act of June 6 (31 Stat. 588, 614) limited lieu selections under the 1897 act to vacant, nonmineral, surveyed public lands subject to homestead entry.

1900. Act of June 6 (31 Stat. 661) authorized the Secretary of the Interior to sell timber from the forest reserves without advertisement up to a stumpage value of $100.

1900. Society of American Foresters was organized on November 30.

1901. Act of January 31 (31 Stat. 745) opened all unoccupied public lands containing salt springs or salt deposits to entry under the placer-mining laws.

1901. Act of February 12 (31 Stat. 785) permitted the Indians on the Grand Portage Lake Indian Reservation in Minnesota to cut and dispose of the timber on their several allotments under rules and regulations prescribed by the Secretary of the Interior. Similar acts followed shortly authorizing the sale of timber on other Indian lands.

1901. Act of February 15 (31 Stat. 790) authorized the Secretary of the Interior to grant rights of way through forest reserves for canals and ditches, dams and reservoirs, electrical lines, and other purposes, revocable at the discretion of the Secretary or his successor.

1901. Agricultural Appropriations Act of March 2 (31 Stat. 922, 929) changed the Division of Forestry in the Department of Agriculture to the Bureau of Forestry.

1901. Agricultural Appropriations Act of March 2 (31 Stat. 922, 931) created the Bureau of Soils in place of the Division of Soils, which had been established in the Department of Agriculture in 1895.

1901. Act of March 3 (31 Stat. 1436) extended the provisions of the Free Timber Act of 1878, as amended in 1891 and 1893, to California, Oregon, and Washington.

1901. A Forestry Division was created in the General Land Office in the Department of the Interior under Filibert Roth.

1902. Newlands Act of June 17 (32 Stat. 388) created the "reclamation fund" out of receipts from the sale and disposal of public lands in certain states west of the Mississippi River; authorized the Secretary of the Interior to construct irrigation works and to withdraw irrigable lands from entry; and provided for the homesteading of irrigated lands and their sale at a price estimated to return to the reclamation fund the cost of construction.

1902. Morris Act of June 27 (32 Stat. 400) amended the act of January 14, 1889, by providing that logging on 200,000 acres on the Chippewa Indian Reservation in Minnesota should be done under supervision of the Bureau of Forestry, with 5 per cent of the pine left as seed trees. No cutting was to be done on certain specified lands, including ten sections to be selected by the Forester with the approval of the Secretary of the Interior.

1903. Act of February 14 (32 Stat. 825, 827) establishing the Department of Commerce and Labor incorporated in it the Commission of Fish and Fisheries, thereafter known as the Bureau of Fisheries.

1903. Second Public Lands Commission, consisting of W. A. Richards, F. H. Newell, and Gifford Pinchot, was appointed by President Roosevelt and submitted reports in 1904 and 1905. These reports recommended classification of the public lands, particularly with reference to their agricultural possibilities; opening of arable lands within the forest reserves to homesteading; repeal of the lieu-land provision in the act of 1897 and of the Timber and Stone Act of 1878; modification of the Desert Land Act and of the commutation provision of the Homestead Act; provision for the sale of timber on the unreserved public domain; and legislation authorizing the President to establish grazing reserves.

1903. President Roosevelt set aside Pelican Island, Florida, as the first Federal wildlife refuge.

1904. Kinkaid Act of April 28 (33 Stat. 547) increased the size of homesteads in western Nebraska to 640 acres of nonirrigable land and required the construction of permanent improvements to the extent of not less than $1.25 per acre.

1905. American Forest Congress met January 2 to 6 in Washington.

1905. Act of January 24 (33 Stat. 614) authorized the President to set aside areas in the Wichita Forest Reserve, Oklahoma, for the protection of game animals and birds. The refuge was transferred to the Bureau of Biological Survey by Presidential Proclamation in 1936 (50 Stat. 1797).

1905. Act of February 1 (33 Stat. 628) (1) transferred the administration of the forest reserves from the Secretary of the Interior to the Secretary of Agriculture; (2) covered all receipts from the forest reserves for a period of five years into a special fund to be available, until expended, as the Secretary of Agriculture might direct, for the protection, administration, improvement, and extension of the reserves; (3) provided that forest supervisors and rangers should be selected, when practicable, from the states or territories in which the reserves were located; (4) authorized the export of pulpwood and wood pulp from Alaska; and (5) granted rights of way for dams, ditches, and flumes across the reserves for various purposes under regulations prescribed by the Secretary of the Interior and subject to state laws.

1905. Act of February 6 (33 Stat. 700) authorized the arrest by any officer of the United States, without process, of any person taken in the act of violating the regulations relating to forest reserves and national parks.

1905. Agricultural Appropriations Act of March 3 (33 Stat. 861, 873) permitted timber on forest reserves to be exported from the state or territory (including Alaska) in which cut except in the Black Hills (South Dakota) and Idaho. This provision was made general in 1913.

1905. Act of March 3 (33 Stat. 861, 872–873) changed the name of the Bureau of Forestry to Forest Service, effective July 1. It also repeated the provisions of the act of February 6, 1905, authorizing forest and park officers to arrest without process any person taken in the act of violating the laws and regulations relating to forest reserves and national parks.

1905. Act of March 3 (33 Stat. 861, 877) changed the name of the Division of Biological Survey to Bureau of Biological Survey.

1905. Act of March 3 (33 Stat. 1264) repealed the lieu-land provision of the act of 1897 but permitted the perfecting of valid selections already made.

1906. President Roosevelt began withdrawal of coal and oil lands for purposes of examination and classification.

1906. Beginning January 1, a charge was made for the first time for grazing on the forest reserves.

1906. Act of April 16 (34 Stat. 116) authorized the Secretary of the Interior to lease for a period of not more than ten years any surplus power developed in connection with an irrigation project, giving preference to municipal purposes.

1906. American Antiquities Act of June 8 (34 Stat. 225) forbade anyone, without proper authority, to appropriate, excavate, injure, or destroy any historic or prehistoric ruin or monument or any object of antiquity on lands owned or controlled by the government of the United States. It also authorized the President to establish by proclamation national monuments for the preservation of features of historic, prehistoric, and scientific interest, under administration of the department already having jurisdiction over the land in question. The area reserved must be as small as compatible with the proper care and manage ment of the objects to be preserved.

1906. Forest Homestead Act of June 11 (34 Stat. 233) authorized the Secretary of Agriculture to open for entry, through the Secretary of the Interior, forest-reserve lands chiefly valuable for agriculture which were not needed for public purposes and which in his judgment might be occupied without injury to the forest. Each tract was to be surveyed by metes and bounds and must not exceed 160 acres in area or 1 mile in length. Commutation was not allowed.

1906. Joint Resolution of June 11 (34 Stat. 831) accepted re-cession by California of the lands in the Yosemite Valley and the Mariposa Big Tree Grove granted to it in 1864 for use as a state park.

1906. Agricultural Appropriations Act of June 30 (34 Stat. 669, 684) provided that 10 per cent of all money received from the forest reserves during any fiscal year, including 1906, was to be turned over to the states or territories for the benefit of the public schools and public roads of the counties in which the reserves were located, but not to the extent of more than 40 per cent of their income from other sources. It also forbade unrestricted spending after June 30, 1908, from the special fund set up in 1905.

1906–1907. Senate and House of Representatives passed separate but similar resolutions requesting the Bureau of Corporations to investigate the lumber industry.

1907. Act of March 4 (34 Stat. 1256, 1269) changed "forest reserves" to "national forests"; permitted the export of national-forest timber from the state or territory in which cut, except from the Black Hills National Forest, South Dakota; forbade the further creation or enlargement of national forests except by act of Congress in Washington, Oregon, Montana, Idaho, Wyoming, and Colorado; abolished the special fund established in 1905 but increased Forest Service appropriations by $1,000,000; required the Forest Service to submit to Congress annually a classified and detailed report of receipts and estimate of expenditures; and raised the Forester's salary from $3,500 to $5,000.

1907. Agricultural Appropriations Act of March 4 (34 Stat. 1256, 1281) appropriated $25,000 for survey of lands in the Appalachian and White Mountains in connection with their proposed purchase for national forests.

1907. On March 14, President Roosevelt appointed the Inland Waterways Commission, with Representative Burton of Ohio as chairman.

1907. Public Lands Convention at Denver in June gave occasion for much criticism of the Forest Service.

1908. Cooperative agreement of January 22 between the Secretary of the Interior and the Secretary of Agriculture gave the Forest Service supervision over the handling of timber on Indian reservations.

1908. Act of March 28 (35 Stat. 51) authorized the cutting of not more than 20 million board feet of mature timber per year on the Menominee Indian Reservation in Wisconsin, including the construction of sawmills for converting the timber into lumber.

1908. Joint Resolution of April 30 (35 Stat. 571) authorized the Attorney-General to start proceedings looking to the forfeiture of the lands granted to aid in the construction of the Oregon and California Railroad and the Coos Bay Wagon Road.

1908. First Conference of Governors, called by President Roosevelt, met May 13 to 15 in Washington, D.C. A National Conservation Commission was appointed, with Pinchot as chairman.

1908. Agricultural Appropriations Act of May 23 (35 Stat. 251, 259) directed such officials of the Forest Service as might be designated by the Secretary of Agriculture to aid in the enforcement of state laws relating to stock, forest-fire control, and fish and game protection, and to aid other Federal bureaus in the performance of their duties.

1908. Act of May 23 (35 Stat. 251, 260) increased the payment to the states for the benefit of county schools and roads to 25 per cent of the gross receipts from national forests, eliminated the 40 per cent limitation, and made the legislation permanent.

1908. Act of May 23 (35 Stat. 251, 267) reserved 12,800 acres (increased in 1909 to 20,000 acres) to establish a permanent National Bison Range in the Flathead Indian Reservation in Montana.

1908. Act of May 23 (35 Stat. 268) created the Minnesota National Forest out of lands covered by the Morris Act of 1902, with appropriate compensation to the Indians. It also increased from 5 to 10 per cent the amount of merchantable pine timber that must be reserved in future sales outside of the "ten sections," in which the Forester was permitted to use such methods of cutting as he thought wise.

1908. First Federal forest experiment station was established at Fort Valley, near Flagstaff, Arizona.

1908. Present regional organization of the Forest Service was put into effect on December 1.

1909. Western Forestry and Conservation Association was organized.

1909. Treaty of January 11 between the United States and Great Britain (36 Stat. 2448) established the International Joint Commission and provided for the utilization and development of the boundary waters between the United States and Canada.

1909. Report of National Conservation Commission was transmitted to Congress by President Roosevelt on January 22, with a request for an appropriation of at least $50,000 to meet the expenses of the Commission, which Congress denied.

1909. North American Conservation Conference, attended by official representatives of the United States, Canada, Newfoundland, and Mexico, was held on February 18 in Washington.

1909. Enlarged Homestead Act of February 19 (35 Stat. 639) made it possible to acquire homesteads of 320 acres in Arizona, Colorado, Montana, Nevada, New Mexico, Oregon, Utah, Washington, and Wyoming. Idaho (1910), California (1912), North Dakota (1912), Kansas (1915), and South Dakota (1915) were later added to the list. The lands entered must be nonmineral, nonirrigable, and contain no merchantable timber. Commutation was not allowed.

1909. Act of March 3 (35 Stat. 781, 783) authorized the Commissioner of Indian Affairs to manage the timber on Indian reservations. It resulted on July 17, 1909, in termination of the 1908 agreement with the Forest Service, and in February, 1910, in the establishment in the Office of Indian Affairs of an Indian Forest Service (later Forestry Branch of the Indian Service) under Jay P Kinney.

1909. Act of March 3 (35 Stat. 815, 818) created the National Waterways Commission, consisting of twelve members of Congress, to conduct investigations and to make recommendations pertaining to water transportation and the improvement of waterways.

1909. Act of March 3 (35 Stat. 844) authorized persons who in good faith had entered coal lands under the nonmineral laws to obtain patent thereto subject to reservation of the coal to the United States.

1909. Sundry Civil Appropriations Act of March 4 (35 Stat. 945, 1027) prohibited the use of any public money for compensation or expenses of any commission, council, board, or other similar body not authorized by law or the detail of personal services from any Federal agency to such body.

1909. First National Conservation Congress was held at Seattle, Washington, August 26 to 28.

1909. Organization of National Conservation Association, with Charles W. Eliot of Harvard University as president, was announced on September 15.

1910. Ballinger-Pinchot controversy led on January 7 to the dismissal of Gifford Pinchot as Forester and O. W. Price as Associate Forester. They were succeeded by H. S. Graves and A. F. Potter.

1910. Act of April 21 (36 Stat. 326) provided in detail for the protection and utilization of fur seals and other fur-bearing animals in Alaska; declared the Pribilof Islands a special reservation; and forbade the killing of fur seals in the Pacific Ocean.

1910. Act of May 16 (36 Stat. 369) established the Bureau of Mines in the Department of the Interior.

1910. Act of June 22 (36 Stat. 583) authorized entry of coal lands under the agricultural land laws but with retention of mineral rights by the government.

1910. Act of June 25 (36 Stat. 847) authorized the President to withdraw temporarily public lands from entry and reserve them for specified purposes, such withdrawals or reservations to remain in force until revoked by him or by Congress. It also provided that all withdrawn lands shall be open to exploration, occupation, and purchase for all minerals other than coal, oil, gas, and phosphates.

1910. Act of June 25 (36 Stat. 855, 857) extended to Indian reservations the penalties provided by the act of February 24, 1897, for failing to extinguish fires built in or near any forest, timber, or other inflammable material upon the public domain. It also provided for the sale and management of timber on Indian reservations (except in Minnesota and Wisconsin) and on certain Indian allotments, for the benefit of the Indians, under regulations prescribed by the Secretary of the Interior.

1910. Forest Products Laboratory was established at Madison, Wisconsin, in cooperation with the University of Wisconsin.

1911. Treaty of February 7 between the United States and Great Britain (37 Stat. 1538) provided for the protection and preservation of fur seals, including the prohibition of pelagic sealing.

1911. Act of February 21 (36 Stat. 925) authorized the sale of surplus water from an irrigation project for use on lands outside the project.

1911. Weeks Law of March 1 (36 Stat. 961): (1) authorized the enactment of interstate compacts for the conservation of forests and the water supply; (2) appropriated $200,000 to enable the Secretary of Agriculture to cooperate with any state which had provided by law for a system of forest-fire protection; and (3) appropriated 1 million dollars for the fiscal year 1910 and 2 million dollars for each succeeding fiscal year until June 30, 1915, for use in the examination, survey, and acquisition by the government of lands located on the headwaters of navigable streams. It also created a National Forest Reservation Commission to pass upon lands approved for purchase and to fix the price at which purchases shall be made and provided for the protection and administration of acquired lands.

1911. Act of March 4 (36 Stat. 1235, 1253) authorized the head of the department having jurisdiction over public lands, national forests, and reservations of the United States to grant rights of way for transmission, telephone, and telegraph lines for a period not exceeding fifty years.

1911. Supreme Court on May 1 and 3 (220 U.S. 506, 523) held that Congress has the constitutional right (1) to reserve portions of the public domain as national forests; (2) to delegate to the Secretary of Agriculture administrative authority to make rules and regulations for their occupancy and use; and (3) to prescribe penalties for the violation of such regulations. It also confirmed the right of the Secretary of Agriculture to charge fees for grazing permits and ruled that state fencing laws gave no right willfully to drive one's stock upon the land of another.

1911. Convention of July 7 between the United States, Great Britain, Japan, and Russia (37 Stat. 1542) prohibited pelagic sealing in the North Pacific Ocean; authorized the four governments to control the taking of fur seals in the territories under their respective jurisdictions; specified the distribution of the take among the governments concerned; and superseded all provisions of the treaty of February 7, 1911, with Great Britain inconsistent with the present convention. Of the total number of sealskins taken by the United States on the Pribilof Islands, 15 per cent were allotted to Canada and 15 per cent to Japan.

1911–1914. Bureau of Corporations submitted a comprehensive report on the lumber industry in four parts.

1912. Act of March 28 (37 Stat. 77) authorized the Commissioner of the General Land Office, on application of adjoining owners, to sell at public auction at not less than $1.25 per acre tracts containing not more than 160 acres of public land which is mountainous or too rough for cultivation, whether isolated or not.

1912. Act of April 30 (37 Stat. 105) opened coal lands for selection by the states and for sale as isolated tracts, with reservation to the United States of the coal in such lands.

1912. Act of June 6 (37 Stat. 123) reduced to three years the length of residence necessary to obtain patent under the Homestead Act and set up certain minimum cultivation requirements. Commutation was allowed after fourteen months of actual residence.

1912. Agricultural Appropriations Act of August 10 (37 Stat. 269, 287): (1) directed the Secretary of Agriculture to select, classify, and segregate all lands that may be opened to settlement and entry under the homestead laws applicable to national forests; (2) authorized and directed the Secretary to sell timber at actual cost to homestead settlers and farmers for their domestic use; and (3) made 10 per cent of the gross receipts from national forests available for expenditure by the Secretary of Agriculture for the construction of roads and trails within national forests. The latter provision was made permanent by the act of March 4, 1913 (37 Stat. 828, 843).

1912. Act of August 20 (37 Stat. 320) authorized settlement of suit against forty-six defendants who had bought Oregon and California Railroad grant lands by forfeiture of the lands to the government with privilege of repurchase by the defendants at $2.50 per acre. All of these cases were settled by 1919.

1912. Act of August 24 (37 Stat. 497) provided that all lands withdrawn by the President under the act of 1910 should at all times be open to exploration, discovery, occupation, and purchase under the mining laws of the United States, so far as these applied to metalliferous minerals. It also added California to the list of states within which national forests cannot be created or enlarged except by act of Congress.

1913. Agricultural Appropriations Act of March 4 (37 Stat. 828, 839) permitted timber cut on any national forest to be exported from the state or territory in which cut.

1913. Act of March 4 (37 Stat. 828, 847) declared all migratory game and insectivorous birds to be within the custody and protection of the government of the United States and forbade their destruction or capture contrary to regulations prescribed by the Secretary of Agriculture. This provision, commonly referred to as the Weeks-McLean Act, was declared unconstitutional and was superseded in 1918 by the Migratory Bird Treaty Act.

1913. Act of March 4 (37 Stat. 828, 855) authorized the National Forest Reservation Commission to acquire lands subject to rights of way, easements, and reservations which the Secretary of Agriculture believes will not interfere with the use of the lands so encumbered.

1913. Act of March 4 (37 Stat. 1015) authorized the Secretary of the Interior to sell any timber on public lands outside of national forests which had been killed or seriously damaged by fire prior to passage of the act.

1913. Act of September 30 (38 Stat. 113) authorized the President to prescribe the methods of opening to entry public lands thereafter excluded from national forests or released from withdrawals.

1913. Act of December 19 (38 Stat. 242) gave San Francisco the right to construct a reservoir in the Hetch Hetchy Valley in the Yosemite National Park to supply the city with water.

1914. Smith-Lever Act of May 8 (38 Stat. 372) provided for cooperative agricultural extension work between the U.S. Department of Agriculture and the land-grant colleges.

1914. Agricultural Appropriations Act of June 30 (38 Stat. 415, 441) increased from 5 to 25 per cent the payment to states of the gross receipts from lands acquired under the Weeks Act of 1911.

1914. Act of July 17 (38 Stat. 509) authorized the entry and patenting, under the nonmineral land laws, of lands withdrawn, classified, or valuable for phosphate, nitrate, potash, oil, gas, or asphaltic minerals, with reservation to the United States of the mineral deposits.

1914. Act of October 20 (38 Stat. 741) stopped the sale of coal lands in Alaska previously authorized by acts of June 6, 1900 (31 Stat. 658), April 28, 1904 (33 Stat. 525), and May 28, 1908 (35 Stat. 424); directed the reservation of certain lands; and provided for the leasing of unreserved coal lands.

1915. Branch of Research was established in the U.S. Forest Service, with E. H. Clapp in charge.

1915. Agricultural Appropriations Act of March 4 (38 Stat. 1086, 1101) authorized the Secretary of Agriculture to grant permits for summer homes, hotels, stores, or other structures needed for recreation or public convenience in national forests in tracts of not more than 5 acres and for periods of not more than thirty years.

1915. Supreme Court on February 23 in the case of *United States v. Midwest Oil Co.* (236 U.S. 459) affirmed the right of the President to withdraw public lands from entry without specific authorization from Congress.

1915. Supreme Court on June 21 (238 U.S. 393) reversed on technical grounds a 1913 decision of the Federal District Court for Oregon forfeiting to the government the Oregon and California Railroad grant lands, but enjoined their further disposal by the railroad pending action by Congress.

1916. Act of May 18 (39 Stat. 123, 137) provided for the establishment of the Red Lake Indian Forest of about 110,000 acres in the Red Lake Indian Reservation in Minnesota and for its administration by the Secretary of the Interior "in accordance with the principles of scientific forestry."

1916. Chamberlain-Ferris Act of June 9 (39 Stat. 218) revested in the United States title to the unsold lands in the grant to the Oregon and California Railroad Company (Southern Pacific) and provided for their classification and disposition.

1916. Act of July 11 (39 Stat. 355, 358) appropriated 1 million dollars a year for ten years for the construction of roads and trails within or partly within national forests when necessary for the use and development of their resources. Additional appropriations of 3 million dollars a year for the same purpose were made for the fiscal years 1919, 1920, and 1921.

1916. Agricultural Appropriations Act of August 11 (39 Stat. 446, 462) authorized the Secretary of Agriculture to require purchasers of national-forest stumpage to make deposits adequate to cover the cost of disposing of brush and other debris resulting from cutting operations. The proviso authorizing return to the purchaser of any deposit in excess of the amount actually required for the work was repealed by act of April 24, 1950 (64 Stat. 82). The 1916 act also authorized the Secretary, under general regulations prescribed by him, to permit the prospecting, development, and utilization of the mineral resources of lands acquired under the Weeks Act of 1911.

1916. Act of August 11 (39 Stat. 446, 476) authorized the President to establish refuges for the protection of game animals, birds, or fish on any lands purchased under the Weeks Act of 1911.

1916. Convention of August 16 between the United States and Great Britain (39 Stat. 1702) provided for the protection by the United States and Canada of migratory game birds, migratory insectivorous birds, and certain other migratory nongame birds.

1916. Act of August 25 (39 Stat. 535) created the National Park Service in the Department of Interior, defined the purposes for which national parks may be established, and authorized the Secretary of the Interior to make such rules and regulations as he may deem necessary for their proper use and management. Grazing was authorized when in the judgment of the Secretary it will not be detrimental to the primary purpose for which the park, monument, or other reservation was established. Funds for the operation of the National Park Service were first provided by the Deficiency Appropriations Act of April 17, 1917 (40 Stat. 2, 20).

1916. Stockraising Homestead Act of December 29 (39 Stat. 862) authorized the Secretary of the Interior to open for entry under the homestead laws not more than 640 acres per person of public lands the surface of which is chiefly valuable for grazing and raising forage crops, do not contain merchantable timber, are not susceptible of irrigation from any known source of water supply, and are of such a character that 640 acres are reasonably required for the support of a family. Instead of cultivation, the entryman must make permanent improvements to the extent of $1.25 per acre. Commutation was not allowed. All of the coal and other minerals were reserved to the United States and made subject to disposal under the coal and mineral-land laws. Lands containing water holes and other bodies of water needed or used by the public for watering purposes were not to be designated, but might be reserved under the act of 1910 and held open for public use. The Secretary of the Interior could also withdraw from entry lands needed to ensure access by the public to watering places and needed for use in the movement of stock to summer and winter ranges or to shipping points.

1917. United States Supreme Court on April 23 (243 U.S. 549) upheld the constitutionality of the act of June 9, 1916, revesting in the United States title to the lands granted to the Oregon and California Railroad.

1917. Act of October 2 (40 Stat. 297) authorized the Secretary of the Interior to lease public lands containing salts of potassium (potash).

1918. Act of May 31 (40 Stat. 593) authorized the Secretary of the Interior to exchange revested Oregon and California Railroad lands for lands of equal value in private ownership within or contiguous to the former limits of the grant.

1918. Migratory Bird Treaty Act of July 3 (40 Stat. 755) provided for effectuation of the convention of August 16, 1916, with Great Britain and authorized the Secretary of Agriculture, subject to the approval of the President, to promulgate regulations for the protection of the migratory birds covered by the convention.

1919. A movement was started by H. S. Graves, Chief of the Forest Service, and by F. E. Olmsted as president and Gifford Pinchot as chairman of a Committee for the Application of Forestry of the Society of American Foresters, to bring about public control of cutting on private forest lands.

1919. Act of February 26 (40 Stat. 1179) settled the suit against the Southern Oregon Company by providing for reconveyance to the United States of the remaining lands in the 1869 grant to aid in the construction of the Coos Bay Wagon Road, with payment by the government of delinquent taxes and of $232,463.07 to the company. The reconveyed lands were to be classified and disposed of as provided in the act of June 9, 1916, relating to revested O. and C. lands. After reimbursement of the government for these items, 25 per cent of the gross receipts from the reconveyed lands was to be paid to the counties for schools and permanent improvements.

1920. H. S. Graves resigned as Forester and A. F. Potter as Associate Forester. They were succeeded by W. B. Greeley and E. A. Sherman.

1920. Act of February 25 (41 Stat. 437) provided for the leasing of deposits of coal, phosphate, sodium, oil, oil shale, or gas, and authorized the Secretary of the Interior to reserve the right to sell, lease, or otherwise dispose of the surface of lands embraced in such leases if not necessary for the use of the lessee. Lessees pay both an annual rental and a royalty per unit of the mineral removed. Of the amount received, $52\frac{1}{2}$ per cent was allocated to the reclamation fund, $37\frac{1}{2}$ per cent to the states for the construction of roads or the support of education, and 10 per cent to the Treasury of the United States. The act applied to national forests created from the original public domain, but not to national forests acquired under the Weeks Act of 1911, to national parks, to game refuges, or to military or naval reservations.

1920. United States Supreme Court on April 19 in the case of *Missouri v. Holland* (252 U.S. 416) confirmed the constitutionality of the Migratory Bird Treaty Act of July 3, 1918.

1920. *Capper Report* on timber depletion, lumber prices, lumber exports, and concentration of timber ownership was transmitted to the Senate on June 1 in response to Senate Resolution 311.

1920. Act of June 4 (41 Stat. 758) authorized the Secretary of the Interior to sell timber on revested Oregon and California Railroad lands and on reconveyed Coos Bay Wagon Road lands classified as power sites, and to exchange reconveyed Coos Bay Wagon Road lands for lands in private ownership.

1920. Act of June 10 (41 Stat. 1063) created the Federal Power Commission consisting of the Secretary of War, Secretary of the Interior, and Secretary of Agriculture, with authority to issue licenses for a period not exceeding fifty years "for the development and improvement of navigation, and for the development, transmission, and utilization of power across, along, from or in any part of the navigable waters of the United States, or upon any part of the public lands and reservations of the United States (including the Territories), or for the purpose of utilizing the surplus water or water power from any Government dam."

1921. Act of March 3 (41 Stat. 1353) prohibited the issuance of permits, licenses, or leases for the development of water in existing national parks or national monuments without specific authority of Congress and repealed that part of the Federal Power Act of 1920 authorizing the issuance of such licenses by the Federal Power Commission. Act of August 26, 1935 (49 Stat. 838), by redefining "reservations" so as to exclude national parks and national monuments, made congressional approval necessary for parks and monuments created after as well as before 1921.

1921. President Harding's Proclamation of April 7 (42 Stat. 2238) set aside the week of May 22 to 28 as the first Forest Protection Week.

1921. Act of August 19 (42 Stat. 171) authorized the states of Arizona, California, Colorado, Nevada, New Mexico, Utah, and Wyoming to enter into a compact for the disposition and apportionment of the waters of the Colorado River.

1921. Federal Highway Act of November 9 (42 Stat. 212, 218) started the practice of appropriating funds specifically for the construction of "forest-development roads" and "forest highways." Cooperation with states was authorized but not required.

1921. Establishment of Eastern forest experiment stations began.

1922. General Exchange Act of March 20 (42 Stat. 465) authorized the Secretary of Agriculture (through the Secretary of the Interior) to exchange surveyed, nonmineral land or timber in national forests established from the public domain for privately owned or state land of equal value within national forests in the same state.

1922. Agricultural Appropriations Act of May 11 (42 Stat. 507, 520) made the first appropriation ($10,000) for the improvement of public camp grounds in national forests, with special reference to protection of the public health and prevention of forest fires.

1922. Act of September 20 (42 Stat. 857) authorized the Secretary of the Interior to protect timber on lands under his jurisdiction from fire, disease, and insects, either directly or in cooperation with other Departments, states, or private owners.

1923. Act of February 16 (42 Stat. 1258) restored the last of the naval-timber reserves (about 3,000 acres in Louisiana) to the public domain.

1923. Convention of March 2 between the United States and Great Britain (43 Stat. 1841) prohibited fishing for halibut in the North Pacific Ocean between November 16 and February 15 and established an International Fisheries Commission consisting of two representatives of each government to conduct studies and make recommendations for the preservation and development of the halibut fishery.

1923. Act of March 4 (42 Stat. 1445) extended the provisions of the Enlarged Homestead Act of 1909 to homestead entries in national forests under certain conditions.

1924. Rachford report, prepared by the Forest Service as result of a four-year study of the grazing situation, recommended substantial increases in national-forest grazing fees.

1924. Department of the Interior Appropriations Act of June 5 (43 Stat.

390, 415) changed the name of the Reclamation Service to Bureau of Reclamation effective July 1.

1924. Act of June 6 (43 Stat. 464) authorized the Secretary of the Interior to reserve fishing areas in any of the waters of Alaska over which the United States has jurisdiction, and within such areas to establish closed seasons during which fishing can be limited or prohibited as he may prescribe.

1924. Northern Pacific Halibut Act of June 7 (43 Stat. 648) provided for effectuation of the convention of March 2, 1923, with Great Britain.

1924. Act of June 7 (43 Stat. 650) provided for the establishment of the Upper Mississippi River Wildlife and Fish Refuge and authorized an appropriation of $1,500,000 for the acquisition of land within the refuge.

1924. Clarke-McNary Act of June 7 (43 Stat. 653) authorized appropriations to enable the Secretary of Agriculture to cooperate in forest-fire control with states meeting prescribed standards, in the growing and distribution of planting stock to farmers, and in promoting the efficient management of farm wood lots and shelterbelts; authorized the purchase of lands anywhere on the watersheds of navigable streams and for timber production as well as stream-flow protection; authorized acceptance of gifts to be added to the national forests; authorized the Secretary of Agriculture to report to Congress such unreserved public timberlands as in his judgment should be added to the national forests; and authorized the creation of military and naval reserves as national forests, without interference with their use for military and naval purposes.

1924. A National Conference on Utilization of Forest Products, called by the Secretary of Agriculture, was held at Washington, November 19 to 20.

1925. A system of ten-year permits for grazing on Western national forests was put into effect by the Forest Service on January 1.

1925. Alaska Game Law of January 13 (43 Stat. 739) created the Alaska Game Commission of five members and authorized the Secretary of Agriculture, upon consultation with or recommendation from the Commission, to adopt regulations governing the taking of game animals, land fur-bearing animals, game birds, and nongame birds, including the issuance of hunting and trapping licenses.

1925. Act of February 28 (43 Stat. 1090) amended the General Exchange Act of 1922 to permit either party to an exchange to make reservations of timber, minerals, or easements, the values of which shall be considered in determining the values of the exchanged lands, provided that such reservations shall be subject to the tax laws of the states concerned.

1925. Act of March 3 (43 Stat. 1127) included watersheds from which water is secured for domestic use or irrigation among the lands on which the Federal government can cooperate with states in control of forest fires under Section 2 of the Clarke-McNary Act of 1924.

1925. Act of March 3 (43 Stat. 1132) authorized the acceptance of contributions to constitute a special fund for the reforestation, administration, or protection of lands within or near national forests. It also increased the appraised value of national-forest timber that can be sold without advertisement from $100 to $500.

1925. Act of March 3 (43 Stat. 1215) authorized the exchange of land or timber for land within the exterior boundaries of national forests acquired under the Weeks Act of 1911 or the Clarke-McNary Act of 1924, on an equal-value basis.

1926. Act of April 12 (44 Stat. 242) authorized the export of timber lawfully cut on any national forest, or on the public lands in Alaska, from the state or territory in which cut, if the supply of timber for local use will not be endangered thereby.

1926. Joint Resolution of April 13 (44 Stat. 250) authorized the Secretary of Agriculture to cooperate with the territories and other possessions in studies of forest taxation and forest insurance and in reforestation and wood-lot management, under the Clarke-McNary Act of 1924, on the same terms as with the states.

1926. Act of April 17 (44 Stat. 301) authorized the Secretary of the Interior to lease public lands containing sulfur in Louisiana under the general provisions of the Mineral Leasing Act of 1920. The authorization was extended to lands in New Mexico by act of July 16, 1932 (47 Stat. 701) and to lands wherever situated (including acquired lands) by act of August 7, 1947 (61 Stat. 913).

1926. First World Forestry Congress was held in Rome, Italy, April 29 to May 5.

1926. Act of May 20 (44 Stat. 576) prohibited the interstate transportation of black bass taken or possessed in violation of the laws of the state from which or to which they were shipped. The act was amended on July 2, 1930 (46 Stat. 845), July 30, 1947 (61 Stat. 517), and July 16, 1952 (66 Stat. 736).

1926. Act of May 25 (44 Stat. 636, 647) increased the period for repayment of construction charges in irrigation projects to forty years (from ten years in 1902 and twenty years in 1914); provided that repayment contracts on new projects should be made only with water-users' organizations; and limited to 160 irrigable acres the area in single ownership that might receive water in an irrigation project.

1926. Act of June 14 (44 Stat. 741) authorized the Secretary of the Interior to make available to states, counties, or municipalities, by exchange, sale, or lease, unreserved, nonmineral public lands classified by him as chiefly valuable for recreational purposes.

1926. Act of June 15 (44 Stat. 745) forbade further creation of or additions to national forests in Arizona and New Mexico except by act of Congress.

1926. Act of July 3 (44 Stat. 890) amended the act of March 4, 1913, to permit the sale of fire-killed timber on the public lands irrespective of the date of its destruction.

1926. Act of July 13 (44 Stat. 915) provided for paying to the counties containing revested Oregon and California Railroad lands the equivalent of the taxes that would have been paid from 1915 to 1926, inclusive, if the lands had remained privately owned and taxable. Taxes were to be computed on the basis of 1915 assessed values and the rate of taxation prevailing in each of the counties involved. After 1926, payments were to be continued on the same basis until all charges against the "Oregon and California land-grant fund" (including back and current payments in lieu of taxes) had been liquidated and the fund showed a credit balance available for distribution under the act of June 9, 1916.

1927. Act of January 25 (44 Stat. 1026) permitted the states to acquire title to school sections that are mineral in character, subject to certain restrictions and reservations.

1927. Act of February 7 (44 Stat. 1057) authorized the Secretary of the Interior to lease public lands containing salts of potassium (potash) under the general provisions of the Mineral Leasing Act of 1920 and repealed the prior act of October 2, 1917.

1927. Act of March 4 (44 Stat. 1452) authorized the Secretary of the Interior to establish grazing districts on unreserved public lands in Alaska and to promulgate rules and regulations for their administration. Net receipts from leases, which may run up to twenty years, are paid to the territory for public education and roads.

1927. A Conference on Commercial Forestry was held at Chicago, Ill., November 16 to 17, under the sponsorship of the Chamber of Commerce of the United States.

1928. Act of March 9 (45 Stat. 253) increased to 320 acres the size of isolated tracts that might be sold at public auction at not less than $1.25 per acre, but left at 160 acres the size of mountainous tracts that might be sold on application of adjoining owners.

1928. Act of April 13 (45 Stat. 429) extended the provisions of the act of June 14, 1926, with certain restrictions, to revested Oregon and California Railroad lands and reconveyed Coos Bay Wagon Road lands.

1928. Act of April 23 (45 Stat. 448) authorized the establishment of the Bear River Migratory Bird Refuge in Utah.

1928. McNary-Woodruff Act of April 30 (45 Stat. 468) authorized appropriation of 2 million dollars in 1928–1929, of 3 million dollars in 1929–1930, and of 3 million dollars in 1930–1931 for the purchase of land under the Weeks Act of 1911 and the Clarke-McNary Act of 1924. Not more than 1 million acres of land was to be purchased in any one state primarily for timber production.

1928. W. B. Greeley resigned as Forester on May 1 and was succeeded by R. Y. Stuart.

1928. Act of May 17 (45 Stat. 597) amended the act of June 9, 1916, to require the cutting and removal of any timber sold on revested Oregon and California Railroad lands under such rules and regulations as might be prescribed by the Secretary of the Interior.

1928. McSweeney-McNary Act of May 22 (45 Stat. 699) authorized a comprehensive ten-year program of research in all phases of forestry and range management, including a timber survey, with an annual appropriation amounting to $3,625,000 by the end of the period, and thereafter such amounts as needed to carry out the provisions of the act.

1928. Society of American Foresters appointed a committee on forest policy to review the situation and make recommendations to the society.

1928. Boulder Canyon Project Act of December 21 (45 Stat. 1057) provided for the construction of works for the protection and development of the Colorado River Basin and authorized the states concerned to enter into supplemental compacts for the development of the Colorado River.

1929. Migratory Bird Conservation Act (Norbeck-Andresen Act) of February 18 (45 Stat. 1222) established the Migratory Bird Conservation Commission and authorized a continuing program for the acquisition of migratory-bird reservations, subject to the consent of the state concerned.

1930. Act of April 10 (46 Stat. 153) authorized an appropriation of $50,-000 to enable the President to appoint a commission to study and report on the conservation and administration of the public domain.

1930. On April 15 the Forestry Branch of the Bureau of Indian Affairs was given responsibility for the handling of grazing on range lands in Indian reservations. A grazing policy was put into effect on July 1, 1931.

1930. Convention of May 9 between the United States and Canada for the preservation of the halibut fishery of the North Pacific Ocean (47 Stat. 1872) supplanted the convention of March 2, 1923, between the United States and Great Britain. It changed the closed season on halibut to run from November 1 to February 15 and authorized the International Fisheries Commission to issue regulations for the protection and conservation of the halibut fishery.

1930. Convention of May 26 between the United States and Canada for the protection and promotion of the sockeye salmon fishery of the Fraser River system (50 Stat. 1355) established an International Pacific Salmon Fisheries Commission to conduct investigations and to promulgate regulations governing the taking of sockeye salmon in certain specified waters.

1930. Knutson-Vandenberg Act of June 9 (46 Stat. 527) authorized appropriation of not to exceed $400,000 a year by the fiscal year 1934 for reforestation activities on the national forests and provided that additional charges could be made in timber sales to provide a special fund for reforestation or silvicultural improvement of the cutover area included in the timber sale.

1930. Act of June 23 (46 Stat. 797) provided that the Federal Power Commission should thereafter consist of five salaried commissioners appointed by the President.

1930. Shipstead-Nolan Act of July 10 (46 Stat. 1020) withdrew from entry all public land in parts of northern Cook, Lake, and St. Louis counties, Minnesota; required the Forest Service to conserve for recreational use the natural beauty of all lakes and streams within this region (chiefly in the Superior National Forest); and provided that there should be no further alteration of the natural water level of any lake or stream within the region without further act of Congress.

1930. President Hoover on December 6 appointed a Timber Conservation Board to study the economic problem of overproduction in the forest industries. The Board made a thorough study of the situation, with special reference to overproduction in the lumber industry, but issued no comprehensive report.

1930. A nationwide forest survey was initiated under authority of the McSweeney-McNary Act of 1928.

1931. Society of American Foresters Committee on Forest Policy submitted a comprehensive report, including endorsement of the principle of public control of cutting on private lands. The report was adopted by the society by a large majority in a referendum vote. The majority favored state rather than Federal control.

1931. On January 16 the Committee on the Conservation and Administration of the Public Domain (appointed by President Hoover in 1930) submitted a report recommending that all portions of the unreserved and unappropriated public domain be placed under responsible administration for the conservation of its natural resources; that areas which are chiefly valuable for the production of forage and which can be effectively conserved and administered by the states containing them be granted to the states which will accept them; and that the President be authorized to consolidate the executive agencies dealing with the administration and disposition of the public domain, the administration of national reservations, and the conservation of natural resources.

1931. Joint Resolution of February 20 (46 Stat. 1200) authorized the Secretary of Agriculture to cooperate with the territories in forest-fire protection under the Clarke-McNary Act of 1924 on the same terms as with the states.

1931. Joint Resolution of March 3 (46 Stat. 1516) extended to Puerto Rico the provisions of the Clarke-McNary Act of 1924 relating to cooperative forest-fire protection and to the acquisition of forest lands by purchase and by gift, with limitation to 50,000 acres of the area that might be purchased.

1931. Multilateral international convention of September 24 prohibited the taking or killing of right whales; required full utilization of the carcasses of baleens or whalebone whales; and provided for the communication of statistical information regarding all whaling operations to the International Bureau for Whaling Statistics at Oslo, Norway.

1931. A National Conference on Land Utilization was held at Chicago, Ill., November 19 to 21, at the call of the Secretary of Agriculture and the Association of Land-Grant Colleges and Universities. Forestry received much consideration.

1932. Northern Pacific Halibut Act of May 2 (47 Stat. 142) provided for effectuation of the convention of May 9, 1930, with Canada.

1933. A *National Plan for American Forestry*, known as the *Copeland Report*, was submitted to the Senate on March 27 by the Secretary of Agriculture. It made two main recommendations: a large extension of public ownership and more intensive management of all publicly owned lands.

1933. Act of March 31 (48 Stat. 22) appropriated funds for the dual purpose of relieving unemployment and promoting conservation of natural resources. In addition to other activities it authorized use of the funds for forest research and for acquisition of land by purchase, donation, condemnation, or otherwise. Executive order of April 5 established the Office of Emergency Conservation Work as an independent agency, which was popularly known as the Civilian Conservation Corps.

1933. Agricultural Adjustment Act of May 12 (48 Stat. 31) provided in detail for relieving the acute economic emergency in agriculture.

1933. Federal Emergency Relief Act of May 12 (48 Stat. 55), and subsequent amendments, provided funds for the relief of unemployment which were used in part for forestry and other conservation activities.

1933. Act of May 18 (48 Stat. 58) created the Tennessee Valley Authority, which includes many phases of conservation in its activities.

1933. President Roosevelt by Executive order, under authority of the Reorganization Act of March 3, 1933, placed all national monuments, the National Capital parks, and national military parks under the administration of the Interior Department.

1933. National Industrial Recovery Act of June 16 (48 Stat. 195) attempted to promote economic recovery by a wide variety of measures, including codes of fair competition, an extensive public-works program, and subsistence homesteads. The Code of Fair Competition for the Lumber and Timber Products Industries, approved August 21, led to the adoption (March 23, 1934) of a Forest Conservation Code which required the various divisions of the industry to formulate and enforce rules of forest practice.

1933. On July 20, the Administrator of Public Works appointed the National Planning Board.

1933. Soil Erosion Service was established in the Department of the Interior on August 25 under authority of National Industrial Recovery Act.

1933. R. Y. Stuart, Chief of the Forest Service, died on October 23. He was succeeded by F. A. Silcox on November 15.

1934. Shelterbelt Project (Prairie States Forestry Project) was started with emergency funds administered by the Forest Service.

1934. Act of March 10 (48 Stat. 400) authorized the President, upon recommendation of the Secretary of Agriculture and the Secretary of Commerce and with the approval of Legislatures of the states concerned, to establish fish and game sanctuaries or refuges in national forests.

1934. Coordination Act of March 10 (48 Stat. 401) authorized the Secretary of Agriculture and the Secretary of Commerce to cooperate with Federal, state, and other agencies in developing a nationwide program of wildlife conservation and rehabilitation; to study the effect of water pollution on wildlife and to recommend remedial measures; and to prepare plans for the maintenance of an adequate supply of wildlife on public lands, Indian reservations, and unallotted Indian lands. It also provided for use for wildlife purposes of water impounded by the Bureau of Reclamation or otherwise and for facilitating the migration of fish in connection with the construction of any future dam by the Federal government or under Federal permit.

1934. Migratory Bird Hunting Stamp Act of March 16 (48 Stat. 451) required takers of migratory waterfowl to buy a $1 Federal hunting stamp, good for one year, and made the proceeds available for the acquisition and management of migratory waterfowl refuges and for the conduct of research.

1934. Act of June 14 (48 Stat. 955) provided under certain conditions for the purchase of lands under the Weeks Act in states not already having given their consent to such purchases.

1934. Act of June 18 (48 Stat. 984, 986) directed the Secretary of the Interior to make rules and regulations for managing Indian forestry units on the principle of sustained yield; for restricting the number of livestock grazed on Indian range units to their estimated carrying capacity; and for protecting the range from deterioration, preventing soil erosion, and assuring full utilization of the range.

1934. Taylor Grazing Act of June 28 (48 Stat. 1269) authorized the Secretary of the Interior to establish not more than 80 million acres of grazing districts in the unreserved public domain (exclusive of Alaska) and to make rules and regulations for their occupancy and use. The act contained specific provisions with respect to mineral resources, hunting and fishing, homestead entry, and the lease of isolated tracts to owners of contiguous lands. Receipts were allocated as follows: 25 per cent, when appropriated by Congress, for the construction, purchase, or maintenance of range improvements; 50 per cent to the counties in which the districts are located; and 25 per cent to the United States Treasury. The act also increased to 760 acres the size of isolated tracts that could be offered for sale and authorized the Secretary of the Interior to sell not more than 160 acres of land that is mountainous or too rough for cultivation, whether isolated or not, to adjoining owners.

1934. Executive order of June 30 created the Quetico-Superior Committee to advise with Federal and other agencies concerning the wilderness sanctuary in the Rainy Lake and Pigeon River watersheds in Minnesota.

1934. Executive order of June 30 replaced the National Planning Board and the Committee on National Land Problems by the National Resources Board.

1935. Connally Hot Oil Act of February 22 (49 Stat. 30) prohibited the interstate transportation of petroleum produced, transported, or withdrawn from storage in excess of the amounts permitted in the state of origin. The act, which was to have expired in 1942, was made permanent by act of June 22, 1942 (56 Stat. 381).

1935. Soil Conservation Act of April 27 (49 Stat. 163) declared it to be the policy of Congress to provide permanently for the control and prevention of soil erosion, delegated all activities relating to soil erosion to the Secretary of Agriculture, and established the Soil Conservation Service in the Department of Agriculture. The latter succeeded the Soil Erosion Service which had been set up in the Department of the Interior in 1933 under the National Industrial Recovery Act and transferred to the Department of Agriculture by Executive order in March, 1935.

1935. Resettlement Administration was established by Executive order of April 30. In 1936 it was transferred to the Department of Agriculture, and in 1937 it was changed to the Farm Security Administration.

1935. Supreme Court on May 27 (295 U.S. 495) invalidated the National Industrial Recovery Act because it involved an unconstitutional delegation of legislative power, exceeded the power of Congress to regulate interstate commerce, and invaded the powers reserved exclusively to the states. The Court's action automatically nullified the lumber-industry code and the rules of forest practice adopted thereunder.

1935. Executive order of June 7 replaced the National Resources Board by the National Resources Committee.

1935. Act of June 15 (49 Stat. 378, 382) authorized the addition to wildlife refuges of land acquired by exchange of (1) land, timber, or other materials in wildlife refuges or (2) of unreserved nonmineral public lands, in both cases on an equal-value basis. It also authorized payment to the counties, for the benefit of schools and roads, of 25 per cent of the gross receipts from wild-life refuges.

1935. Bankhead-Jones Act of June 29 (49 Stat. 436) authorized an annual appropriation increasing from 1 to 5 million dollars for conduct by the Secretary of Agriculture and by the agricultural experiment stations of research into laws and principles underlying basic problems of agriculture in its broadest aspects. The act also authorized appropriations for the further development of co-operative agricultural extension work and the more complete endowment and support of land-grant colleges.

1935. A Forest Service report entitled *National Pulp and Paper Require-ments in Relation to Forest Conservation (Hale Report)* was transmitted to the Senate on July 12 in response to Senate Resolution 205.

1935. Act of August 21 (49 Stat. 666) authorized the Secretary of the In-terior to acquire and administer historic sites and buildings and established an Advisory Board on National Parks, Historic Sites, Buildings, and Monuments.

1935. Act of August 26 (49 Stat. 866) authorized the appropriation of re-ceipts from the Uinta and Wasatch National Forests in Utah for the purchase of lands therein. Later acts contained similar provisions with respect to the Cache National Forest in Utah (52 Stat. 347), the San Bernardino and Cleve-land National Forests in California (52 Stat. 699), the Nevada and Toiyabe National Forests in Nevada (52 Stat. 1205), the Ozark and Ouachita National Forests in Arkansas (54 Stat. 46), the Angeles National Forest in California (54 Stat. 299), and the Sequoia National Forest in California (54 Stat. 402).

1935. Fulmer Act of August 29 (49 Stat. 963) authorized an appropriation of 5 million dollars for the purchase by the Federal government of lands to be administered as state forests under plans of management satisfactory to the Secretary of Agriculture. Congress has never appropriated funds to put the act into operation.

1936. United States Supreme Court on January 6 (297 U.S. 1) declared unconstitutional the agricultural-adjustments parts of the Agricultural Adjust-ment Act of 1935, dealing chiefly with acreage allotments, benefit payments, and processing taxes, on the ground that they invaded powers reserved to the states.

1936. Convention of February 7 between the United States and Mexico (50 Stat. 1311) provided for the protection by the United States and Mexico of migratory game and nongame birds and for the control of transportation be-tween the two countries of migratory birds and game mammals, dead or alive.

1936. Soil Conservation and Domestic Allotment Act of February 29 (49 Stat. 1148) attempted to attain the objectives of the Agricultural Adjustment Act of 1933 by authorizing the Secretary of Agriculture to make benefit pay-ments to farmers as a soil-conservation measure. "Parity payments" were also authorized in order to make the purchasing power of the farmer comparable to that existing in 1909 to 1914.

1936. A comprehensive report by the Forest Service entitled *The Western Range* was transmitted to the Senate on April 28 in response to Senate Resolu-tion 289.

1936. Whaling Treaty Act of May 1 (49 Stat. 1246) provided for effectuation of the multilateral convention of 1931 for the regulation of whaling, including authorization of the Secretary of the Treasury and the Secretary of Commerce to make the necessary regulations for the control of whaling.

1936. Act of June 15 (49 Stat. 1515) amended the McSweeney-McNary Act of 1928 to authorize the establishment of the Great Plains Forest Experiment Station.

1936. On June 16 the Division of Forestry and Grazing in the Bureau of Indian Affairs was given charge of all matters relating to wildlife management on Indian reservations.

1936. Act of June 20 (49 Stat. 1555) provided for effectuation of the convention of February 7, 1936, with Mexico and authorized the Secretary of Agriculture, subject to approval by the President, to promulgate regulations to that end.

1936. Flood Control Act of June 22 (49 Stat. 1570) recognized the fact that flood control on navigable waters or their tributaries is a proper activity of the Federal government, in cooperation with the states and their political subdivisions. It provided that thereafter Federal investigations and improvements of rivers and other waterways for flood control and allied purposes should be under the jurisdiction of the War Department, and Federal investigations of watersheds and measures for runoff and water-flow retardation and soil-erosion prevention on watersheds under the jurisdiction of the Department of Agriculture. The act authorized interstate flood-control compacts and authorized a long list of projects for prosecution by the Army Engineers.

Amendments in 1937 (50 Stat. 876) and 1938 (52 Stat. 1215) authorized additional surveys and examinations at specific localities and directed the Secretary of Agriculture to make runoff and erosion surveys on all watersheds specified for flood-control surveys by the Secretary of War. The 1937 act also authorized the Secretary of Agriculture to impose such conditions as he might deem necessary in prosecuting measures for retarding runoff and preventing erosion on non-Federal lands.

1936. Act of June 23 (49 Stat. 1894) authorized and directed the National Park Service to make a comprehensive study, other than on lands under the jurisdiction of the Department of Agriculture, of the public park, parkway, and recreational-area programs of the United States and of the several states and political subdivisions thereof, and of the lands chiefly valuable as such areas, and to cooperate with the states and their political subdivisions in planning such areas. It also authorized the states to enter into interstate compacts for the establishment and development of park, parkway, and recreational areas, subject to the approval of the state legislatures and of Congress.

1936. Act of June 26 (49 Stat. 1976) increased the maximum allowable area of grazing districts to 142 million acres. Exchange of lands with states was authorized on either an equal-value or equal-area basis. The President was authorized, with the advice and consent of the Senate, to select a Director of Grazing; and the Secretary of the Interior was authorized to appoint such assistant directors and other employees as necessary to administer the act, provided that every appointee must have been for one year a bona fide citizen or resident of the state in which he is to serve.

1936. Second World Forestry Congress was held in Budapest, Hungary, September 10 to 14.

1936. An "upstream engineering" conference called by President Roosevelt was held in Washington, September 22 to 23, to emphasize the importance of this phase of flood and erosion control.

1936. Presidential Proclamation of November 27 (50 Stat. 1797) abolished the Wichita National Forest in Oklahoma and placed the Wichita Mountains Wildlife Refuge under the administration of the Bureau of Biological Survey.

1937. Convention of January 29 between the United States and Canada (50 Stat. 1351) strengthened the authority of the International Fisheries Commission to issue regulations for the protection and conservation of the halibut fishery in the North Pacific Ocean.

1937. Act of February 11 (50 Stat. 19) created a Disaster Loan Corporation to make such loans as it may determine to be necessary because of floods or other catastrophes. Funds were provided by the Corporation in connection with the Northeastern timber-salvage work of the Forest Service. It was dissolved and its functions transferred to the Reconstruction Finance Corporation on June 30, 1945 (59 Stat. 310).

1937. Cooperative Farm Forestry Act (Norris-Doxey Act) of May 18 (50 Stat. 188) authorized an annual appropriation of $2,500,000 for the promotion of farm forestry in cooperation with the states. The first appropriation ($300,-000) was for the fiscal year 1940.

1937. Act of June 28 (50 Stat. 319) established the Civilian Conservation Corps as the official successor to the Emergency Conservation Work; provided in detail for its administration; authorized the use of ten hours a week for educational and vocational training on a voluntary basis; and extended its life to June 30, 1940. In 1939 (53 Stat. 1253), the C.C.C. was continued through June 30, 1943. Subsequent acts extended its life through June 30, 1944.

1937. Northern Pacific Halibut Act of June 28 (50 Stat. 325) provided for effectuation of the convention of January 29, 1937, between the United States and Canada.

1937. Bankhead-Jones Farm Tenant Act of July 22 (50 Stat. 522) provided for loans to farm tenants, for rehabilitation loans, and for the retirement and rehabilitation of submarginal agricultural lands. Acquired lands could be sold, exchanged, leased, or otherwise disposed of, under specified conditions, one of which was the reservation of an undivided three-fourths interest in all coal, oil, gas, and other minerals. The Secretary was also authorized to cooperate with Federal, state, and other public agencies in developing plans for a program of land conservation and land utilization.

1937. Water Facilities Act of August 28 (50 Stat. 869) provided for the development by the Secretary of Agriculture of facilities for water storage and utilization in the arid and semiarid regions.

1937. Act of August 28 (50 Stat. 874) provided for reclassification of the lands of the revested Oregon and California Railroad and the reconveyed Coos Bay Wagon Road grants and for sustained-yield management by the Secretary of Interior of those classified as timberlands. The Secretary was authorized to establish sustained-yield forest units for the support of dependent communities and local industries and to make cooperative agreements with other Federal agencies, with state agencies, and with private forest owners to secure coordinated administration. Lands chiefly valuable for agriculture could be opened to homestead entry or sale under the terms of the Taylor Grazing Act of 1934. Receipts from O. and C. lands were to be distributed as follows: 50 per cent to the counties immediately concerned, an additional 25 per cent to the counties permanently after satisfying reimbursable Federal charges against the lands, and 25 per cent for administration in such amounts as appropriated by Congress.

1937. Wildlife Restoration Act (Pittman-Robertson Act) of September 2 (50 Stat. 917) authorized the setting apart of the tax on firearms, shells, and cartridges in the "Federal aid to wildlife-restoration fund" to be used for cooperation with the states in approved wildlife-restoration projects up to 75 per cent of the total cost of the projects. Each cooperating state must pass legislation for the conservation of wildlife, including a prohibition against the diversion of license fees paid by hunters for any other purpose than the administration of its fish and game department.

1938. Agricultural Adjustment Act of February 16 (52 Stat. 31) provided in great detail for benefit and parity payments to farmers, stressed the idea of the "ever-normal granary," and inaugurated Federal crop insurance for wheat through the creation of the Federal Crop Insurance Corporation under control of the Secretary of Agriculture. The act also provided for the establishment by the Department of Agriculture of four regional laboratories for the conduct of investigations relating to the industrial use of farm products. These were located at Philadelphia, Pa., Peoria, Ill., New Orleans, La., and Albany, Calif.

1938. Act of May 11 (52 Stat. 345) authorized the Secretary of Commerce to establish salmon-cultural stations in the Columbia River Basin; to conduct investigations; and to install devices for improving feeding and spawning conditions in order to protect migratory fish from irrigation projects.

1938. Act of June 1 (52 Stat. 609) authorized the Secretary of the Interior to sell or lease not more than 5 acres of certain public lands, outside of Alaska, which he may classify as chiefly valuable as home, cabin, health, convalescent, recreational, or business sites, subject to a reservation to the United States of all oil, gas, and other mineral deposits. Regulations under the act provide for leases of not more than five years.

1938. Concurrent Resolution of June 14 (52 Stat. 1452) created a Joint Congressional Committee on Forestry to study the present and prospective situation with respect to the forest land of the United States and to make a report and recommendations by April 1, 1939. The time limit was later extended to April 1, 1941. The report was presented March 24, 1941.

1938. Act of June 23 (52 Stat. 1033) authorized the Secretary of the Interior to lease at rates determined by him any state, county, or private land chiefly valuable for grazing within the exterior boundaries of a grazing district. Such leases are to run for not more than ten years, and the fees paid for grazing privileges on the leased lands shall not be less than the rental paid by the

United States for them. All moneys received in the administration of leased lands are made available, when appropriated by Congress, for the leasing of lands under this act, and shall not be distributed to the states as are other receipts.

1938. Act of June 29 (52 Stat. 1241) established the Olympic National Park and authorized the President, after eight months, to add national-forest or other lands to the park, provided the total area of the park shall not exceed 898,292 acres.

1939. Reorganization Plan No. I of April 25 (53 Stat. 1423), approved by Congress on June 7 to take effect July 1 (53 Stat. 813), established the National Resources Planning Board in the Executive Office of the President by transfer and consolidation of the National Resources Committee and the Federal Employment Stabilization Office in the Department of Commerce.

1939. Reorganization Plan No. II of May 9 (53 Stat. 1431, 1433) transferred the Bureau of Fisheries from the Department of Commerce, and the Bureau of Biological Survey from the Department of Agriculture, to the Department of the Interior, and made the Secretary of the Interior chairman of the Migratory Bird Conservation Commission.

1939. Department of the Interior Appropriations Act of May 10 (53 Stat. 685, 692) made the first appropriation ($37,500) for the prevention and suppression of fires on the public domain in Alaska.

1939. Act of May 24 (53 Stat. 753) provided for payments to Coos County and Douglas County, Oregon, in lieu of taxes on reconveyed Coos Bay Wagon Road lands on the basis of the appraised value of the lands and the current rate of taxation, provided that payments during any ten-year period shall not exceed 75 per cent of the receipts from the reconveyed lands. Not more than 25 per cent of the receipts was made available, when appropriated by Congress, for administration of the lands.

1939. Act of July 14 (53 Stat. 1002) provided for the establishment within each grazing district of an advisory board of five to twelve local stockmen elected by the users of the range but appointed by the Secretary of the Interior, who may also on his own initiative appoint one wildlife member on each board. Each advisory board shall offer advice on applications for grazing permits and on all other matters affecting the administration of the Taylor Grazing Act. Except in an emergency the Secretary of the Interior shall request the advice of the advisory board prior to the promulgation of any rules and regulations affecting the district.

1939. Act of July 20 (53 Stat. 1071) restored to the President authority to establish national forests in Montana.

1939. Act of July 31 (53 Stat. 1144) authorized the Secretary of the Interior to exchange revested Oregon and California Railroad lands and reconveyed Coos Bay Wagon Road lands for lands of equal value in private, state, or county ownership within or contiguous to the former limits of the grants; and repealed the previous acts relating to such exchanges.

1939. Reclamation Project Act of August 4 (53 Stat. 1187) effected various reforms in existing reclamation legislation; authorized the sale of electric power or lease of power privileges for periods of not more than forty years, at such rates as would cover an appropriate share of the cost of operation and maintenance as well as the construction investment, with preference to municipalities and other public corporations or agencies; and specified the basis of payment for the various kinds of benefits provided by multipurpose reclamation projects.

1939. Act of August 11 (53 Stat. 1418) authorized the Secretary of the Interior to undertake construction of water conservation and utilization projects in the Great Plains and arid and semiarid areas of the United States in cooperation with the Department of Agriculture.

1939. F. A. Silcox, Chief of the Forest Service, died on December 20. E. H. Clapp served as Acting Forester until 1943.

1939. Office of Director of Forestry was established in the Department of the Interior to coordinate the forestry work of the Department.

1940. Office of Land Utilization was established in the Office of the Secretary of the Interior to coordinate the conservation activities of the Department.

1940. *Forest Outings*, a comprehensive report on forest recreation, was published by the Forest Service.

1940. Reorganization Plan No. III of April 2 (54 stat. 1231, 1232) consolidated the Bureau of Fisheries and the Bureau of Biological Survey into the Fish and Wildlife Service.

1940. Lea Act of April 26 (54 Stat. 168) provided for Federal cooperation in the protection of forest lands from white pine blister rust, irrespective of ownership, provided that on state or private lands Federal expenditures must be at least matched by state or local authorities or by individuals or organizations.

1940. Act of May 28 (54 Stat. 224) authorized the President, on the basis of a cooperative agreement between the Secretary of Agriculture and the municipality concerned, to withdraw national-forest lands from which a municipality obtains its water supply from all forms of location, entry, or appropriation. The Secretary of Agriculture may prescribe such rules and regulations as he considers necessary for adequate protection of the watershed.

1940. Transportation Act of September 18 (54 Stat. 898, 954) authorized the payment of full commercial tariff rates for the transportation of persons or property for the United States (except military or naval property or members of military or naval forces traveling on official duty) to land-grant railroads which within one year would waive all further claims under their grants. Lands already patented, certified for patent, or sold to innocent purchasers were not affected.

1940. Convention of October 12 between the United States and other American republics (56 Stat. 1354) committed the signatory powers to take appropriate steps for the protection of nature and the preservation of wildlife in their respective countries.

1940. Supreme Court on December 16 in the case of *United States v. Appalachian Electric Power Company* (311 U.S. 377) held that a waterway constitutes "navigable water of the United States" if it can be made available for navigation by the construction of improvements, whether such improvements have actually been made or even authorized; and that a navigable water of the United States does not lose its character because its use for interstate commerce has lessened or ceased. The Court also stated that navigation is only a part of interstate commerce and that "flood protection, watershed development, recovery of the cost of improvements through utilization of power are likewise parts of commerce control."

1941. *Forest Lands of the United States,* report of the Joint Committee on Forestry established in 1938, was submitted to Congress on March 24.

1941. Supreme Court on June 2 in the case of *Oklahoma v. Atkinson Company* (313 U.S. 508) stated that "it is clear that Congress may exercise its control over the non-navigable stretches of a river in order to preserve or promote commerce on the navigable portions" and added that "the power of flood control extends to the tributaries of navigable streams."

1941. Act of November 15 (55 Stat. 763) extended to all lands owned by, leased by, or under the jurisdiction of the United States, including Indian lands and lands in process of acquisition, the penalties (somewhat modified) for setting and for failing to extinguish fires in or near any timber, underbrush, grass, or other inflammable material.

1942. Executive order of April 12 delegated to the Secretary of the Interior authority to make withdrawals and restorations of public lands.

1942. Act of May 4 (56 Stat. 267) approved the Atlantic States Marine Fisheries Compact.

1942. Provisional Fur Seal Agreement with Canada (58 Stat. 1379), effective June 1, 1942, virtually continued the Fur Seal Convention of 1911 so far as the United States and Canada are concerned, except that thereafter Canada was to receive 20 per cent instead of 15 per cent of the take at the Pribilof Islands.

1942–1943. Acts of July 2, 1942 (56 Stat. 562, 569) and July 12, 1943 (57 Stat. 494, 499) provided for liquidation of the C.C.C. as quickly as possible but not later than June 30, 1944. Liquidation of most of the personnel was accomplished by August 15, 1942.

1943. Forest Products Research Society was organized on January 3.

1943. L. F. Watts was appointed Chief of the Forest Service on January 8.

1943. Act of July 26 (57 Stat. 169, 170) abolished the National Resources Planning Board effective August 31, except for such administrative action as required to wind up its affairs by January 1, 1944, and forbade performance of the Board's functions by any other agency except as thereafter provided by law.

1943. The American Forestry Association undertook a Forest Resources Appraisal, a three-year inventory of the nation's forest resources.

1944. Treaty of February 3 between the United States and Mexico (59 Stat. 1219) established the International Boundary and Water Commission and provided for the allocation of the flow of the Rio Grande River and the Colorado River, the construction of dams, and other purposes.

1944. Act of February 26 (58 Stat. 100) gave effect to the Provisional Fur Seal Agreement of 1942 with Canada.

1944. Sustained-Yield Forest Management Act of March 29 (58 Stat. 132) authorized the Secretary of Agriculture and/or the Secretary of the Interior to establish cooperative sustained-yield units consisting of Federal forest land and private forest land or Federal sustained-yield units consisting only of

Federal forest land, when in their judgment the maintenance of stable communities is primarily dependent upon Federal stumpage and when such maintenance cannot be secured through usual timber-sale procedures. Provision is made for the sale of Federal stumpage to cooperating landowners or to responsible purchasers within communities dependent on Federal stumpage, without competitive bidding at prices not less than the appraised value of the timber.

1944. Act of May 5 (58 Stat. 216) amended the Clarke-McNary Act of 1924 by authorizing annual increases in the appropriation for cooperative forest-fire protection with the states and for studies of tax laws and forest-fire insurance up to a maximum of 9 million dollars for the fiscal year 1948 and thereafter.

1944. Act of May 31 (58 Stat. 265) authorized an annual appropriation of $750,000 to complete the initial survey of forest resources inaugurated by the McSweeney-McNary Act of 1928, with the stipulation that total appropriations for this purpose should not exceed $6,500,000. An additional appropriation of $250,000 annually was authorized to keep the survey current.

1944. Department of Agriculture Organic Act of September 21 (58 Stat. 734, 736), among many other administrative provisions, authorized the Secretary of Agriculture to pay rewards for information leading to arrest and conviction for violating laws and regulations relating to fires in or near national forests or for the unlawful taking of, or injury to, government property. It also authorized an annual expenditure during the existing emergency of not more than 1 million dollars for cooperative forest-fire protection under the Clarke-McNary Act of 1924 without requiring an equal expenditure by state and private owners.

1944. Act of September 27 (58 Stat. 745) authorized the Secretary of the Interior to dispose of sand, stone, gravel, vegetation, and timber or other forest products on unreserved public lands during the period of hostilities. This authority terminated on December 31, 1946, but was restored by the permanent and more comprehensive act of July 31, 1947.

1944. Federal-Aid Highway Act of December 20 (58 Stat. 838, 842) included authorization of annual appropriations of $25,000,000 for forest highways and $12,500,000 for forest roads and trails during each of the first three postwar fiscal years.

1944. Flood Control Act of December 22 (58 Stat. 887, 889) provided that thereafter Federal investigations and improvements of rivers and waterways for flood control and allied purposes should be under the jurisdiction of the War Department, and that Federal investigations of watersheds and measures for runoff and water-flow retardation and soil-erosion prevention on watersheds should be under the jurisdiction of the Secretary of Agriculture.

1945. Act of July 14 (59 Stat. 467) amended the Small Tract Act of 1938 to permit employees of the Department of the Interior stationed in Alaska to purchase or lease one small tract, except business sites.

1945. Presidential Proclamation of September 28 (59 Stat. 884) stated that the United States regards the natural resources in the continental shelf as subject to its jurisdiction and control without thereby affecting the free and unimpeded navigation of the high seas above the continental shelf.

1945. Presidential Proclamation of September 28 (59 Stat. 885) stated that the United States regards it as proper, without affecting the freedom of navigation, to establish conservation zones in parts of the high seas contiguous to its coasts, in which fishing activities shall be subject to the regulation and control of the United States, either alone or in cooperation with other nations.

1945. The United States on October 16 signed the Constitution of the Food and Agriculture Organization of the United Nations (60 Stat. 1886). "Agriculture" was defined as including fisheries, marine products, forestry, and primary forest products.

1946. Reorganization Plan No. 3 of May 16 (60 Stat. 1097, 1099) transferred to the Secretary of the Interior, subject to the approval of the Secretary of Agriculture, and to such conditions as he may specify, the jurisdiction formerly exercised by the latter over the development of mineral resources on lands acquired under the Weeks Act of 1911 and various emergency appropriations. It also consolidated the General Land Office and the Grazing Service to form the Bureau of Land Management in the Department of the Interior.

1946. Act of July 24 (60 Stat. 656) amended the Wildlife Restoration Act of 1937 by limiting the apportionment of funds to any one state to not less than ½ per cent and not more than 5 per cent of the total amount apportioned, and by permitting the use of not more than 25 per cent of the Federal apportionment for maintenance of completed wildlife-restoration projects.

1946. Joint Resolution of August 8 (60 Stat. 930) directed the Fish and Wildlife Service to prosecute investigations, experiments, and a vigorous program for the elimination of the sea lamprey from the Great Lakes.

1946. Farmers' Home Administration Act of August 14 (60 Stat. 1062) provided for a Farmers' Home Administration to replace the Farm Security Administration and to assume certain functions of the Farm Credit Administration and the National Housing Agency in order "to simplify and improve credit services to farmers and promote farm ownership."

1946. Act of August 14 (60 Stat. 1080) strengthened the Coordination Act of 1934 by authorizing the Secretary of the Interior, through the Fish and Wildlife Service, to provide assistance to, and cooperate with, Federal, state, and public or private agencies and organizations in the development, protection, and rehabilitation of wildlife resources of the United States.

1946. Agricultural Research and Marketing Act (Hope-Flannagan Act) of August 14 (60 Stat. 1082) provided for further research into basic laws and principles relating to agriculture and for improving and facilitating the marketing and distribution of agricultural products. It also authorized the Secretary of Agriculture to appoint a national advisory committee of eleven members and such other committees as he deemed appropriate to assist in effectuating specific research and service programs.

1946. An American Forest Congress was held October 9 to 11 in Washington, D.C., under the sponsorship of the American Forestry Association. The directors of the association subsequently drafted a detailed program which was overwhelmingly adopted early in 1947 by a referendum vote of the membership.

1946. Convention of December 2 between the United States and fourteen other governments (62 Stat. 1716) approved a schedule for the regulation of whaling and established an International Whaling Commission with authority to conduct investigations and to amend the schedule of regulations.

1946. On December 9, the Secretary of the Interior delegated to the Commissioner of Indian Affairs authority to approve sales of timber from Indian lands up to 40,000 M board feet and to adjust stumpage prices on these sales.

1946. The Forest Service, in December, issued the first of six *Reappraisal Reports,* based on its nationwide reappraisal project conducted in 1945 and 1946.

1946. Local research centers were organized for the first time under the Southern and the Southeastern Forest Experiment Stations.

1947. Forest Pest Control Act of June 25 (61 Stat. 177) declared it to be the policy of the government to protect all forest lands irrespective of ownership from destructive forest insect pests and diseases. It authorized the Secretary of Agriculture either directly or in cooperation with other Federal agencies, state and local agencies, and private concerns and individuals to conduct surveys to detect infestations and to determine and carry out control measures against incipient, potential, or emergency outbreaks.

1947. Act of July 7 (61 Stat. 246) established the Commission on Organization of the Executive Branch of the Government (Hoover Commission) to make recommendations to promote economy, efficiency, and improved service in the executive branch of the government.

1947. Act of July 24 (61 Stat. 419) approved the Pacific Marine Fisheries Compact between California, Oregon, and Washington.

1947. Sockeye Salmon Fishery Act of July 29 (61 Stat. 511) provided for effectuation of the convention of May 26, 1930.

1947. Act of July 30 (61 Stat. 630) increased the size of isolated tracts that might be offered for sale to 1,520 acres and of mountainous tracts to 760 acres.

1947. Act of July 31 (61 Stat. 681) authorized the Secretary of the Interior to dispose of sand, stone, gravel, clay, timber, and other materials on public lands exclusive of national forests, national parks, national monuments, and Indian lands. Material exceeding $1,000 in appraised value must be sold at public auction. Receipts are disposed of in the same manner as receipts from the sale of public lands.

1947. First issue of *Unasylva* was published by the Division of Forestry and Forest Products in the Food and Agriculture Organization of the United Nations.

1947. Interior Department Appropriations Act of August 6 (61 Stat. 790) amended the Taylor Grazing Act of 1934 to authorize the Secretary of the Interior in fixing fees for the grazing of livestock in grazing districts to "take into account the extent to which such districts yield public benefits over and above those accruing to the users of the forage for livestock purposes." Such fees were thereafter to consist of (1) a grazing fee, 12½ per cent of which is distributed to the states for the benefit of the counties in which the grazing districts are located, and (2) a range-improvement fee which, when appropriated by Congress, is available for the construction, purchase, or maintenance of range improvements. Of the receipts from public lands not in grazing districts which are leased for grazing under Section 15 of the Taylor Grazing Act, 25 per cent is available, when appropriated by Congress, for range improvements, and 50 per cent is distributed to the states for the benefit of the counties. Of the receipts from grazing districts on ceded Indian lands, 33⅓ per cent is distributed to the states for the benefit of the counties, and the remaining 66⅔ per cent is deposited to the credit of the Indians.

1947. Mineral Leasing Act for Acquired Lands of August 7 (61 Stat. 913) authorized the Secretary of the Interior to lease acquired lands containing deposits of coal, phosphate, oil, oil shale, gas, sodium, potassium, and sulfur under the provisions of the mineral leasing laws, with the consent of the head of the department having jurisdiction over the lands and subject to such conditions as he may prescribe.

1947. On September 14, the Commissioner of Indian Affairs authorized the regional offices to approve sales of timber up to 15,000 M board feet.

1948. Act of February 10 (62 Stat. 19) provided that whoever, without lawful authority or permission, shall go upon any national-forest land while it is closed to the public by a regulation of the Secretary of Agriculture made pursuant to law, shall be subject to fine and imprisonment.

1948. Act of March 30 (62 Stat. 100) permitted the use and occupancy of national-forest lands in Alaska for purposes of residence, recreation, public convenience, education, industry, agriculture, and commerce for periods not exceeding thirty years and in tracts not exceeding 80 acres. Lands so leased are not subject to disposal or leasing under the mining laws.

1948. Act of April 8 (62 Stat. 162) provided for the reopening of the revested Oregon and California Railroad lands and the reconveyed Coos Bay Wagon Road lands, except power sites, to exploration, location, entry, and disposition under the general mining laws.

1948. Secretary of Agriculture in May established the National Forest Board of Review, the name of which was changed in 1950 to National Forest Advisory Council.

1948. Act of May 19 (62 Stat. 240) authorized transfer of certain real property controlled but no longer needed by Federal agencies (1) to the states for wildlife-conservation purposes other than for migratory birds or (2) to the Secretary of the Interior if the property has particular value in carrying out the national migratory-bird management program.

1948. Act of June 22 (62 Stat. 568) authorized appropriations not to exceed a total of $500,000 for the purpose of acquiring certain specified lands in Cook, Lake, and St. Louis Counties in the Superior National Forest, Minnesota, the development or exploitation of which might impair the unique qualities and natural features of the remaining wilderness canoe country. It also directed payment to the counties, in lieu of the usual 25 per cent of gross receipts, of 0.75 per cent of the fair appraised value of the land in the area covered by the act, as determined by the Secretary of Agriculture at ten-year intervals.

1948. Water Pollution Control Act (Taft-Barkley Act) of June 30 (62 Stat. 1155) provided for technical and financial cooperation by the Federal government with states and municipalities in the formulation and execution of programs for the abatement of stream pollution. Necessary appropriations were authorized for the five-year period ending June 30, 1953.

1948. An Inter-American Conference on Conservation of Renewable Natural Resources was held in Denver, Colorado, September 7 to 20.

1949. Convention of January 25 with Mexico (T.I.A.S. 2094) provided for the establishment of an International Commission for the Scientific Investigation of Tuna.

1949. First of the reports of the Commission on Organization of the Executive Branch of the Government (Hoover Commission) was sent to Congress on February 5.

1949. Convention of February 8 between the United States and ten other countries (T.I.A.S. 2089) established the International Commission for the Northwest Atlantic Fisheries, with authority to conduct investigations and to promulgate regulations for the taking of fish in the Northwest Atlantic Ocean.

1949. Joint Resolution of May 19 (63 Stat. 70) approved the Gulf States Marine Fisheries Compact.

1949. Convention of May 31 with Costa Rica (T.I.A.S. 2044) provided for the establishment of an Inter-American Tropical Tuna Commission.

1949. Act of June 25 (63 Stat. 271) approved the Northeastern Interstate Forest Fire Protection Compact for the purpose of promoting effective prevention and control of forest fires in the Northeastern United States.

1949. Act of June 25 (63 Stat. 271) increased to 1 million dollars a year the authorized appropriation for the conduct of the nationwide forest survey provided for by the McSweeney-McNary Act of 1928, as amended in 1944, with a limitation of 11 million dollars on total expenditures, and increased to $1,500,000 a year the authorized appropriation for keeping the survey current.

1949. Third International Forestry Congress was held at Helsinki, Finland, July 10 to 20.

1949. Act of August 12 (63 Stat. 599) increased to $2 the price of the hunting stamp required for the taking of migratory waterfowl under the act of March 16, 1934.

1949. United Nations Scientific Conference on the Conservation and Utilization of Resources was held at Lake Success, N.Y., August 17 to September 6. An International Technical Conference on the Protection of Nature was held simultaneously.

1949. Act of August 30 (63 Stat. 679) provided for the sale at public auction of public land in Alaska not within national parks, monuments, forests, Indian lands, or military reservations which have been classified by the Secretary of the Interior as suitable for industrial or commercial purposes, including the construction of housing, in tracts not to exceed 160 acres, to any bidder who furnishes satisfactory proof that he has bona fide intentions and the means to develop the tract for the intended use.

1949. Anderson-Mansfield Reforestation and Revegetation Act of October 11 (63 Stat. 762) authorized a schedule of appropriations for the reforestation and revegetation of the forest and range lands of the national forests. "It is the declared policy of the Congress to accelerate and provide a continuing basis for the needed reforestation and revegetation of national-forest lands and other lands under administration or control of the Forest Service."

1949. Act of October 26 (63 Stat. 909) amended the Clarke-McNary Act of 1924 by authorizing (1) annual increases in the appropriation for cooperative forest-fire protection with the states up to a maximum of $20,000,000 for the fiscal year 1955 and thereafter; (2) annual increases in the appropriation for cooperation with the states in providing planting stock for farmers and others up to a maximum of $2,500,000 for the fiscal year 1953 and thereafter; and (3) an annual appropriation of $500,000 for cooperation with the land-grant colleges or other suitable state agencies in educating farmers in the management of forest lands and in harvesting, utilizing, and marketing the products thereof.

1949. United States Supreme Court on November 7 (338 U.S. 863) upheld the decision of the Washington Supreme Court affirming the constitutionality of the Washington law of 1945 providing for the control of cutting on privately owned forest lands.

1949. Southern Regional Education Compact was ratified by the Legislatures of ten states. Coordination of teaching and research in forestry has been attempted under the compact.

1950. Reorganization Plan No. 3 of March 13 (64 Stat. 1262) transferred to the Secretary of the Interior, with two exceptions, all functions of all agencies and employees of the Department; authorized the Secretary to effect such organization of the Department as he deemed appropriate; and added an assistant secretary and an administrative assistant secretary to the Department.

1950. Granger-Thye Act of April 24 (64 Stat. 82), among many other provisions "to facilitate and simplify the work of the Forest Service," broadened the authority granted the Secretary of Agriculture by the act of March 3, 1925, to accept contributions for administration, protection, improvement, reforestation, and other work on non-Federal lands within or near national forests; provided for sales and exchanges of nursery stock with public agencies; authorized the lease, protection, and management of public and private range land intermingled with or adjacent to national-forest land; made available, when appropriated by Congress, an amount equivalent to 2 cents per animal-month for sheep and 10 cents per animal-month for other kinds of livestock under permit on a national forest for range improvements on that forest; provided for the organization of local advisory boards on petition of a majority of the grazing permittees on a national forest; authorized the Secretary of Agriculture to issue permits for the grazing of livestock on national forests for periods not exceeding ten years and renewals thereof; and repealed the provision of the Weeks Act of 1911 limiting contributions to counties to 40 per cent of their income from other sources.

1950. Act of May 10 (64 Stat. 149) established the National Science Foundation "to promote the progress of science" and for other purposes.

1950. Whaling Convention Act of August 9 (64 Stat. 421) provided for effectuation of the convention of December 2, 1946, for the regulation of whaling.

1950. Fish Restoration and Management Act (Dingell-Johnson Act) of

August 9 (64 Stat. 430) authorized the annual appropriation of an amount equivalent to the revenue from the tax on fishing rods, creels, reels, and artificial lures, baits, and flies, to be used for cooperation with the states in fish restoration and management projects up to 75 per cent of the total cost of the projects.

1950. Cooperative Forest Management Act of August 25 (64 Stat. 473) authorized an annual appropriation of $2,500,000 to enable the Secretary of Agriculture to cooperate with state foresters in providing technical services to private forest landowners and operators and to processors of primary forest products. The Cooperative Farm Forestry Act of 1937 was repealed effective June 30, 1951.

1950. Tuna Conventions Act of September 7 (64 Stat. 777) provided for effectuation of the conventions of January 25, 1949, with Mexico and of May 31, 1949, with Costa Rica.

1950. Northwest Atlantic Fisheries Act of September 27 (64 Stat. 1067) provided for effectuation of the international convention of February 8, 1949. The Secretary of the Interior, through the Fish and Wildlife Service, was authorized to administer and enforce all of the provisions of the convention.

1950. *A Water Policy for the American People,* the first volume of a three-volume report of the President's Water Resources Policy Commission (Cooke Commission), was transmitted to the President on December 11.

1950. President Truman's "Point Four" proposed a cooperative program for aid in the development of economically undeveloped areas of the world.

1950. The Society of American Foresters celebrated its Golden Anniversary and published a comprehensive history of forestry in the United States during the previous fifty years.

1952. Convention of May 9 with Canada and Japan (T.I.A.S. 2786) established the North Pacific Fisheries Commission with authority to conduct investigations and make recommendations aimed at protecting the interests of the three countries concerned in the fisheries of the North Pacific Ocean. The Commission has no regulatory powers.

1952. Act of May 13 (66 Stat. 71) authorized participation in the Northeastern Interstate Forest Fire Protection Compact of 1949 of any Province of Canada contiguous to any state which is a party to the compact.

1952. Act of May 27 (66 Stat. 95) increased to $2,000 the appraised value of national-forest stumpage that can be sold without advertisement.

1952. *Resources for Freedom,* report of the President's Materials Policy Commission (Paley Commission), was transmitted to the President on June 2.

1952. Interior Department Appropriations Act of July 9 (66 Stat. 445, 447) made the appropriation for access roads to O. and C. lands deductible from the 75 per cent of gross receipts payable to the counties.

1952. Act of July 17 (66 Stat. 755) extended the financial authorizations approved by the Water Pollution Control Act of 1948 to June 30, 1956.

1952. Sixth International Grasslands Congress was held at State College, Pa., August 17 to 23.

1952. Secretary of Agriculture appointed a **Forest Research Advisory Com-**

mittee under authority of the Agricultural Research and Marketing Act of August 14, 1946.

1953. Convention of March 2 with Canada (T.I.A.S. 2900) strengthened the powers of the International Fisheries Commission to regulate the catch of halibut in the North Pacific Ocean.

1953. Reorganization Plan No. 2 of March 25 (67 Stat. 633) transferred to the Secretary of Agriculture, with certain specified exceptions, all functions of all agencies and employees of the Department; authorized the Secretary to effect such organization of the Department as he deemed appropriate; and added two assistant secretaries and an administrative assistant secretary to the Department.

1953. Submerged Lands Act of May 22 (67 Stat. 29) confirmed and established the titles of the states to lands beneath navigable waters within state boundaries and to the natural resources within such lands and waters; provided for the use and control of such lands and resources; and confirmed the jurisdiction and control of the United States over the natural resources of the sea bed of the continental shelf seaward of state boundaries.

1953. Act of June 6 (67 Stat. 45) gave congressional approval to the Connecticut River Flood Control Compact between the states of Massachusetts, Connecticut, New Hampshire, and Vermont.

1953. L. F. Watts resigned as Chief of the Forest Service and was succeeded on July 1 by R. E. McArdle.

1953. Act of July 10 (67 Stat. 142) established a Commission on Organization of the Executive Branch of the Government (Second Hoover Commission), which dealt with some aspects of Federal natural-resource activities.

1953. Act of July 10 (67 Stat. 145) established a Commission on Intergovernmental Relations (Kestnbaum Commission), one of the committees of which dealt with Federal–state relations in the field of natural resources.

1953. Agricultural Appropriations Act of July 28 (67 Stat. 205, 214) appropriated 5 million dollars to conduct studies and carry out preventive measures for the protection of watersheds under the provisions of the Soil Conservation Act of 1935.

1953. Act of August 7 (67 Stat. 462) provided for the jurisdiction of the United States over the submerged lands of the outer continental shelf and authorized the Secretary of the Interior to lease such lands for certain purposes.

1953. Act of August 8 (67 Stat. 489) extended the forest survey to the territories and possessions of the United States.

1953. Act of August 8 (67 Stat. 494) amended the Northern Pacific Halibut Act of 1937 to provide for effectuation of the convention with Canada signed March 2, 1953.

1953. Act of August 13 (67 Stat. 559) established a national Advisory Committee on Weather Control to study and evaluate public and private experiments in weather modification.

1953. Act of August 15 (67 Stat. 613) amended the Federal Reserve Act to authorize national banks to make loans secured by first liens up to 40 per cent of their appraised value "upon forest tracts which are properly managed in all respects." Loans may not be made for more than two years, except that they may be made for ten years under a mortgage providing for their amortization at a rate of not less than 10 per cent a year.

1953. Fourth American Forest Congress was held October 29 to 31 in Washington, D.C., under the sponsorship of the American Forestry Association. Following the Congress "A Program for American Forestry" was formulated by the directors and in 1954 was overwhelmingly approved by the membership of the association.

1953. Forest Conservation Society of America was founded on November 30 at Washington, D.C.

1953. Panama adhered to the convention of May 31, 1949, between Costa Rica and the United States and thus became a member of the Inter-American Tropical Tuna Commission.

1954. President Eisenhower on May 26 established a Cabinet Committee on Water Resources Policy consisting of the Secretaries of the Interior (chairman), Agriculture, Commerce, Army, and Health, Education, and Welfare, and the chairman of the Federal Power Commission.

1954. Act of May 28 (68 Stat. 151) removed the limitation of 142 million acres on the total area that might be included in grazing districts.

1954. Act of June 8 (68 Stat. 239) amended the Small Tract Act of 1938 to permit the sale or lease of small tracts chiefly valuable for residence, recreation, business, or community site purposes, to individuals, associations, corporations, states, municipalities, or other governmental subdivisions, if such sale or lease will not unreasonably interfere with the use of water for grazing purposes or unduly impair the protection of watershed areas. It also permitted the leasing of revested Oregon and California Railroad lands and of reconveyed Coos Bay Wagon Road lands except for business purposes.

1954. Act of June 17 (68 Stat. 250) provided for the orderly termination of Federal supervision over the members and property (including forests and range lands) of the Menominee Indian Tribe in Wisconsin.

1954. Act of June 24 (68 Stat. 270) declared the controverted Oregon and California Railroad lands in the indemnity strip to be O. and C. lands, which shall continue to be administered as national-forest lands, and the receipts from which shall be disposed of as provided in the act of August 28, 1937. In order to facilitate administration and accounting, the Secretary of Agriculture was authorized to designate in each county an area of national-forest land of substantially equal value, revenues from which shall be disposed of under the 1937 act. The Secretary of the Interior and the Secretary of Agriculture were also directed to block up national-forest and intermingled and adjacent O. and C. lands, exclusive of those in the indemnity strip, by exchange of administrative jurisdiction on approximately an equal-value (and so far as practicable an equal-area) basis.

1954. Act of July 27 (68 Stat. 563) approved the Southeastern Interstate Forest Fire Protection Compact "to promote effective prevention and control of forest fires in the Southeastern region of the United States." It applied to Alabama, Florida, Georgia, Kentucky, Mississippi, North Carolina, South Carolina, Tennessee, Virginia, and West Virginia, and to any contiguous state on approval by the legislature of each member state.

1954. Watershed Protection and Flood Prevention Act of August 4 (68 Stat. 666) authorized the Secretary of Agriculture, under specified conditions, to cooperate with states and local organizations for the purpose of preventing erosion, floodwater, and sediment damages and of furthering the conservation, development, utilization, and disposal of water. It repealed the authority granted the Secretary under the Flood Control Act of 1936 to make preliminary examinations and surveys and to prosecute certain works of improvement on watersheds, but preserved his authority to prosecute the eleven projects authorized by the Flood Control Act of 1944 and to prosecute emergency measures under the 1938 act.

1954. North Pacific Fisheries Act of August 12 (68 Stat. 698) provided for effectuation of the convention of May 9, 1952, with Canada and Japan.

1954. Act of August 13 (68 Stat. 718) provided for orderly termination of Federal supervision over the Klamath Indian Tribe of Oregon.

1954. Act of August 17 (68 Stat. 734) deleted the phrase "in the arid and semiarid areas of the United States" from the act of August 28, 1937, providing Federal aid in the development of facilities for water storage and utilization. It also provided that no further construction work should be done by the Secretary of Agriculture but authorized him to make loans for the purpose of attaining the objectives of the act.

1954. Internal Revenue Code of August 16 (68A Stat. 67) authorized farmers in computing income taxes to deduct expenditures for soil or water conservation or for the prevention of erosion, up to 25 per cent of gross income.

1954. Act of August 24 (68 Stat. 783) approved the South Central Interstate Forest Fire Protection Compact. It applied to the states of Arkansas, Louisiana, Mississippi, Oklahoma, and Texas, and to any contiguous state on approval by the legislature of each member state.

1954. Act of September 3 (68 Stat. 1146) authorized the issuance by Federal agencies of permits, leases, or easements to states or local governmental bodies, for periods not to exceed thirty years, on lands within their respective jurisdictions. The act applied to "public lands and national forests, except national parks and monuments."

1954. A National Watershed Congress was held in Washington, D.C., December 6 to 7.

1955. Act of July 14 (P. L. 159) authorized the Surgeon General of the Public Health Service to cooperate with other agencies in providing research and technical assistance relating to the control of air pollution.

1955. Act of July 23 (P. L. 167) amended the Materials Disposal Act of July 31, 1947, by adding common pumice, pumicite, and cinders to the materials specified in that act, and authorized the disposal of all such materials on both unreserved and reserved public lands except national parks, national monuments, and Indian lands by the secretary of the department having jurisdiction over the lands in question. It also provided that on unpatented claims hereafter located the United States shall have the right to dispose of the timber and other nonmineral surface resources, provided that such disposal shall not endanger or materially interfere with mining operations; and it established a procedure whereby the right to the use of timber and other surface resources on existing, inactive mining claims may be canceled or waived.

1955. Act of July 28 (P. L. 185) authorized the extension of the Interstate Oil Compact, first approved in 1935, for a period of four years from September 1, 1955.

1955. Act of August 1 (P. L. 206) repealed the provisions of the Timber and Stone Act of June 3, 1878, as amended, providing for the sale of public lands chiefly valuable for timber or stone.

1955. A second National Watershed Conference was held in Washington, D.C., December 5 to 6.

1956. Soil Bank Act of May 28 (P. L. 540) established an acreage reserve program and a conservation reserve program. It also authorized the Secretary of Agriculture to cooperate with states in tree-planting programs.

1956. Great Lakes Fishery Act of June 4 (P. L. 557) provided for effectuation of the convention of September 10, 1954, with Canada.

1956. Act of June 22 (P. L. 607) extended the area in the Superior National Forest authorized for purchase by the act of June 22, 1948, and increased the authorization for such purchase by $2,000,000.

1956. Act of July 9 (P. L. 660) amended and strengthened the Water Pollution Control Act of June 30, 1948.

1956. Act of July 25 (P. L. 790) approved the Middle Atlantic Interstate Forest Fire Protection Compact. It applied to the states of Delaware, Maryland, New Jersey, Pennsylvania, Virginia, and West Virginia.

1956. Act of August 2 (P. L. 925) authorized the establishment of the Virgin Islands National Park.

1956. Small Reclamation Projects Act of August 6 (P. L. 984) authorized loans and grants to states for the construction of reclamation projects costing not more than $5,000,000, or under certain circumstances not more than $10,000,000.

1956. Fish and Wildlife Act of August 8 (P. L. 1024) reorganized the Fish and Wildlife Service under an Assistant Secretary for Fish and Wildlife in the Department of the Interior.

APPENDIX 3

Selected Bibliography

GENERAL

Albion, Robert G.: "Forests and Sea Power," Harvard University Press, Cambridge, Mass., 1926.

Allen, Edward T.: The Application and Possibilities of the Federal Forest Reserve Policy, *Soc. Amer. Foresters Proc.*, 1:41–52, 1905.

Allen, Shirley W.: "An Introduction to American Forestry," McGraw-Hill Book Company, Inc., New York, 1950.

———: "Conserving Natural Resources—Principles and Practice in a Democracy," McGraw-Hill Book Company, Inc., New York, 1955.

Cameron, Jenks: "The National Park Service—Its History, Activities and Organization," D. Appleton-Century-Crofts, Inc., New York, 1922.

———: American Forest Influences—Sea Power, *Amer. Forests*, 32:707–711, 767–768, 1926.

———: President Adams' Acorns, *Amer. Forests*, 34:131–134, 1928.

———: An Anchor to Forestward, *Amer. Forests*, 34:199–201, 235, 1928.

———: Who Killed Santa Rosa? *Amer. Forests*, 34:263–266, 312, 1928.

———: "The Development of Governmental Forest Control in the United States," Johns Hopkins Press, Baltimore, 1928.

———: "The Bureau of Biological Survey—Its History, Activities and Organization," Johns Hopkins Press, Baltimore, 1929.

Chapline, W. R.: Range Management History and Philosophy, *Jour. Forestry*, 49:634–638, 1951.

Chapman, H. H.: The Responsibilities of the Profession in the Present Situation, *Jour. Forestry*, 33:204–210, 1935.

———: Some Legal and Economic Aspects of the O. and C. Controversy, *Jour. Forestry*, 43:566–568, 1945.

———: The Cure for the O. and C. Situation, *Jour. Forestry*, 43:569–574, 1945.

Clapp, Earle H.: Federal Forest Policies of the Future, *Jour. Forestry*, 39:80–83, 1941.

———: Public Forest Regulation, *Jour. Forestry*, 47:527–530, 1949.

Clepper, Henry: The Journal of Forestry—An Historical Summary of the First Fifty Years, *Jour. Forestry*, 50:899–912, 1952.

Coffman, John D.: Forestry in the Department of the Interior, *Jour. Forestry,* 39:84–91, 1941.

Compton, Wilson: "The Organization of the Lumber Industry," American Lumberman, Chicago, 1916.

———: Private Enterprise Offers Better Opportunity for Progress in Forestry than Nationalization, *Jour. Forestry,* 42:81–88, 1944.

Conference of Lumber and Timber Products Industries with Public Agencies on Forest Conservation, *Jour. Forestry,* 32:275–307, 1934.

Connery, Robert H.: "Governmental Problems in Wildlife Conservation," Columbia University Press, New York, 1935.

Conover, Milton: "The General Land Office—Its History, Activities and Organization," Johns Hopkins Press, Baltimore, 1923.

Cox, Herbert J.: On the Other Side of Certain Public Lands, *Jour. Forestry,* 43:315–321, 1945.

Dana, Samuel T.: Certain Public Lands, *Jour. Forestry,* 42:703–704, 1944.

———: The Growth of Forestry in the Past Half Century, *Jour. Forestry,* 49:86–92, 1951.

———: "Forest Policy in the United States," University of British Columbia, Vancouver, B.C., 1953.

——— (ed.): "History of Activities in the Field of Natural Resources, University of Michigan," University of Michigan Press, Ann Arbor, Mich., 1953.

Defebaugh, James E.: "History of the Lumber Industry of America," 2 vols., American Lumberman, Chicago, 1906–1907.

Fernow, Bernhard E.: "Economics of Forestry," T. Y. Crowell & Company, New York, 1902.

———: "A Brief History of Forestry in Europe, the United States and Other Countries," University Press, Toronto, Ont., 1911.

Fitch, Edwin M.: "The Tariff on Lumber," Tariff Research Committee, Madison, Wis., 1936.

Fritz, Emanuel: A Proposal for Reorganizing and Realigning Federal Forest, Forage, Park, and Game Lands, *Jour. Forestry,* 44:278–281, 1946.

Gillett, Charles A.: The Forest Industries Public Action Program, *Jour. Forestry,* 44:724–726, 1946.

Granger, Christopher M.: Mining Claims on the National Forests—It's Time to Take Another Look, *Jour. Forestry,* 50:355–358, 1952.

Graves, Henry S. (Chairman): Standardization of Instruction in Forestry, Report of the Committee of the Conference of Forest Schools, *Forestry Quart.,* 10:341–394, 1912.

Graves, Henry S.: Private Forestry, *Jour. Forestry,* 17:113–121, 1919.

———: A National Lumber and Forest Policy, *Jour. Forestry,* 17:351–363, 1919.

———: A Policy of Forestry for the Nation, *Jour. Forestry,* 17:901–910, 1919.

——— and Cedric H. Guise: "Forest Education," Yale University Press, New Haven, Conn., 1932.

Greeley, William B.: Self-Government in Forestry, *Jour. Forestry,* 18:103–105, 1920.

———: "Forests and Men," Doubleday & Company, Inc., New York, 1951.

———: "Forest Policy," McGraw-Hill Book Company, Inc., New York, 1953.

Hall, William L.: Progress and Problems of Private Forest Practice, *Jour. Forestry*, 37:110–113, 1939.

———: The Society of American Foresters—Its Contributions to Our National Economy, *Jour. Forestry*, 49:86–92, 1951.

Hawes, Austin F.: A Chapter in American Forest History—The Association of Eastern Foresters, *Jour. Forestry*, 25:325–337, 1927.

———: Forty Years of State Forestry, *Jour. Forestry*, 39:95–99, 1941.

Herbert, P. A.: A Forest Policy for the Nation, *Jour. Forestry*, 28:806–812, 1930.

———: A Foreign Forest Policy for the Nation, *Jour. Forestry*, 42:631–636, 1944.

Heyward, Frank, Jr.: Industrial Self-Regulation, *Jour. Forestry*, 39:231–235, 1941.

Hibbard, Benjamin H.: "History of the Public Land Policies," The Macmillan Company, New York, 1924.

Hill, Robert Tudor: The Public Domain and Democracy, *Studies Hist. Econ. Pub. Law*, vol. 38, no. 1, Columbia University, New York, 1910.

Holt, William S.: "The Office of the Chief of Engineers of the Army—Its Non-Military History, Activities and Organization," Johns Hopkins Press, Baltimore, 1923.

Hopkins, Howard: A Nation-wide Forestry Program, *Jour. Forestry*, 39:523–529, 1941.

Hosmer, Ralph S.: The Society of American Foresters—An Historical Summary, *Jour. Forestry*, 38:837–854, 1940.

———: The National Forestry Program Committee, *Jour. Forestry*, 45:627–645, 1947.

———: The Society of American Foresters—An Historical Summary, *Jour. Forestry*, 48:756–777, 1950.

———: "Forestry at Cornell—A Retrospect of Proposals, Developments, and Accomplishments in the Teaching of Professional Forestry at Cornell University," Cornell University, Ithaca, N.Y., 1950.

Hotchkiss, George W.: "History of the Lumber and Forest Industry of the Northwest," G. W. Hotchkiss & Company, Chicago, 1898.

Howard, William G.: New York's Forest Practice Act, *Jour. Forestry*, 45:405–407, 1947.

Ise, John: "The United States Forest Policy," Yale University Press, New Haven, Conn., 1920.

Kauffman, Erle: "The Conservation Yearbook," Washington, 1954.

Kaylor, Joseph F.: Maryland's Forest Conservancy Districts Law, *Jour. Forestry*, 42:352–354, 1944.

Kendall, Harry T.: The Lumbermen's Attitude toward Forestry, *Cut-Over Lands*, 2:20–23, 1919; reprinted in *Jour. Forestry*, 17:647–649, 1919.

Kinney, Jay P: "Forest Legislation in America Prior to March 4, 1789," Cornell University, Ithaca, N.Y., 1916.

———: Forest Policy in Indian Timberlands, *Jour. Forestry*, 25:430–436, 1927.

————: "A Continent Lost—A Civilization Won: Indian Land Tenure in America," Johns Hopkins Press, Baltimore, 1937.

————: "Indian Forest and Range: A History of the Administration and Conservation of the Redman's Heritage," Forestry Enterprises, Washington, 1950.

Kirkland, Burt P.: The Democracy of National Control, *Jour. Forestry*, 18:448–450, 1920.

Korstian, Clarence F.: Forestry on Private Lands in the United States, *Duke Univ. Forestry Bul.* 8, 1944.

Lambert, John H., Jr.: Massachusetts' Forest Cutting Practices Law, *Jour. Forestry*, 42:799–804, 1944.

Lillard, Richard G.: "The Great Forest," Alfred A. Knopf, Inc., New York, 1947.

Martin, Clyde S.: History and Influence of the Western Forestry and Conservation Association on Cooperative Forestry in the West, *Jour. Forestry*, 43:165–169, 1945.

Mason, A. T.: "Bureaucracy Convicts Itself," The Viking Press, Inc., New York, 1941.

Mason, David T.: "Forests for the Future," Forest Products History Foundation, St. Paul, Minn., 1952.

McGlothlin, William J.: Toward a Regional Program of Forestry Training and Research in the South, *Jour. Forestry*, 50:195–197, 1952.

Murphy, Louis S.: State versus Federal Competence, *Jour. Forestry*, 46:6–15, 1948.

Nelson, DeWitt: Progressive Forest Management in California, *Jour. Forestry*, 50:259–261, 1952.

Olmsted, Frederick E.: The Work Ahead, *Jour. Forestry*, 17:227–235, 1919.

Ostrander, George N.: The History of the New York Forest Preserve, *Jour. Forestry*, 40:301–304, 1942.

Peffer, E. Louise: "The Closing of the Public Domain," Stanford University Press, Stanford, Calif., 1951.

Pinchot, Gifford: The Lines Are Drawn, *Jour. Forestry*, 17:899–900, 1919.

————: The Public Good Comes First, *Jour. Forestry*, 39:208–212, 1941.

————: "Breaking New Ground," Harcourt, Brace and Company, Inc., New York, 1947.

Powell, Fred W.: "The Bureau of Plant Industry—Its History, Activities and Organization," Johns Hopkins Press, Baltimore, 1927.

Puter, S. A. D.: "Looters of the Public Domain," Portland Printing House, Portland, Ore., 1908.

Reynolds, Harris A.: The Place of the Forestry Association in Conservation, *Jour. Forestry*, 39:120–122, 1941.

Robbins, Roy M.: "Our Landed Heritage," Princeton University Press, Princeton, N.J., 1942; Peter Smith, New York, 1950.

Rodgers, Andrew Denny, III: "Bernhard Eduard Fernow: A Story of North American Forestry," Princeton University Press, Princeton, N.J., 1951.

Rogers, N. S.: Objectives and Accomplishments of the Oregon Forest Laws, *Jour. Forestry*, 42:480–482, 1944.

Roosevelt, Theodore: Forestry and Foresters, *Soc. Amer. Foresters Proc.*, 1:3–9, 1905.

Rupp, Alfred E.: History of Land Purchase in Pennsylvania, *Jour. Forestry*, 22:490–497, 1924.

Saunderson, Mont H.: "Western Land and Water Use," University of Oklahoma Press, Norman, Okla., 1950.

Schenck, Carl Alwin: "The Biltmore Story," American Forest History Foundation, Minnesota Historical Society, St. Paul, Minn., 1955.

Shankland, Robert: "Steve Mather of the National Parks," Alfred A. Knopf, Inc., New York, 1951.

Silcox, F. A.: Foresters Must Choose, *Jour. Forestry*, 33:198–204, 1935.

———: Forestry—A Public and Private Responsibility, *Jour. Forestry*, 33:460–468, 1935.

———: A Federal Plan for Forest Regulation within the Democratic Pattern, *Jour. Forestry*, 37:116–119, 1939.

Smith, Darrell H.: "The Forest Service—Its History, Activities and Organization," Brookings Institution, Washington, 1930.

Smith, Glen A.: The Attack on the Forest Service Grazing Policy, *Jour. Forestry*, 24:136–140, 1926.

Smith, Herbert A.: Forest Education before 1898, *Jour. Forestry*, 32:684–689, 1934.

Spring, Samuel N. (ed.): "The First Half Century of the Yale School of Forestry," Yale University, New Haven, Conn., 1950.

Stahl, Rose Mildred: The Ballinger-Pinchot Controversy, *Studies in History*, vol. XI, no. 2, Smith College, Northampton, Mass., 1926.

Stoddart, Laurence A., and Arthur D. Smith: "Range Management," McGraw-Hill Book Company, Inc., New York, 1943.

Tilden, Freeman: "The National Parks—What They Mean to You and Me," Alfred A. Knopf, Inc., New York, 1951.

Toumey, James W.: Second National Conference on Education in Forestry, *Jour. Forestry*, 19:167–172, 1921.

U.S. Code Annotated, Title 16, "Conservation," West Publishing Company, St. Paul, Minn., Edward Thompson Company, Brooklyn, N.Y., 1928, and supplements.

U.S. Code Annotated, Title 43, "Public Lands," West Publishing Company, St. Paul, Minn., Edward Thompson Company, Brooklyn, N.Y., 1928, and supplements.

Watson, Russell: Historic Sources of Congressional Trouble in Conservation, *Jour. Forestry*, 22:480–489, 1924.

Watts, Lyle F.: Comprehensive Forest Policy Indispensable, *Jour. Forestry*, 41:783–788, 1943.

———: A Forest Program to Help Sustain Private Enterprise, *Jour. Forestry*, 42:81–88, 1944.

———: Where Are the Goal Posts? *Jour. Forestry*, 42:159–163, 1944.

———: Regulation Is No Dilemma, *Jour. Forestry*, 42:416–419, 1944.

Weber, Gustavus A.: "The Bureau of Entomology—Its History, Activities and Organization," Brookings Institution, Washington, 1930.

Weyerhaeuser, J. P., Jr.: Answering the Threat of Forest Depletion, *Jour. Forestry,* 47:524–526, 1949.

Wirt, George H.: A Half Century of Forestry in Pennsylvania, *Jour. Forestry,* 41:730–734, 1943.

Woods, John B.: Forty Years of Private Forest Ownership, *Jour. Forestry,* 39: 106–110, 1941.

Yard, Robert Sterling: "Our Federal Lands," Charles Scribner's Sons, New York, 1928.

ORGANIZATIONAL

American Forestry Association: "A Program for American Forestry, 1947."

——: "A Program for American Forestry, 1954."

Proc. Amer. Forest Cong., 1905, American Forestry Association, Washington, 1905.

Proc. Amer. Forest Cong., 1946, American Forestry Association, Washington, 1947.

Proc. 4th Amer. Forest Cong., 1953, American Forestry Association, Washington, 1953.

Proc. Amer. Forestry Assoc., 1890–1897.

Proc. Amer. Forestry Cong., 1882–1889.

Butler, Ovid (comp. and ed.): "American Conservation—in Picture and Story," American Forestry Association, Washington, 1941.

Butler, Ovid (Chairman): "The Progress of Forestry, 1945 to 1950," Report of Special Committee, American Forestry Association, Washington, 1951.

Chamber of Commerce of the United States: "Policy Declarations on Natural Resources," Washington, 1955.

Chapman, H. H., and Henry Schmitz: Grazing vs. Forestry, Majority Report of the Committee on Grazing, Society of American Foresters, *Jour. Forestry,* 24:378–394, 1926.

Charter Members of the Society of American Foresters, *Jour. Forestry,* 48:753–755, 1950.

Clapp, Earle H. (Chairman): "A National Program of Forest Research," Report of a Special Committee of the Washington Section of the Society of American Foresters, American Tree Association, Washington, 1926.

Conference on Commercial Forestry, "Report," Chamber of Commerce of the United States, Washington, 1927.

Forest Industries Council: "Growing Trees in a Free Country," 1953.

Graves, Henry S. (Chairman): "Problems and Progress of Forestry in the United States," Report of the Joint Committee on Forestry, National Research Council and the Society of American Foresters, Washington, 1947.

Joint Committee on Recreational Survey of Federal Lands of the American Forestry Association and the National Parks Association: "Recreational Resources of Federal Lands," National Conference on Outdoor Recreation, Washington, 1928.

Kaufert, Frank H., and William H. Cummings: "Forestry and Related Research in North America," Society of American Foresters, Washington, 1955.

National Association of Manufacturers: "Industry Believes," pp. 35–41, New York, 1954.

National Lumber Manufacturers Association: Statement of Forestry Policy, Jour. Forestry, 27:776–781, 1929.

Past and Present Officers of the Society of American Foresters, Jour. Forestry, 48:778–780, 1950.

Resources for the Future: "The Nation Looks at Its Resources: Report of the Mid-Century Conference on Resources for the Future," Washington, 1954.

Sampson, A. W.: Grazing vs. Forestry, Minority Report of the Committee on Grazing, Society of American Foresters, Jour. Forestry, 24:395–405, 1926.

Society of American Foresters, Committee for the Application of Forestry (Gifford Pinchot, Chairman): Forest Devastation—a National Danger and a Plan to Meet It, Jour. Forestry, 17:911–945, 1919.

Society of American Foresters, Appalachian Section: "Cumulated Index for Proceedings of the Society of American Foresters, vols. 1–11 (May, 1905–October, 1916); Forestry Quarterly, vols. 1–14 (October, 1902–December, 1916); Journal of Forestry, vols. 15–27 (January, 1917–December, 1929)," Society of American Foresters, Washington, 1930.

———: "Second Cumulated Index for Journal of Forestry, vols. 28–37 (January, 1930–December, 1950)," Society of American Foresters, Washington, 1940.

———: "Third Cumulated Index for Journal of Forestry, vols. 38–48 (January, 1940–December, 1950)," Society of American Foresters, Washington, 1952.

Stanford Research Institute: "America's Demand for Wood, 1929–1975," Weyerhaeuser Timber Company, Tacoma, Wash., 1954.

Winters, Robert K. (ed.): "Fifty Years of Forestry in the U.S.A.," Society of American Foresters, Washington, 1950.

GOVERNMENTAL [1]

Commission on Intergovernmental Relations (Meyer Kestnbaum, Chairman): "Report" and Committee Reports on Natural Resources and Conservation, Federal Aid to Agriculture, and Twenty-five Federal Grant-in-Aid Programs, 1955.

Commission on Organization of the Executive Branch of the Government (Herbert Hoover, Chairman): "Reports on the Department of Agriculture and the Department of the Interior" and Appendices K, L, M, and Q, 1949.

———: "Reports on Real Property Management and on Water Resources and Power" and Task Force Reports on same subjects, 1955.

Conf. Governors U.S. Proc., 1909.

Curran, C. E., and C. Edward Behre: "National Pulp and Paper Requirements in Relation to Forest Conservation" ("Hale Report"), S. Doc. 175, 74th Cong., 1st Sess., 1935.

Donaldson, Thomas: "The Public Domain," H.R. Misc. Doc. 45, 47th Cong., 2d Sess., part 4, 1884.

---

[1] All of the publications in this group were printed by the Government Printing Office, Washington.

Egleston, Nathaniel H.: "Report upon Forestry," vol. IV, 1884.

Fernow, Bernhard E.: "Report upon Forestry Investigations, 1877–1898," H. Doc. 181, 55th Cong., 3rd Sess., 1899.

Greeley, William B.: "Some Public and Economic Aspects of the Lumber Industry," Report 114, Office of the Secretary of Agriculture, 1917.

Hough, Franklin B.: "Report upon Forestry," vols. I–III, 1878, 1880, 1882.

Hynning, Clifford J.: "State Conservation of Resources," 1939.

Ickes, Harold L.: "Not Guilty—An Official Inquiry into the Charges Made by Glavis and Pinchot against Richard A. Ballinger, Secretary of the Interior, 1909–1911, 1940."

Inland Waterways Commission (Theodore E. Burton, Chairman): "Preliminary Report," S. Doc. 325, 60th Cong., 1st Sess., 1908.

Irion, Harry (comp.): The Principal Laws Relating to the Establishment and Administration of the National Forests and to Other Forest Service Activities, *U.S. Dept. Agr. Handbook* 20, 1951.

McArdle, Richard E.: "Timber Resources for America's Future (A Summary of the Timber Resource Review)," Forest Service, U. S. Dept. Agric., 1955.

Munns, Edward N.: A Selected Bibliography of North American Forestry, *U.S. Dept. Agr. Misc. Pub.* 364, 2 vols., 1940.

National Academy of Sciences, Forestry Commission: "Forest Policy for the Forest Lands of the United States," S. Doc. 105, 55th Cong., 1st Sess., 1897.

Natl. Conf. on Land Utilization Proc., 1932.

National Conference on Utilization of Forest Products "Report," 1925.

National Conservation Commission (Gifford Pinchot, Chairman): "Report," 3 vols., S. Doc. 676, 60th Cong., 2d Sess., 1909.

National Waterways Commission (Theodore E. Burton, Chairman): "Preliminary Report," S. Doc. 301, 61st Cong., 2d Sess., 1910.

————: "Final Report," S. Doc. 469, 62d Cong., 2d Sess., 1912.

President's Materials Policy Commission (William S. Paley, Chairman): "Resources for Freedom," 5 vols., 1952.

President's Water Resources Policy Commission (Morris L. Cooke, Chairman): "A Water Policy for the American People," 3 vols., 1950.

Sargent, Charles S.: "Report on the Forests of North America (Exclusive of Mexico)," Tenth Census (1880), 1884.

U.S. Congress, Joint Committee to Investigate the Department of the Interior and the Bureau of Forestry: "Report," 13 vols., S. Doc. 719, 61st Cong., 3d Sess., 1911.

————, Senate Select Committee on Reforestation (Charles L. McNary, Chairman): "Reforestation," S. Report 28, 68th Cong., 1st Sess., 1924.

————, Joint Committee on Forestry (John H. Bankhead II, Chairman): "Forest Lands of the United States," S. Doc. 32, 77th Cong., 1st Sess., 1941.

U.S. Department of Agriculture: "Report of the Secretary," 1862 to date.

————, Forest Service: "Report of the Chief," 1887 to date.

————: "Timber Depletion, Lumber Prices, Lumber Exports, and Concentration of Timber Ownership" ("Capper Report"), Report on S. Res. 311, 66th Cong., 2d Sess., 1920.

————: "A National Plan for American Forestry" ("Copeland Report"), 2 vols., S. Doc. 12, 73d Cong., 1st Sess., 1933.

————: "The Western Range," S. Doc. 199, 74th Cong., 2d Sess., 1936.

————: Reappraisal Reports 1–6, 1946–1947.

————: Forests and National Prosperity—A Reappraisal of the Forest Situation in the United States, *U.S. Dept. Agr. Misc. Pub.* 668, 1948.

U.S. Department of Commerce and Labor, Bureau of Corporations: "The Lumber Industry," 3 vols., 1913–1914.

U.S. Department of the Interior: "Report of the Secretary," 1849 to date.

————: "Forest Conservation on Lands Administered by the Department of the Interior," 1940.

————: "Forestry on Indian Lands," 1940.

————: "Land Management in the Department of the Interior," 1946.

————: A Century of Conservation 1849–1949, *Conservation Bul.* 39, 1950.

————: "Rebuilding the Federal Range," 1951.

————: "Years of Progress, 1945–1952."

————, Bureau of Land Management: "Brief Notes on the Public Domain," by Irving Senzel, 1950.

————: "Graphic Notes on the Public Domain," 1950.

————: Land of the Free, *Conservation Bul.* 40, 1951.

————, General Land Office: "Report of the Commissioner," 1849–1946.

————: "Land of the Free," 1940.

U.S. National Security Resources Board: "The Objectives of United States Materials Resources Policy and Suggested Initial Steps in Their Accomplishment," report by Jack Gorrie, Chairman, 1952.

U.S. Public Land Commission (J. A. Williamson, Chairman): "Preliminary Report," H. R. Exec. Doc. 46, 46th Cong., 2d Sess., 1880.

U.S. Public Lands Commission (W. A. Richards, Chairman): "Reports," S. Doc. 189, 58th Cong., 3d Sess., 1905.

Index

Page numbers from 372 on refer to items in the Chronological Summary, Appendix 2

435